NONLINEAR MODEL-BASED IMAGE/VIDEO PROCESSING AND ANALYSIS

NONLINEAR MODEL-BASED IMAGE/VIDEO PROCESSING AND ANALYSIS

Edited by

C. Kotropoulos and I. Pitas

A Wiley Interscience Publication
JOHN WILEY & SONS, INC.
New York • Chichester • Weinheim • Brisbane • Singapore • Toronto

For ordering and customer service, call 1-800-CALL-WILEY

Library of Congress Cataloging in Publication Data is available:

C. Kotropoulos and I. Pitas
   Nonlinear Model-Based Image/Video
   Processing and Analysis
ISBN 0-471-37735-X

10 9 8 7 6 5 4 3 2 1

To my mother Πέλλα and late father Λεόντιος (C.K.)

Contributors

HANS BURKHARDT, Albert-Ludwigs-Universität Freiburg, Computer Science Department, Institute for Pattern Recognition and Image Processing, 79085 Freiburg, Germany

SERGIO CARRATO, Department of Electronics, University of Trieste, Italy

ROBERT CASTAGNO, Swiss Federal Institute of Technology, Lausanne, Switzerland

ANDREA CAVALLARO, Swiss Federal Institute of Technology, Lausanne, Switzerland

ETIENNE DECENCIERE FERRANDIERE, Center of Mathematical Morphology, Ecole des Mines de Paris, Fontainebleau, France

TOURADJ EBRAHIMI, Swiss Federal Institute of Technology, Lausanne, Switzerland

MONCEF GABBOUJ, Signal Processing Laboratory, Tampere University of Technology, Tampere, Finland

CONSTANTINE KOTROPOULOS, Aristotle University of Thessaloniki, Department of Informatics, Artificial Intelligence and Information Analysis Laboratory, Box 451 Thessaloniki 540 06, Greece

STEPHEN MARSHALL, University of Strathclyde, Department of Electronic and Electrical Engineering, 204 George Street, Glasgow G1 1XW, Scotland, United Kingdom

STEFANO MARSI, Department of Electronics, University of Trieste, Italy

DOINA PETRESCU, Signal Processing Laboratory, Tampere University of Technology, Tampere, Finland

IOANNIS PITAS, Aristotle University of Thessaloniki, Department of Informatics, Artificial Intelligence and Information Analysis Laboratory, Box 451 Thessaloniki 540 06, Greece

GIOVANNI RAMPONI, Department of Electronics, University of Trieste, Italy

DRUTI SHAH, University of Strathclyde, Department of Electronic and Electrical Engineering, 204 George Street, Glasgow G1 1XW, Scotland, United Kingdom

SVEN SIGGELKOW, Albert-Ludwigs-Universität Freiburg, Computer Science Department, Institute for Pattern Recognition and Image Processing, 79085 Freiburg, Germany

IOAN TĂBUŞ, Signal Processing Laboratory, Tampere University of Technology, Tampere, Finland

FRANCESCO ZILIANI, Swiss Federal Institute of Technology, Lausanne, Switzerland

Contents

Preface

This book is addressed to senior undergraduate and postgraduate students, engineers, and scientists interested in the fields of digital image processing, computer vision, digital video processing, visual information retrieval, and multimedia applications. It is assumed that the reader is familiar with basic concepts of digital signal processing, digital image processing, and pattern recognition, for which a number of very good textbooks exist.

The book focuses on certain issues related to still image and image sequence (i.e., digital video) processing and analysis. In particular, the book is devoted to image sequence filtering with applications to digital video restoration, noise removal, image sequence interpolation, lossless image compression; image sequence analysis with applications to video manipulation and traffic surveillance; invariant features with applications to visual information retrieval; and image models for facial feature tracking with applications to lip reading. The techniques are all nonlinear in the aforementioned tasks.

This book reflects the research undertaken within the ESPRIT LTR-20229 project "NOBLESSE" (Nonlinear Model-Based Analysis and Description of Images for Multimedia Applications) and its predecessor ESPRIT-BRA project "NAT" (Nonlinear and Adaptive Techniques in Digital Image Processing, Analysis, and Computer Vision) both funded by the European Commission. Considerable effort is also made to present as many results by other researchers as possible so that an up to date state-of-the-art account of the most important advances in each topic is provided.

The editors wish to express their appreciation to the contributing authors for the high-quality illustrative materials they have provided with their manuscript. We are obliged to the Institute of Electrical and Electronic Engineers (IEEE), the Society of Photo-Optical Instrumentation Engineers (SPIE), and Kluwer Academic Publishers for granting the contributing authors permission to use material published in their journals in this book as detailed in specific citations in the text. Special acknowledgment is made to Mr. Theoharis Kosmidis for his assistance in preparing the final camera-ready manuscript. Finally we want to thank Mr. George Telecki, Ms. Rosalyn Farkas, Ms. Sara Paracka, and Ms. Cassandra Craig at Wiley for their patience and technical support.

CONSTANTINE KOTROPOULOS
IOANNIS PITAS

Thessaloniki, January 2001

1 Introduction to Nonlinear Model-Based Image/Video Processing and Analysis

C. KOTROPOULOS and I. PITAS

Aristotle University of Thessaloniki
Department of Informatics
Artificial Intelligence and Information Analysis Laboratory
Box 451, Thessaloniki 540 06, Greece

This book is devoted to the manipulation of still images and image sequences, namely video, on a computer. Such a manipulation has been an established research activity for more than 30 years. However, digital image and video processing equipment has been widely accessible to the general public only recently. A major contributing factor has been the massive admission of nonacademic users to the world wide web (WWW) [43]. Manipulating digital imagery is prevalent on the WWW. Indeed, all images and image sequences on the WWW are in a digital form. Moreover most images in press and advertisement have been transformed to a digital form somewhere between the shot taking and the printing. Images seen on television or those acquired by satellites or through medical imaging modalities have often been processed or transmitted digitally [72].

Digital image processing and analysis are the basic subjects of the book. The interest in digital image processing methods stems from two practical needs: the improvement of pictorial information for human interpretation, and scene processing for autonomous machine perception [28]. Through the book, a progressive transition is performed from still images to image sequences, that is, from the processing of a single video frame to the processing of a sequence of video frames by exploiting the spatiotemporal redundancy between successive video frames. Both monochrome (i.e., gray scale) and color images are treated in the book. Although in some applications (e.g., X ray, ultrasonic imaging, etc.) the images are inherently monochrome, in many others (surveillance of vehicles, content-based information retrieval, object extraction and tracking–e.g., face detection and tracking–industrial applications) color plays a crucial role and is an indispensable information source.

We will reserve the term digital *image processing* for referring to image enhancement (e.g., image filtering), image restoration, and image compression. From this point of view, digital image processing can be considered as a branch of the more general field of digital signal processing. *Image analysis* encompasses image segmentation, edge detection, shape representation (e.g., mathematical morphology for shape representation), recognition, and interpretation. It is seen that image analysis involves a decision-making step. Consequently, image analysis is influenced by pattern recognition. Computer vision is a superset of image analysis aiming at machine perception and artificial vision.

In the early development of image processing, linear filters were the primary tools. Their mathematical simplicity, the existence of some desirable properties (e.g., the superposition principle) in addition to their satisfactory performance in many applications made them quite popular. However, linear filters have poor performance in the presence of noise that is not additive or whose statistics deviate from Gaussianity as well as when system nonlinearities are encountered [64]. Many other problems, such as the mean square estimation [61, p. 175], the maximum entropy criterion are inherently nonlinear. In pattern recognition, the Bayes classifier results in principle to a nonlinear classifier [25]. Other classifiers, such as the perceptron (two-layer or multilayer), the radial basis function neural networks, the decision trees are also nonlinear [85]. Support vector machines [94] are nonlinear systems as well. Nonlinear systems are able to discriminate between equivalence classes as opposed to linear systems that can discriminate between spectral components [9]. For such reasons nonlinear techniques have become indispensable tools in filtering, estimation, detection and classification problems which are frequently met in digital image processing and computer vision tasks. This explains why the term *"nonlinear"* appears in the title of the book. The continuing interest on nonlinear signal/image processing is manifested by the success of a series of dedicated workshops and special issues on this topic [53, 54, 56, 58, 1].

Let us now argue on the two remaining words, *"model-based"*, in the title of the book. The interpretation of this term is twofold. It can be read as either the ad hoc selection of the system chosen to accomplish a specific task or as an object description. The first interpretation is valid for Chapters 2 to 5. From this point of view, a number of models are reviewed, such as Boolean and stack filters (Chapter 2), rational systems (Chapter 3), morphological filters (Chapter 4), and order statistic filters of varying complexity (Chapter 5). The second interpretation is met in Chapters 6 to 8. For example, image regions are the object models in Chapter 6 on video segmentation. Invariants are the models for content-based retrieval in Chapter 7. Templates, active contours, wireframes, and appearance-based models [87] are fitted to the data in order to describe the face or the facial characteristics in Chapter 8. Clearly the two interpretations converge, since the system selection specifies the properties of the data that should be preserved in the system output. For example, L-filters aim at rejecting outliers in their output, whereas $L\ell$-filters and their extensions aim at reducing tone interference in addition to outlier rejection. It is well known that averages of polynomials computed on a local support, as those studied in Chapter 7, are invariant under the group of Euclidean motion. Accordingly we can consider them as polynomial systems whose rational forms are investigated in Chapter 3. Moreover if the polynomial expansions are replaced by permutation lattice ones, the permutation order statistic lattices result that are reviewed in Chapter 5. However, we need to point out that model-based approaches are not panacea for image processing and computer vision problems. For example, the quest for a generalized model-based image coding system would be greatly aided by the ability to code scenes without object recognition, even if only partial coding was achieved [76]. Three important advantages accrue from the use of low-level primitives (e.g., corner features): generality, opportunism, and graceful degradation. It should be emphasized that the combination of both top-down (i.e., model-based) and bottom-up approaches is the strategy that guarantees a higher success in most practical problems.

The book is not intended as an introductory text on image processing and computer vision. It assumes that the reader is already familiar with the basic concepts for which a number of excellent recent monographs and textbooks in image processing [67, 28, 36, 42, 96] and computer vision [30, 38, 92, 19] are cited. For beginners and practitioners a number of textbooks including either source code [49, 70, 17, 63] or run-time libraries [3, 35, 93, 63] are also available. The book enriches the already existed literature on nonlinear digital filtering [64, 2, 20] by covering topics that are not included in it. Moreover it offers complementary material with respect to the textbook [84], the monographs [43, 7] and the edited book [39]. The outline of the book is as follows.

The subject of Chapter 2 is the optimal design of *Boolean* and *stack filters* and their applications in image processing. The theory of stack and Boolean filters provides a common framework that unifies order statistic filters and morphological filters. The problem of the optimal design of Boolean and stack filters under the mean absolute error criterion is addressed in a training approach, where a representative target image and a noise corrupted version of this image are available for filter design. Novel applications of optimal Boolean and stack filters to *noise suppression, edge detection* and *lossless image compression* are described. The chapter includes a detailed list of references. Besides this application on lossless image compression, the book does not consider in depth the problem of image and video compression. Readers interested in these topics may consult [91, 84, 16, 34, 4, 68, 86, 73].

Chapter 3 is devoted to *rational filters*. Rational filters are expressed as the ratio of two polynomials. The input/output relationship of a nonlinear processing system can be modeled as a rational filter. They have been introduced as alternatives to the discrete Volterra filters [52] aiming at reducing the number of parameters when the order of the nonlinearity is high. Applications of rational filters in *noise smoothing, image interpolation and contrast enhancement* are presented. A unique feature of this chapter is the discussion on the *realization* of rational filters. Very large scale integration (VLSI) realizations on both application-specific integrated circuits (ASIC) and field-programmable gate arrays (FPGA) are described. Moreover implementations using digital signal processors (DSPs) are considered. Readers interested in very long instruction word (VLIW) architectures may also consult the tutorials [23, 69]. Another tutorial on FPGA can be found in [95]. General readings on DSPs are [21, 24].

Mathematical morphology has been very useful for 2-D and 3-D image processing and analysis [74, 51, 33, 80]. Chapter 4 investigates the application of mathematical morphology to *image sequence restoration*. Algorithms are developed for treating the most common defects in old movies, namely the flicker, the jitter, the vertical scratches, and the local random defects. The latter point is studied in more detail. Openings and closings by reconstruction have proved efficient tools for the detection of local random defects, such as dirt and blotches. Closely related to the subject of this chapter is the monograph [43] where techniques based on 2-D and 3-D autoregressive models and Bayesian inference are presented for image sequence modeling, motion estimation, missing data reconstruction, and line scratch detection and removal. The need for image restoration algorithms exists also at the output of any decoder of a compressed image or video bitstream, because in every image and video transmission system information is lost due to channel errors. Channel errors appear either as random bit or erasure errors. The former corrupt random patterns of bits of the transmitted signal. The latter result in the loss of continuous segments of bits of the transmitted signal. Since the information is typically predictively coded, the aforementioned errors propagate through a decoded sequence over space/time. Furthermore any lossy image and video compression system introduces artifacts. Algorithms for removing the artifacts in compressed images, compensating for channel errors and preventing their propagation in video streams are described in [39].

Adaptive order statistic filters are reviewed in Chapter 5. This chapter can be considered as a continuation of Chapters 2 and 4. Order statistic filters use the concept of data ordering [65]. The motivation behind using data ordering is the following. If outliers corrupt the observed signal/image, these will be located in either the lower or higher data ranks. Accordingly, if we are able to appropriately weigh the ordered samples, then we can eliminate or reduce the effect of outliers. There is now a plethora of nonlinear filters based on data ordering. Three major classes of order statistic filters: the L-filter, the $L\ell$-filter, and the ranked-order filters. These are the fundamental filter structures which are the nonlinear models under discussion. The first two fundamental filter models have the form of a linear combination of the observations, and they exploit either the rank or the combined rank and location information that is inherent in the observations. Although these filter structures have been proved highly effective in suppressing the noise, in processing of signals with nonstationary mean levels and with abrupt changes, they lead unavoidably to blurring. Such signal characteristics are frequent in image processing, where important visual cues provided by the edges and fine image details should be preserved. In this chapter we are interested in adaptive order statistic filter designs. Two broad classes of adaptive filters are considered: (1) adaptive filters whose coefficients are determined by iterative algorithms (e.g., least mean squares, recursive least squares) for the minimization of the mean squared error between the filter output and the desired response; (2) signal-adaptive filters.

We will confine ourselves to the use of the least mean squares (LMS) algorithm [98, 32] in the design of adaptive nonlinear filters of the first class. The motivation behind using LMS is its ability to cope with nonstationary and/or time varying signals [50, 75, 27]. LMS has been successfully applied to channel equalization, echo cancellation and prediction. All these problems can be described by a common linear noisy model, that relates two signals $x(n)$ and $s(n)$ according to $s(n) = \mathcal{F}(x(n)) + v(n)$, where $v(n)$ is the noise [50]. Note that in the special case of prediction $x(n) = s(n-1)$. We call $x(n)$ and $s(n)$ input and reference, respectively. If $\mathcal{F}$ is assumed to be a linear filter, an adaptive estimate of the noise $v(n)$, $\hat{v}(n) = s(n) - \mathcal{H}_{n-1}(x(n))$, can be obtained to control adaptively $\mathcal{H}_{n-1}$ so that $\hat{v}(n)$ becomes as small as possible. If the noise probability density function deviates from the Gaussian distribution, or in the presence of acquisition or transmission errors (i.e., in the presence of outliers) the linear model fails. This is not the case with the adaptive LMS L- and $L\ell$-filters that are reviewed in this chapter. The performance of the above-mentioned filter structures in noise smoothing is studied for both still images and video sequences. Besides the fundamental structures we review powerful extensions and their generalizations.

Signal-adaptive filters are filters that change their smoothing properties at each image pixel according to the local image content. Most of these filters change their coefficients close to the image edges or close to impulses, so that their performance becomes similar to that of the median filter. Their inclusion in this chapter serves as a bridge between the filters that stem from the L- and $L\ell$-filters and those stemming from the ranked-order filters. The fundamental filter in this class is the *signal-adaptive median* filter. We describe a signal-dependent adaptive L-filter structure that employs two adaptive L-filters, one applied to homogeneous regions and a second filter applied close to image edges. The adaptation of the window in the signal-adaptive median filter by employing morphological operations is also presented.

Applications of adaptive order statistics to *noise filtering* for both gray scale and color still images and image sequences are presented. Noise filtering is a generic application that may refer to the compensation for transmission errors, the compensation for channel errors in the presence of differential pulse code modulation decoders [81], speckle filtering aiming at increasing lesion detection in ultrasonic images [44], vector rational interpolation of erroneously received motion fields in MPEG-2 coded video streams for error-concealment purposes [89]. An application of permutation order statistic lattices in inverse halftoning is also reported [40]. Readers interested in color image processing may also consult [88, 72, 66].

Video segmentation and tracking of video objects are the subjects of Chapter 6. The starting point is the segmentation of a video frame to regions and subsequently, the tracking of these image regions in the upcoming video frames. Region extraction is a fundamental image analysis operation [67, 28, 36, 3] or low-level computer vision [38, 92]. Beginners may find useful to get acquainted with image segmentation by consulting a textbook that is accompanied either with code or run-time libraries such as [35, 93, 63]. The importance of image regions has increased in content-based coding (MPEG-4) and description (MPEG-7) standards [31, 7, 12, 13, 11].

For a discussion of the basic principles of model-based coding the interested reader may consult [84]. It is well known that image regions do not possess semantic content per se. Semantically meaningful objects are obtained by grouping regions through user interaction. The region extraction uses multiple features, such as color, texture, motion information by using the components of optical flow [84, 26, 82] and position. Feature normalization and weighting schemes are incorporated. A *constrained fuzzy C-means algorithm* is developed for clustering the feature vectors. Two important applications are described, namely *video editing* [29] and *automatic traffic surveillance*.

Invariants are properties (i.e., functions) of geometric configurations which do not change under a certain class of transformations [19, 87]. For instance, the length of a line segment does not change if the segment is rotated and translated. We claim that length is invariant under rigid motion. To briefly introduce the concept of invariants to the reader, let us consider the problem of object recognition. This problem implies that image data are compared to a model database, where each model is simply an object description. When a model is found to correspond to a subset of the data, we claim that the model matches the data. The matched model reveals the object identity. One of the key problems in object recognition from luminance images is that the appearance of an object depends on imaging conditions. If we were able to identify shape properties that do not change with imaging conditions, this problem would be solved. That is, we aim at defining invariant functions of some image properties (e.g., contours), which yield sufficiently different values for all shapes of interest.Vectors of invariant functions are used to index the models. Invariants are known only for some shape classes, and defining useful invariants for general object shapes is not easy at all.

Scalar, algebraic invariants of the geometric imaging transformation (the so-called transformation groups) are considered in Chapter 7. The systematic construction of invariants is reviewed in this chapter. Three principal approaches, namely the normalization approach, the differential approach and the integral approach are discussed. Invariants that are functions of object contours are constructed by normalization. It is demonstrated that an affine normalization can provide a solution up to undetermined scaling, translation, and rotation. The latter transformations can be eliminated by using Fourier descriptors. The elements of many transformation groups can be parameterized by a suitable vector. Invariants then can be derived by employing Lie algebra. This is the so-called differential approach. The chapter also discusses thoroughly the construction of invariants by integrating over the group of translations and the group of Euclidean motion. The invariants are now functions of the gray scale images. Group averages of monomials are shown to be invariant under translations. Averages of polynomials computed on a local support are shown to be invariant under the group of Euclidean motion. Histograms of locally invariant features are proved to preserve the invariance property. However, the histograms are not continuous functions of feature values. By replacing them with their fuzzy variants, we can obtain clustered classes in the feature space. *Content-based image retrieval appli-*

cations using either color or combined color-texture histograms of invariant features are presented.

Content-based image retrieval is related to a number of applications, ranging from art galleries and museum archives, to picture/photograph archiving [62] and communications, to criminal investigation, to medical and geographic databases, and so on. It is the melting pot of many research disciplines, such as image/video processing and analysis, pattern recognition, computer vision, multimedia data modeling [78], multimedia indexing, human-machine interface, data visualization, to mention a few. A very good monograph on this topic has recently been released [7]. A very small sample of references related to content-based image/video retrieval is provided to stimulate to further reading [22, 37, 79, 15, 41, 90]. A detailed list of references can be found in [7].

Chapter 8 discusses the closely related problems of *face detection, facial feature localization, and tracking.* Special emphasis is given to *lower lip tracking* and *face recognition.* The objective is to review models that can integrate audiovisual information. The description starts with intensity models for face detection and face recognition. Face recognition has been an active research topic the current decade. It is in the cross section of the research disciplines on human-centered interfaces [5, 55, 57, 60, 77], biometric person authentication [6, 59, 99, 97, 46, 45, 47, 83], content-based image retrieval [62], and so on. Two tutorial papers on face recognition [71, 10] and several special issues on journals, such as [18], appeared during the past decade. Both holistic and analytic approaches are reviewed. Wireframe models and active contour models [8] are two proven powerful models for face and facial feature tracking. The research related to these models is outlined. Statistical techniques based on maximizing the posterior probability of a deformed template given an observed image and heuristic methods for model fitting are quoted. The problem of lower lip extraction and tracking is studied in more detail. Readers interested in lip reading, speech-driven face animation, lip synchronization, and audio-to-visual mapping may also consult [14, 48].

The applications discussed in Chapters 6 to 8 are fundamental ones under the so-called field of *multimedia signal processing* [12, 13]. This new field can be conceived as the place where signal processing and telecommunications meet computer vision, computer graphics, and document processing. The former disciplines have been traditionally explored in electrical engineering curricula whereas the latter ones have grown within computer science societies. The integration and the interaction of different data streams, such as text, images, graphics, speech, audio, video, animation, handwriting, and data files create many challenging research opportunities whose scope goes beyond signal compression and coding.

REFERENCES

1. G.R. Arce, P. Maragos, Y. Neuvo, and I. Pitas, eds. *IEEE Trans. Image Processing*. Special Issue on Nonlinear Image Processing. June 1996.

2. J. Astola and P. Kuosmanen. *Fundamentals of Nonlinear Digital Filtering*. CRC Press, Boca Raton, FL, 1997.

3. G.A. Baxes. *Digital Image Processing: Principles and Applications*. Wiley, New York, 1994.

4. V. Bhaskaran and K. Konstantinides. *Image and Video Compression Standards: Algorithms and Architectures*. Kluwer Academic, Norwell, MA, 1995.

5. M. Bichsel, ed. *Proc. First Int. Conf. Automatic Face and Gesture Recognition*. Zurich, Austria, 1995. Multimedia Laboratory, Dept. of Computer Science, University of Zurich.

6. J. Bigün, G. Chollet, and G. Borgefors, eds. *Audio- and Video-Based Biometric Person Authentication*, Springer-Verlag, Berlin, 1997.

7. A. Del Bimbo. *Visual Information Retrieval*. Morgan Kaufmann, San Francisco, CA, 1999.

8. A. Blake and M. Isard. *Active Contours: The Application of Techniques from Graphics, Vision, Control Theory and Statistics to Visual Tracking of Shapes in Motion*. Springer-Verlag, New York, 1999.

9. H. Burkhardt and S. Siggelkow. Chapter 7, this volume.

10. R. Chellappa, C.L. Wilson, and S. Sirohey. Human and machine recognition of faces: A survey. *Proc. IEEE*, 83(5): 705-740, May 1995.

11. H.H. Chen, T. Ebrahimi, G. Rajan, C. Horne, P.K. Doenges, and L. Chiariglione, eds. *IEEE Trans. Circuits and Systems for Video Technology*. Special Issue on Synthetic/Natural Hybrid Video Coding. March 1999.

12. T. Chen, K.J. Ray Liu, and A. Murat Tekalp, eds. *Proc. IEEE*. Special Issue on Multimedia Signal Processing–Part 1. May 1998.

13. T. Chen, K.J. Ray Liu, and A. Murat Tekalp, eds. *Proc. IEEE*. Special Issue on Multimedia Signal Processing–Part 2. June 1998.

14. T. Chen and R.R. Rao. Audio-visual integration in multimodal communication. *Proc. IEEE*, 86(5): 837-852, May 1998.

15. M. Christel, S. Stevens, T. Kanade, M. Mauldin, R. Reddy, and H. Wactlar. Techniques for the creation and exploration of digital video libraries. In B. Furht, ed., *Multimedia Tools and Applications*. Kluwer Academic, Norwell, MA, 1996, pp. 283-327.

16. R.J. Clarke. *Digital Compression of Still Images and Video*. Academic Press, London, 1995.

17. R. Crane. *A Simplified Approach to Image Processing. Classical and Modern Techniques in C*. Prentice Hall PTR, Upper Saddle River, NJ, 1997.

18. J. Daugman, ed. *IEEE Trans. Pattern Analysis and Machine Intelligence*. Special Issue Face and Gesture Recognition. July 1997.

19. E.R. Davies. *Machine Vision*. Academic Press, San Diego, CA, 1997.

20. E.R. Dougherty and J. Astola. *Nonlinear Filters for Image Processing*. SPIE Optical Engineering Press, Bellingham, WA, 1999.

21. J. Eyre and J. Bier. The evolution of DSP processors: From early architectures to the latest developments. *IEEE Signal Processing Magazine*, 17(2): 43-56, March 2000.

22. C. Faloutsos. *Searching Multimedia Databases by Content*. Kluwer Academic, Norwell, MA, 1996.

23. P. Faraboschi, G. Desoli, and J.A. Fisher. The latest word in digital and media processing. *IEEE Signal Processing Magazine*, 15(2): 59-107, March 1998.

24. J. Fridman. Sub-word parallelism in DSP: Applying the TigerSHARC arhitecture. *IEEE Signal Processing Magazine*, 17(2): 27-36, March 2000.

25. K. Fukunaga. *Introduction to Statistical Pattern Recognition*. Academic Press, San Diego, CA, 1990.

26. B. Furht, J. Greenberg, and R. Westwater. *Motion Estimation Algorithms for Video Compression*. Kluwer Academic, Norwell, MA, 1997.

27. G.-O. Glentis, K. Berberidis, and S. Theodoridis. Efficient least squares adaptive algorithms for FIR transversal filtering. *IEEE Signal Processing Magazine*, 16(4): 13-41, July 1999.

28. R.C. Gonzalez and R.E. Woods. *Digital Image Processing*. Addison-Wesley, Reading, MA, 1992.

29. A. Hampapur, R. Jain, and T.E. Weymouth. Production model based digital video segmentation. In B. Furht, ed., *Multimedia Tools and Applications*. Kluwer Academic, Norwell, MA, 1996, pp. 111-153.

30. R. Haralick and L. Shapiro. *Computer and Robot Vision*, Vols. I-II. Addison-Wesley, Reading, MA, 1992.

31. B.G. Haskell, A. Puri, and A.N. Netravali. *Digital Video: An Introduction to MPEG-2*. Kluwer Academic, Norwell, MA, 1997.

32. S. Haykin. *Adaptive Filter Theory*, 3rd ed. Prentice Hall, Englewood Cliffs, NJ, 1995.

33. H.J.A.M. Heijmans and J.B.T.M. Roerdink, eds. *Mathematical Morphology and Its Applications to Image and Signal Processing*. Kluwer Academic, Dordrecht, 1998.

34. G. Held and T.R. Marshall. *Data and Image Compression*. Wiley, Winchester, England, 1995.

35. B. Jahne. *Digital Image Processing*. Springer-Verlag, Berlin, 1997.

36. A.K. Jain. *Fundamentals of Digital Image Processing*. Prentice Hall, Englewood Cliffs, NJ, 1989.

37. R. Jain. Infoscopes: Multimedia information systems. In B. Furht, ed., *Multimedia Systems and Techniques*. Kluwer Academic, Norwell, MA, 1996, pp. 217-253.

38. R. Jain, R. Kasturi, and B.G. Schunk. *Machine Vision*. McGraw-Hill, New York, 1995.

39. A.K. Katsaggelos and N.P. Galatsanos, eds. *Signal Recovery Techniques for Image and Video Compression and Transmission*. Kluwer Academic, Dordrecht, 1998.

40. Y.T. Kim, G.R. Arce, and N. Grabowski. Inverse halftoning using binary permutation filters. *IEEE Trans. Image Processing*, 4(9): 1296-1311, September 1995.

41. Y. Kiyoki, T. Kitagawa, and T. Hayama. A metadatabase system for semantic image search by a mathematical model of meaning. In A. Sheth and W. Klas, eds., *Multimedia Data Management: Using Metadata to Integrate and Apply Digital Media*. McGraw-Hill, New York, 1998, pp. 191-222.

42. R. Klette and P. Zamperoni. *Handbook of Image Processing Operators*. Wiley, Chichester, England, 1996.

43. A.C. Kokaram. *Motion Picture Restoration: Digital Algorithms for Artefact Suppression in Degraded Motion Picture Film and Video*. Springer-Verlag, London, 1998.

44. C. Kotropoulos and I. Pitas. Optimum nonlinear signal detection and estimation in the presence of ultrasonic speckle. *Ultrasonic Imaging*, 14(3): 249-275, July 1992.

45. C. Kotropoulos, A. Tefas, and I. Pitas. Frontal face authentication using discriminating grids with morphological feature vectors. *IEEE Trans. Multimedia*, 2(1): 14-26, March 2000.

46. C. Kotropoulos, A. Tefas, and I. Pitas. Frontal face authentication using morphological elastic graph matching. *IEEE Trans. Image Processing*, 4(9): 555-560, 2000.

47. C. Kotropoulos, A. Tefas, and I. Pitas. Morphological elastic graph matching applied to frontal face authentication under well-controlled and real conditions. *Pattern Recognition*, 33(12): 31-43, 2000.

48. S.-Y. Kung and J.-N. Hwang. Neural networks for intelligent multimedia processing. *Proc. IEEE*, 86(6): 1244-1272, June 1998.

49. C.A. Lindley. *Practical Image Processing in C: Acquisition, Manipulation, Storage*. Wiley, New York, 1991.

50. O. Macchi. *Adaptive Processing: The Least Mean Squares Approach with Applications in Transmission*. Wiley, Chichester, England, 1995.

51. P. Maragos, R.W. Schafer, and M.A. Butt, eds. *Mathematical Morphology and Its Applications to Image and Signal Processing*. Kluwer Academic, Norwell, MA, 1996.

52. V.J. Mathews and G. Sicuranza. *Polynomial Signal Processing*. Wiley, New York, 2000.

53. *Proc. 1993 Winter Workshop Nonlinear Digital Signal Processing*, January 18-20 1993. Tampere, Finland.

54. *Proc. 1995 IEEE Workshop Nonlinear Signal and Image Processing*, Vols. I-II, June 20-22 1995. Neos Marmaras, Halkidiki, Greece.

55. *Proc. Second Int. Conf. Automatic Face and Gesture Recognition*, Los Amitos, CA, 1996. IEEE Computer Society Press.

56. *CDROM Proc. 1997 IEEE-EURASIP Workshop Nonlinear Signal and Image Processing*, September 8-10 1997. Mackinac Island, MI.

57. *Proc. Third Int. Conf. Automatic Face and Gesture Recognition*, Los Amitos, CA, April 14-16 1998. IEEE Computer Society Press.

58. *Proc. 1999 IEEE-EURASIP Workshop Nonlinear Signal and Image Processing*, June 20-23 1999. Antalya, Turkey.

59. *Proc. Second Int. Conf. Audio- and Video-Based Biometric Person Authentication*, March 22-23 1999. Washington, DC.

60. *Proc. Fourth IEEE Int. Conf. Automatic Face and Gesture Recognition*, Los Amitos, CA, March 26-30 2000. IEEE Computer Society Press. Grenoble, France.

61. A. Papoulis. *Probabilities, Random Variables and Stochastic Processes*, 3rd ed. McGraw-Hill, New York, 1991.

62. A. Pentland, R.W. Picard, and S. Sclaroff. Photobook: Content-based manipulation of image databases. In B. Furht, ed., *Multimedia Tools and Applications*. Kluwer Academic, Norwell, MA, 1996, pp. 43-80.

63. I. Pitas. *Digital Image Processing Algorithms and Applications*. Wiley, New York, 2000.

64. I. Pitas and A.N. Venetsanopoulos. *Nonlinear Digital Filters: Principles and Applications*. Kluwer Academic, Norwell, MA, 1990.

65. I. Pitas and A.N. Venetsanopoulos. Order statistics in digital image processing. *Proc. IEEE*, 80(12): 1893-1921, December 1992.

66. K.N. Plataniotis and A.N. Venetsanopoulos. *Color Image Processing and Applications*. Springer-Verlag, Berlin, 2000.

67. W.K. Pratt. *Digital Image Processing*. Wiley, New York, 1991.

68. K.R. Rao and J.J. Hwang. *Techniques and Standards for Image, Video, and Audio Coding*. Prentice Hall, Upper Saddle River, NJ, 1996.

69. S. Rathnam and G. Slavenburg. Processing the new world on interactive media: The Trimedia VLIW CPU architecture. *IEEE Signal Processing Magazine*, 15(2): 108-117, March 1998.

70. J.C. Russ. *The Image Processing Handbook*. CRC Press, Boca Raton, FL, 1992.

71. A. Samal and P.A. Iyengar. Automatic recognition and analysis of human faces and facial expressions: A survey. *Pattern Recognition*, 25(1): 65-77, 1992.

72. S.J. Sangwine and R.E.N. Horne, eds. *The Colour Image Processing Handbook*. Chapman and Hall, London, 1998.

73. G.M. Schuster and A.K. Katsaggelos. *Rate-Distortion Based Video Compression: Optimal Video Frame Compression and Object Boundary Encoding*. Kluwer Academic, Norwell, MA, 1997.

74. J. Serra and P. Soille, eds. *Mathematical Morphology Its Applications To Image Processing*. Kluwer Academic, Dordrecht, 1994.

75. W. Sethares. The least mean squares family. In N. Kalouptsidis and S. Theodoridis, eds., *Adaptive System Identification and Signal Processing Algorithms*. Prentice Hall, London, 1993, pp. 84-122.

76. L.S. Shapiro. *Affine Analysis of Image Sequences*. Cambridge University Press, Cambridge, U.K., 1995.

77. R. Sharma, V.I. Pavlović, and T.S. Huang. Toward multimodal human-computer interface. *Proc. IEEE*, 86(5): 853-869, May 1998.

78. A. Sheth and W. Klas, eds. *Multimedia Data Management: Using Metadata to Integrate and Apply Digital Media*. McGraw-Hill, New York, 1998.

79. S.W. Smoliar and H. Zhang. Video indexing and retrieval. In B. Furht, ed., *Multimedia Systems and Techniques*. Kluwer Academic, Norwell, MA, 1996, pp. 293-322.

80. P. Soille. *Morphological Image Analysis*. Springer-Verlag, Berlin, 1999.

81. X. Song, T. Viero, and Y. Neuvo. Interframe DPCM with robust median-based predictors for transmission of image sequences over noisy channels. *IEEE Trans. Image Processing*, 5(1): 16-32, January 1996.

82. C. Stiller and J. Konrad. Estimating motion in image sequences. *IEEE Signal Processing Magazine*, 16(4): 70-91, July 1999.

83. A. Tefas, C. Kotropoulos, and I. Pitas. Face authentication by using elastic graph matching and support vector machines. In *Proc. 2000 IEEE Int. Conf. Acoustics, Speech, and Signal Processing*, Vol. IV, 2000, pp. 2409-2412.

84. A.M. Tekalp. *Digital Video Processing*. Prentice Hall PTR, Upper Saddle River, NJ, 1995.

85. S. Theodoridis and K. Koutroumbas. *Pattern Recognition*. Academic Press, San Diego, CA, 1999.

86. L. Torres and M. Kunt, eds. *Video Coding: The Second Generation Approach*. Kluwer Academic, Norwell, MA, 1996.

87. E. Trucco and A. Verri. *Introductory Techniques for 3-D Computer Vision*. Prentice Hall, Upper Saddle River, NJ, 1998.

88. H.J. Trusell, J. Allebach, M.D. Fairchild, B. Funt., and P.W. Wong, eds. *IEEE Trans. Image Processing*. Special Issue Color Imaging. July 1997.

89. S. Tsekeridou, F. Alaya Cheikh, M. Gabbouj, and I. Pitas. Motion field estimation by vector rational interpolation for error concealment purposes. In *Proc. 1999 IEEE Int. Conf. Acoustics, Speech, and Signal Processing*, Vol. 6, 1999, pp. 3397-3400.

90. S. Tsekeridou and I. Pitas. Audio-visual content analysis for content-based video indexing. In *Proc. IEEE Int. Conf. Multimedia Computing and Systems*, Vol. 1, 1999, pp. 667-672.

91. G. Tziritas and C. Labit. *Motion Analysis for Image Sequence Coding*. Elsevier, Amsterdam, 1994.

92. S. Ullman. *High-Level Vision: Object Recognition and Visual Cognition*. The MIT Press, Cambridge, MA, 1996.

93. S.E. Umbaugh. *Computer Vision and Image Processing: A Practical Approach Using CVIPtools*. Prentice Hall PTR, Upper Saddle River, NJ, 1998.

94. V. Vapnik. *Statistical Learning Theory*. Wiley, New York, 1998.

95. J. Villasenor and B. Hutchings. The flexibility of configurable computing. Proving the hardware for data-intensive real-time processing. *IEEE Signal Processing Magazine*, 15(5): 67-84, September 1998.

96. A.R. Weeks. *Fundamentals of Electronic Image Processing*. SPIE Optical Engineering Press and IEEE Press, Bellingham, Washington-Piscataway, NJ, 1996.

97. J. Weng and D.L. Swets. Face recognition. In A. Jain, R. Bolle, and S. Pankanti, eds., *Biometrics: Personal Identification in Networked Society*. Kluwer Academic, Norwell, MA, 1999, pp. 65-86.

98. B. Widrow and S.D. Stearns. *Adaptive Signal Processing*. Prentice Hall, Englewood Cliffs, NJ, 1985.

99. J. Zhang, Y. Yan, and M. Lades. Face recognition: Eigenface, elastic matching and neural nets. *Proc. IEEE*, 85(9): 1423-1435, September 1997.

2 Optimal Design of Boolean and Stack Filters and Their Application in Image Processing

D. PETRESCU[†], I. TĂBUŞ, and M. GABBOUJ

Signal Processing Laboratory
Tampere Universty of Technology
Tampere, Finland

2.1 INTRODUCTION

Linear filtering has been the dominating tool in solving image processing problems in the past. However, it does not always achieve the desired level of performance in tasks, such as filtering long-tailed noise distributions, preserving or detecting image details. Researchers have been looking for more appropriate solutions to these types of problems within various classes of nonlinear filters.

In the last decades, new classes of nonlinear filters have been developed, providing powerful methods which clearly outperformed the linear filters in specific tasks, such as impulsive noise rejection from images, feature extraction, and shape description. Two of these nonlinear filters classes, which historically stem from different theoretical foundations–the order statistics filters and the morphological filters–are strongly connected in their structural aspects. The recently developed theory of stack and Boolean filters provides a common framework which unifies these classes of filters, as well as other related classes, and enables the selection of the "best" suited class for each particular problem at hand. The reader may consult [1, 15, 18, 27, 28, 30, 38, 41, 42, 65, 74]. The optimal design theory under the mean absolute error (MAE) criterion for selecting the best filter inside the stack filter class was developed in [7, 8, 9, 13, 14, 34, 77, 81]. Most of the cited papers developed the

[†]Currently with Billions Operations per Second Inc., 6340 Quadrangle Dr., Chapel Hill, NC 27514, USA

15

theory based on the availability of noise and signal models. For image processing applications these models are not always available and therefore the optimal solutions are derived for representative target images, using the sum of absolute errors criterion [35, 67, 70]. These theoretical advances enabled the extension of the application area for the new classes of filters. In order to develop new applications, more complex architectures have also been introduced, extending and combining nonlinear filters to suit the application at hand [50].

In this chapter novel extensions and applications of optimal Boolean and stack filters are developed, aiming to solve a number of image processing problems, including image restoration, feature extraction, and lossless compression. New filtering architectures are proposed to combine the beneficial properties of Boolean and stack filters with other types of filters.

In lossless image compression applications, the performance achieved in the prediction stage has major influence over the final compression ratio. Different solutions using median filters were proposed and shown to outperform classical linear predictors [23, 43, 45, 50]. In this chapter optimal Boolean and stack filters, designed using the procedure developed in [70], are used for the prediction task in sequential and progressive compression schemes.

The feature extraction from images is the excellency domain of mathematical morphology. Shape description, edge detection, and the extraction of medial axis from objects were formulated and solved using morphological operators, as can be seen in [6, 17, 24, 31, 40, 64]. Most of these analysis tools perform well only for noise-free images. Different techniques for edge enhancement and edge detection in noisy images were proposed, mainly using order statistics filters, in [20, 39, 57]. However, the techniques based on optimal Boolean and stack filters are more efficient in removing noise and simultaneously preserving image details, as shown in [79]. In this chapter solutions for edge detection from noisy images are developed using the optimal design of a Boolean filter with a reference filter.

Throughout this chapter our aim is to provide new insights into the applications of optimal Boolean and stack filters in image processing, using efficient design tools. The chapter presents in turn the optimal design, extended filtering architectures, and applications of Boolean and stack filters in the prediction stage for lossless image coding and the edge detection in noisy images.

In Section 2.2, the optimal design of Boolean and stack filters is presented. The problem of the optimal design of Boolean and stack filters under the sum of absolute errors criterion is addressed in a *training approach*, when a representative target image and a noise corrupted version of this image are available for filter design. When additional requirements are imposed on the filter behavior, such as the preservation of certain details or regularities, the design problem is solved by slightly changing some stages of the procedure. The *optimal design with a reference Boolean filter* requires that the output of the optimal Boolean filter in specified noisy environments is as close as possible to the output of a *reference* Boolean filter acting on the uncorrupted image. In all the design settings of this chapter, the optimal design procedures consist of two stages. The first stage deals with computing the coefficients representing the cost for deciding the Boolean function of the filter. In the second stage a fast design procedure is used for deciding the Boolean function for the binary vectors [70].

In Section 2.3 different filtering architectures are presented for image restoration applications using optimally designed stack filters. Parallel multilevel and cascade multistage structures are described. A new combined architecture using one layer of median filters whose outputs are fed into an optimal stack filter is presented as a generalization of multistage filtering configurations using median filters [2, 49].

In Section 2.4 optimal Boolean and stack filters are used in another image processing task, namely in the prediction for lossless image compression. The connections between the entropy of prediction errors and the sum of absolute errors criterion are revealed and a design procedure for Boolean predictors under the Error-Entropy criterion is proposed. The performance of Boolean filters used for prediction in simple and combined architectures is experimentally illustrated.

In Section 2.5 Boolean filters designed with a reference filter are used for edge detection in noisy images. Conclusions regarding the performance achieved by optimal Boolean filters in different image processing tasks are included in the last section.

2.2 BOOLEAN AND STACK FILTERS: BASIC STRUCTURES AND OPTIMAL DESIGN

2.2.1 Boolean and Stack Filters: Definitions

A large number of digital filters share a feature known as the *threshold decomposition* property, which relies on the threshold decomposition of integers into a set of binary values, defined by means of the *thresholding* operator. Thresholding an integer value i at level m is denoted by $\mathcal{T}^{(m)}(i)$ and defined in [12] as

$$\mathcal{T}^{(m)}(i) = \begin{cases} 1, & i \geq m \\ 0, & i < m. \end{cases} \tag{2.1}$$

Any integer x between 0 and M can be represented as a sum of M binary values obtained by thresholding x at all integer values $m \in \{1, \ldots, M\}$:

$$x = \sum_{i=1}^{M} \mathcal{T}^{(m)}(x). \tag{2.2}$$

The output of a filter having the threshold decomposition property can be obtained by decomposing the input signal into a set of binary signals using the threshold decomposition, carrying out the filtering operation separately on each binary signal, and then summing up the results. The filters satisfying the threshold decomposition property can be fully specified in the binary domain by a set of Boolean functions

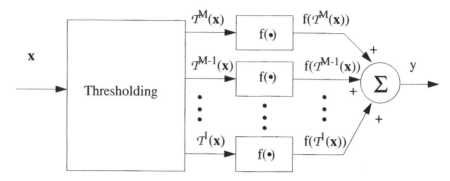

Fig. 2.1   Boolean filtering architecture.

$\{f_m(\cdot)\}_{m=1}^M$, and their output representation is obtained as

$$y = \sum_{m=1}^M f_m(\mathcal{T}^{(m)}(\mathbf{x})), \tag{2.3}$$

where $\mathbf{x}$ is the N-length vector containing the input values in filter's window $\mathcal{W}$, and the threshold operator is applied component-wise to the vector arguments.

Boolean filters, called also threshold Boolean filters in [30], are a class of nonlinear filters possessing the threshold decomposition property. They are specified in the binary domain by a Boolean function $f(\cdot)$, and $f_m(\cdot) = f(\cdot)$ for all values of m, $1 \leq m \leq M$. A Boolean function depending on N logical input variables can be defined either using a logical expression, with logical AND, OR, and negation operations, or using a 2^N-length binary vector representation, which is equivalent to the truth table:

$$\mathbf{f} = [f(\mathbf{b}_1)\, f(\mathbf{b}_2) \dots f(\mathbf{b}_{2^N})]^T, \tag{2.4}$$

where $\mathbf{b}_i$ denotes the N-length binary vector, corresponding to the binary representation of $(i-1)$, i.e., $(i-1) = b_{i_1} 2^0 + b_{i_2} 2^1 + \dots + b_{i_N} 2^{N-1}$.

The output of a Boolean filter is defined as

$$y = BF_{\mathbf{f}}(\mathbf{x}) = \sum_{m=1}^M f(\mathcal{T}^{(m)}(\mathbf{x})), \tag{2.5}$$

where $BF_{\mathbf{f}}$ denotes the Boolean filtering and $\mathbf{f}$ is the vector representation of the filter's Boolean function. The filtering architecture for Boolean filters is presented in Fig. 2.1.

By thresholding, each component of an N-dimensional integer vector $\mathbf{x}$ a set of binary vectors is obtained which obeys the *stacking property* [74]:

$$\mathcal{T}^{(m)}(\mathbf{x}) \leq \mathcal{T}^{(n)}(\mathbf{x}) \; \forall m \geq n \text{ with } m, n \in \{1, \dots, M\}. \tag{2.6}$$

The relation $\mathbf{x} \leq \mathbf{z}$ holds if it is true for every component of the two vectors:

$$\mathbf{x} \leq \mathbf{z} \text{ iff } x_k \leq z_k \ \forall k \in \{1, \ldots N\}. \tag{2.7}$$

It should be noted that a set of maximum $N + 1$ different vectors are obtained by thresholding an N-dimensional vector. These vectors are

$$\{\mathbf{1}, \ \mathcal{T}^{x_{(2)}}(\mathbf{x}), \ \ldots \ \mathcal{T}^{x_{(N)}}(\mathbf{x}), \ \mathbf{0}\}, \tag{2.8}$$

where $x_{(i)}$ denotes the ith order statistic of $\mathbf{x}$ ($x_{(1)} \leq x_{(2)} \leq \cdots \leq x_{(N)}$), and $\mathbf{1}$ and $\mathbf{0}$ denote the all-ones and all-zeros N-dimensional vectors.

For a fast computation of the Boolean filter output, the fact that at most $N + 1$ different binary vectors are obtained in the threshold decomposition can be used [30]. The multilevel output may be computed as

$$y = x_{(1)}f(\mathbf{1}) + \sum_{i=1}^{N-1}(x_{(i+1)} - x_{(i)})f(\mathcal{T}^{(x_{(i+1)})}(\mathbf{x})) + (M - x_{(N)})f(\mathbf{0}). \tag{2.9}$$

Stack filters are a special case of Boolean filters, whose Boolean function satisfies the *stacking property* in [74]:

$$\forall \ \mathbf{x} \leq \mathbf{z} \Rightarrow f(\mathbf{x}) \leq f(\mathbf{z}). \tag{2.10}$$

The Boolean function $\mathbf{f}$ of a stack filter has the property that it is *positive* [41], meaning that it can be represented in the minimum sum of products form using only uncomplemented logical variables. Positive Boolean functions have the property that they commute with thresholding, and the output of the stack filter can be computed directly in the integer domain, using the following equivalences between the logical operators in the binary domain and min and max operations in the integer domain:

$$\min(i, j) = \sum_{m=1}^{M} \mathcal{T}^{(m)}(i) \cdot \mathcal{T}^{(m)}(j),$$

$$\max(i, j) = \sum_{m=1}^{M} \mathcal{T}^{(m)}(i) + \mathcal{T}^{(m)}(j), \tag{2.11}$$

where the product and sum represent the logical *and* and logical *or* operations.

Example

For a Boolean function whose logical expression is $g(\mathbf{x}) = x_1 \cdot x_2 + x_3$, $N = 3$, with x_1, x_2, x_3 logical input variables, the vector representation is $\mathbf{g} = [0\ 0\ 0\ 1\ 1\ 1\ 1\ 1]^T$. The filter is a stack filter, and its output in the integer domain for an integer input vector $\mathbf{x}$ can be computed as $y = BF\mathbf{g}(\mathbf{x}) = \max\{\min\{x_1, x_2\}, x_3\}$.

Further insight into the relation between Boolean and stack filters was provided in [30], where it is shown that any Boolean filter with N-dimensional processing window can be represented as a linear combination of up to $N + 1$ stack filters. An efficient spectral algorithm for representing any Boolean function as a linear combination of positive Boolean functions is proposed in [5]. Statistical analyses of stack filters are performed in [25, 27], and their deterministic properties are studied in [1, 15, 16]. The stack filter class includes a large number of filters with very different behavior, such as the weighted-order statistics filters in [78], the morphological filters with flat structuring elements in [18, 41], the soft morphological filters in [26, 28].

2.2.2   Optimal Design of Boolean and Stack Filters

Optimal design techniques based on the minimization of the mean absolute error (MAE) criterion are developed in [8, 9, 14, 15, 34] for selecting the best filter inside the stack filter class. The MAE criterion was used mainly because the stacking property allows the conversion of the criterion into a sum of absolute errors at each level of the threshold decomposition.

In image filtering applications, the optimal stack filter is designed using a "representative" image and minimizing a criterion that is the sum of absolute errors between the input and a target image [36, 70]. For the optimal design of Boolean filters the sum of absolute error criterion is used at all threshold levels [30]. The design problem for such a filter is specified as follows:

Given

- the deterministic image $\{s(l)\}_{l=1}^{L}$ with pixel values $s(l) \in \{0, 1, \ldots, M\}$, where pixel locations are indexed by $l = 1, \ldots, L$,

- the input signal as a set of vectors $\{\mathbf{x}(l)\}_{l=1}^{L}$, each vector $\mathbf{x}(l)$ containing the values of the pixels processed to yield the output at the pixel indexed by l,

find the Boolean filter $BF_{\mathbf{f}}$, which minimizes, at all threshold levels, the sum of absolute error criterion:

$$J^b = \frac{1}{L} \sum_{l=1}^{L} \sum_{m=1}^{M} |\mathcal{T}^{(m)}(s(l)) - f(\mathcal{T}^{(m)}(\mathbf{x}(l)))|. \qquad (2.12)$$

When the solution is sought inside the stack filter class, this criterion is equivalent to the sum of absolute errors at the integer level, defined as

$$J = \frac{1}{L} \sum_{l=1}^{L} |s(l) - BF_{\mathbf{f}}(\mathbf{x}(l))| = \frac{1}{L} \sum_{l=1}^{L} |\sum_{m=1}^{M} (\mathcal{T}^{(m)}(s(l)) - f(\mathcal{T}^{(m)}(\mathbf{x}(l))))|. \qquad (2.13)$$

Then J is equal to J^b for stack filters since the thresholding and modulus operations commute for positive Boolean functions [8, 70].

The method presented in the following was introduced in [67, 70]. It consists of deriving an optimal solution after preprocessing the available data to extract the needed statistics. The optimal design procedure is organized in two stages. In the first stage, coefficients representing the cost for deciding the filter's Boolean function for the binary vectors are computed from the initial data set. In the second stage, a fast design procedure [70] is used for deciding the Boolean function for the binary vectors.

After substituting the values of $\mathcal{T}^{(m)}(s(l))$ from Eq. (2.1) into Eq. (2.12), the criterion J^b is further processed as

$$
\begin{aligned}
J^b &= \frac{1}{L}\sum_{l=1}^{L}\left(\sum_{m=1}^{s(l)}(1 - f(\mathcal{T}^{(m)}(\mathbf{x}(l)))) + \sum_{m=s(l)+1}^{M} f(\mathcal{T}^{(m)}(\mathbf{x}(l))))\right) \\
&= \frac{1}{L}\sum_{l=1}^{L}\left(s(l) - \sum_{m=1}^{s(l)} f(\mathcal{T}^{(m)}(\mathbf{x}(l)))) + \sum_{m=s(l)+1}^{M} f(\mathcal{T}^{(m)}(\mathbf{x}(l))))\right) (2.14)
\end{aligned}
$$

Alternatively, an equivalent form can be obtained in which J^b is represented as a linear combination of the values of the Boolean function f for all 2^N possible input vectors [67], as in [70]:

$$
J^b = C_0 + \sum_{i=1}^{2^N} c(\mathbf{b}_i) f(\mathbf{b}_i), \tag{2.15}
$$

The coefficient C_0 in this equation is the sample mean value of the target signal,

$$
C_0 = \frac{1}{L}\sum_{l=1}^{L} s(l). \tag{2.16}
$$

Let $\mathcal{M}_0(\mathbf{b}_i)$ be the set of all pairs (l, m) for which $\mathcal{T}^{(m)}(s(l)) = 0$, and $\mathcal{T}^{(m)}(\mathbf{x}(l)) = \mathbf{b}_i$, and $\mathcal{M}_1(\mathbf{b}_i)$ be the set of all pairs (l, m) for which $\mathcal{T}^{(m)}(s(l)) = 1$ and $\mathcal{T}^{(m)}(\mathbf{x}(l)) = \mathbf{b}_i$. The coefficients $\{c(\mathbf{b}_i)\}_{i=1}^{2^N}$, referred to as *cost coefficients*, are computed as joint statistics from the target and input signals using

$$
c(\mathbf{b}_i) = \frac{1}{L}(N_0(\mathbf{b}_i) - N_1(\mathbf{b}_i)) \tag{2.17}
$$

where

$$
N_0(\mathbf{b}_i) = Card(\mathcal{M}_0(\mathbf{b}_i)) \tag{2.18}
$$

$$
N_1(\mathbf{b}_i) = Card(\mathcal{M}_1(\mathbf{b}_i)). \tag{2.19}
$$

A Boolean function $\mathbf{f}^{b*}$ minimizing J^b may be easily found [30, 70]. The values for the Boolean function for the 2^N vectors $\mathbf{b}_i \in \{0, 1\}^N$ are assigned as follows:

$$f^{b*}(\mathbf{b}_i) = \begin{cases} 1 & \text{if } c(\mathbf{b}_i) < 0 \\ 0 & \text{if } c(\mathbf{b}_i) > 0 \\ (\text{don't care}) & \text{if } c((\mathbf{b}_i) = 0. \end{cases} \quad (2.20)$$

The "don't care" positions can be set to either 1 or 0, without affecting the performance cost, and therefore they are assigned taking into account different criteria, e.g., the lowest complexity of implementation of the resulting $\mathbf{f}^{b*}$. If $\mathbf{f}^{b*}$ is positive, then the optimal filter is a stack filter with Boolean function $\mathbf{f}^{b*}$. In this case the optimal Boolean and stack filters are identical. If $\mathbf{f}^{b*}$ is not positive, J and J^b are not equivalent, and the optimal stack and Boolean filters will be different.

If the solution is sought inside the stack filters class, the stacking constraint must be imposed. One way to impose this constraint during minimization is to solve a linear programming problem with $O(N \cdot 2^N)$ variables and constraints [8] (N being the size of the filter window). To avoid the complexity of linear programming, fast procedures are proposed in [70, 81]. In [81] a suboptimal procedure using a compare-and-select algorithm is presented while, a projection procedure into the class of stack filters is developed in [70]. The projection of $\mathbf{f}^{b*}$ into a positive Boolean function $\mathbf{f}^{+*}$ is performed iteratively. In the first step, the positive function $\mathbf{f}^+$ is initialized with the positive function having either all ones or all zeros of its vector representation completely contained in $\mathbf{f}^{b*}$ representation. At each iteration $\mathbf{f}^+$ is modified by adding or deleting one positive term in the minimum sum of products representation as long as the value of the criterion decreases. Linear programming is only an option in the last stage. This procedure, called FASTAF in [70], is extremely fast for small windows, such as 3×3, because $\mathbf{f}^{b*}$ is often already positive. However, for larger window sizes, this advantage disappears.

The Boolean filter $\mathbf{f}^b$ obtained minimizing J^b leads to a lower value of the criterion than that achieved by the optimal (with respect to J) stack filter [30].

An adaptive approach to optimal stack filtering was proposed in [36], and a fast algorithm was derived based on it in [35]. The procedure acts similarly to LMS algorithms, training the filter at each new pair of (input, target) data such as to preserve the stacking property and to advance toward the optimal solution. Nonadaptive and adaptive constrained least mean absolute algorithms were developed for the estimation of stack filters, based on their representation by polynomial separating functions and through the linearization of the unit step function in the design criterion [77]. Many of the solutions derived for the optimal stack filters design problem use linear programming techniques [8, 14].

2.2.3 Optimal Design With a Reference Boolean Filter

Different image processing tasks can be performed, in the noise free case, using Boolean or stack filters. A typical example is edge detection [57, 79]. Edges can be detected by applying a local gradient and then thresholding the result. A simple nonlinear gradient image is obtained by replacing each pixel value with the difference between the local maximum and the local minimum in [58]. The gradient image can be obtained as the output of a Boolean filter or, equivalently, as the difference of the outputs of two stack filters. When noisy images are processed, the local maximum and local minimum are strongly affected by noise, resulting in lack of continuity and false detection of edges (if impulsive noise is present, and it affects a fraction p of image pixels, then in the gradient image, up to $N \cdot p$ of the pixels may be affected, where N is the size of the processing mask). One way to reduce this drawback is to design a Boolean filter able to process the noisy image, such that the filtering result is close to the gradient image obtained in the noise-free case. Additionally, the design solution must converge to the Boolean filter producing the local gradient, in the absence of noise. The design problem for finding such a Boolean filter under the sum of absolute errors criterion at all threshold levels is specified as follows:

Given

- the deterministic image $\{s(l)\}_{l=1}^{L}$ with integer values $s(l) \in \{0, 1, \ldots, M\}$,

- the Boolean function $\mathbf{g}$,

- the input signal as a set of vectors $\{\mathbf{x}(l)\}_{l=1}^{L}$

find the Boolean filter $\mathbf{f_g}$, which minimizes the criterion:

$$J_g^b = \frac{1}{L} \sum_{l=1}^{L} \sum_{m=1}^{M} |g(\mathcal{T}^{(m)}(\mathbf{s}(l))) - f_g(\mathcal{T}^{(m)}\mathbf{x}(l)))|, \qquad (2.21)$$

where $\mathbf{x}(l) = \mathbf{s}(l) + \mathbf{e}(l)$ denotes the vector containing the values processed to yield the output at position l, obtained by adding the noise values in $\mathbf{e}$ to the signal values $\mathbf{s}$ inside the filter mask.

The problem is solved following a similar approach described in Section 2.2.2. The only changes occur in the computation of the cost coefficients, where at each threshold value m, $\mathcal{T}^{(m)}(s)$ is replaced by $g(\mathcal{T}^{(m)}(s))$.

If $\mathbf{g}$ is a positive Boolean function, then the criterion in Eq. (2.21) is equivalent to that in Eq. (2.12) presented in Section 2.2.2, when the target signal is $\{p(l)\}_{l=1}^{L} = \{BF_{\mathbf{g}}(\mathbf{s}(l))\}_{l=1}^{L}$. If $\mathbf{g}$ is not positive, the above equivalences do not hold. The criterion in Eq. (2.21) is not identical to that in Eq. (2.12) because, according to the threshold decomposition and Boolean function definitions, $g(\mathcal{T}^{(m)}(\mathbf{s}))$ may differ from $\mathcal{T}^{(m)}(p)$, when $p = BF_{\mathbf{g}}(\mathbf{s})$. The criterion in Eq. (2.21) introduced in this section has the advantage that, $\mathbf{f} = \mathbf{g}$ in the noise-free case. This is not always the case if the criterion in Eq. (2.12) from Section 2.2.2 is used, as illustrated in the following example.

Example

Consider the Boolean function $x_1 \bar{x}_2$, of two input variables, represented in the vector form as $\mathbf{g}=[0\ 0\ 1\ 0]^T$. Take the input vector, in the noise-free case, $\mathbf{s} = \mathbf{x} = [7\ 5]^T$, and $M = 15$. In this case $p = BF_\mathbf{g}(\mathbf{s}) = 2$. The computation of the cost coefficients using the criteria in Eq. (2.21) and Eq. (2.12) is presented in parallel:

$$\sum_{m=1}^{M} |g(\mathcal{T}^{(m)}(\mathbf{s})) - f(\mathcal{T}^{(m)}(\mathbf{x}))| \qquad \sum_{m=1}^{M} |\mathcal{T}^{(m)}(p) - f(\mathcal{T}^{(m)}(\mathbf{x}))|$$

$$\begin{aligned} c([1\ 1]) &= 5 & c([1\ 1]) &= 1 \\ c([1\ 0]) &= -2 & c([1\ 0]) &= 2 \\ c([0\ 0]) &= 8 & c([0\ 0]) &= 8 \end{aligned}$$

The Boolean functions obtained in the design procedure for the two cases are as follows:

$$\begin{aligned} f_g([1\ 1]) &= 0 & f_g([1\ 1]) &= 0 \\ f_g([1\ 0]) &= 1 & f_g([1\ 0]) &= 0 \\ f_g([0\ 0]) &= 0 & f_g([0\ 0]) &= 0 \end{aligned}$$

and for the binary input vectors contained in the threshold decomposition for this particular case, $f_g(\mathbf{b}_i)=g(\mathbf{b}_i)$ only for the left side, where the criterion in Eq. (2.21) is used.

The optimal design with a reference Boolean function is successfully used for edge detection and skeletonization of noise corrupted images in [51, 52, 79].

2.2.4 Optimal Design Under Structural Constraints

The optimal Boolean and stack filter design problem presented in the previous section was concerned only with the aspect of restoring degraded images. However, this may often result in a loss or degradation of some features or details, to which the human eye is sensitive, and whose preservation should be considered as important as the elimination of noise. Noise attenuation and detail preservation are generally conflicting aspects, and the problem of designing a filter able to achieve "best" noise reduction and at the same time preserve certain desired signal structures is known in the literature under the name of *optimal design under structural constraints* [9, 14].

Structural constraints can be imposed in the design of nonlinear filters in two different ways. One can transform these constraints to the binary domain by threshold decomposition, and constrain the output values of filter's Boolean function for the thresholded set of vectors. This approach is used for stack filters in [9, 14] and for weighted median filters in [37, 76]. The optimization of stack filters under rank selection and structural constraints in a probabilistic approach is presented in [25]. In [80] the structural constraints are imposed in the binary domain, modifying the Boolean function of a standard median filter such that it fulfills the constraints.

Alternatively, the structural constraints can be applied directly in the multilevel domain. This can be achieved using a *mixing criterion* [69] that combines the two parts of the problem: noise rejection and preservation of geometrical structures (*features*)

$$J_{mix} = \lambda J_f + (1 - \lambda)J_n, \tag{2.22}$$

where J_n denotes the criterion for the noise rejection part of the problem, computed for the target deterministic image $\{s_n(l)\}_{l=1}^{L_n}$ [using Eq. (2.12)]. J_f is the criterion for the features preservation part of the problem, computed for the feature data set $\{s_f(l)\}_{l=1}^{L_f}$. $\lambda \in [0, 1]$ is a mixing factor that allows different weights for the penalties of errors in the two sides of the problem.

The mixing criterion can be equivalently represented using mixing coefficients [69]:

$$J_{mix}(\lambda) = C_{0mix} + \sum_{i=1}^{2^N} c_{mix}(\mathbf{b}_i)f(\mathbf{b}_i), \tag{2.23}$$

where

$$
\begin{aligned}
C_{0mix} &= \lambda C_{0f} + (1 - \lambda)C_{0n}, \\
c_{mix}(\mathbf{b}_i) &= \lambda c_f(\mathbf{b}_i) + (1 - \lambda)c_n(\mathbf{b}_i), \quad \forall i \in \{1, \ldots, 2^N\}. \tag{2.24}
\end{aligned}
$$

The computation of the cost coefficients c_f, c_n is performed as in Section 2.2.2.

When λ varies between 0 and 1, a finite number of different solutions are obtained. The solution may change when $c_{mix}(\mathbf{b}_i)$ changes sign, and this happens only when $c_f(\mathbf{b}_i)$ and $c_n(\mathbf{b}_i)$ have different signs and λ crosses the value:

$$\lambda_{change}(i) = \frac{c_n(\mathbf{b}_i)}{c_n(\mathbf{b}_i) - c_f(\mathbf{b}_i)}. \tag{2.25}$$

The solution emphasizing the most the effect of the structural constraints will be obtained for $\lambda \in [\max_i\{\lambda_{change}(i)\}, 1]$.

2.3 FILTERING ARCHITECTURES FOR IMAGE RESTORATION USING BOOLEAN AND STACK FILTERS

Optimal stack filters are successfully used in image restoration mainly when images are contaminated by impulsive noise [15]. The performance of optimal stack filters improves when the size of the processing mask increases [70]. However, even by using fast design procedures, the processing mask size for optimal stack filters is

(a) Cascade multistage filtering architecture

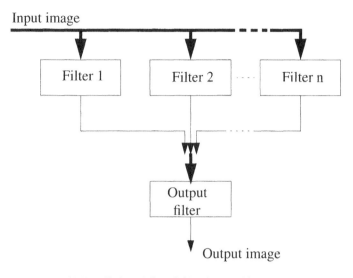

(b) Parallel multilevel filtering architecture

Fig. 2.2 Combined filtering architectures: (a) cascade and (b) parallel filtering architectures. Bold arrows correspond to a processing mask and thin arrows correspond to one pixel.

limited (e.g., no more than 16 input variables), due to memory requirement and computation time which grow exponentially with the number of input variables.

One possibility for improving filtering performance while still using small processing masks, is to utilize filtering architectures combining several filters to yield the final output. Two types of combined architectures with optimal stack filters have been proposed: *cascade multistage optimal stack filters* [68, 82] and *parallel multilevel filtering architectures using optimal stack filters* [55, 66, 84]. The corresponding architectures are presented in Fig. 2.2. The filters in these architectures often have different processing masks, oriented along directions at which the human visual system is most sensitive to details, such as horizontal and vertical directions.

2.3.1 Cascade Multistage Optimal Stack Filters

The **cascade multistage filtering architecture** connects several filtering stages in a sequential order, such that each filtering stage can be performed only after the preceding one is completed as in Fig. 2.2(a). This filtering architecture is equivalent to a single filtering stage using a processing mask obtained by morphologically dilating

(a)   (b)

Fig. 2.3 Equivalent processing masks of combined filtering architectures: (a) cascade multistage filtering architecture, and (b) parallel multilevel filtering architecture.

the masks at each filtering stage. An example of the composition of processing masks is shown in Fig. 2.3(a).

The optimal design for the overall cascade multistage stack filtering architecture is equivalent to the design of a filter with the equivalent processing mask; see Fig. 2.3(a). It is computationally intractable for large window sizes. However, this particular filtering structure allows the design of an optimal stack filter for each stage, using a data set containing the ideal image and the result of filtering the noise corrupted image in the previous stages. This architecture was successfully used in image restoration as shown in [68, 82].

2.3.2   Parallel Multilevel Filtering Architectures Using Optimal Stack Filters

In a **parallel multilevel filtering architecture** each filtering level consists in several filtering processes carried out in parallel. The outputs are collected to form the processing mask for the next level. The overall filtering architecture is equivalent to one filter having a processing mask equal to the union of the masks of the parallel filters. A parallel two-level filtering architecture is sketched in Fig. 2.2(b) and an example of the composition of processing masks is shown in Fig. 2.3(b).

Many nonlinear filters introduced by different authors can be viewed as particular cases of the parallel multilevel filtering architecture. When the filters, Filter 1, . . ., Filter n, in Fig. 2.2(b) are median filters with small processing masks, and the output filter is a median filter, the overall filter was shown to achieve better detail preservation than the median filter directly acting on the equivalent processing mask. Such architectures were introduced in [49] and analyzed in [2]. Other publications propose modifications of multilevel median architectures to improve their performance [61, 73].

Another class of parallel multilevel filtering architectures uses linear FIR filters (with various processing masks) in the first level of Fig. 2.3(b) and a nonlinear filter (median, stack filter, Boolean filter) as the output filter. The basic FIR-median hybrid structure in [21] can be represented as in Fig. 2.3(b), where Filter 1 and Filter 3 are averaging filters, Filter 2 is the identity filter and the output filter is the median filter. These filters are used for preserving details along averaging masks. Several FIR-median filters, with differently oriented masks, can be used in a tree structure [49] to make the filters more responsive to orientations of fine details. The FIR stack hybrid filter, introduced in [84], further generalizes the FIR median hybrid filters. The first level consists in averaging filters with different masks, and the output filter is a stack filter. When the stack filter is optimally designed, the performance obtained in image restoration applications is superior to that of FIR median hybrid filters. An even more general FIR hybrid structure is the FIR-Boolean hybrid filter, proposed in [54]. A Boolean filter is used to process the outputs of FIR linear filters transforming the image in parallel. This structure was successfully used in image prediction for lossless coding as will be shown in Section 2.4.

A parallel multilevel stack filtering architecture, generalizing the multilevel median filters, was proposed in [66]. All filters in the structure in Fig. 2.3(b) are stack filters. It was shown that this type of architecture, involving optimal stack filters for small masks, is able to outperform multilevel median filtering architectures.

Parallel multilevel architectures achieve good detail preservation when details are oriented along the processing mask of one of the filters. In the architectures presented above the processing masks were fixed, nonadaptive, and the filtering performance may be nonsatisfactory for images containing mostly details oriented in other directions than those spanned by the selected processing masks.

A different type of parallel multilevel architecture, the $\mathcal{L}$–M–S filter in [55], designed for image restoration applications, was proposed to reduce the drawback of using fixed processing masks. The $\mathcal{L}$–M–S filter class aims to generalize previous multilevel median filtering architectures, combining the good properties of median filters for small processing masks with the fine tuning properties of stack filters for larger processing masks. N median filters using different masks, chosen from a library of processing masks $\mathcal{L}$, are applied in parallel to the input image. A stack filter processes the outputs of the median filters, yielding the overall output. This architecture is motivated by the following properties observed when analyzing the behavior of previous multilevel filtering architectures:

- The filtering performance of multilevel median filters [2] can be improved if the output of the median filters acting in parallel over the image are processed by an optimal stack filter.

- The optimal stack filter for 3-pixels processing masks, having different orientations, is in most cases the median filter.

- Median or optimal stack filters for one-directional processing masks achieve the highest filtering performance if the orientation of the processing mask is parallel to the dominant edges in the image.

The $\mathcal{L}$–M–S filters are stack filters with an equivalent processing mask resulted from the union of the masks selected for the median filters. The $\mathcal{L}$–M–S filter architecture is presented in Fig. 2.4. The optimal $\mathcal{L}$–M–S filter in a class (defined

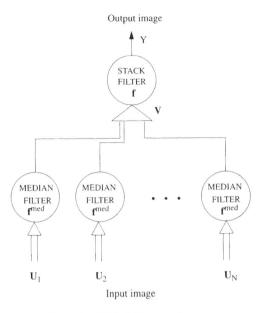

Output image

Fig. 2.4 $\mathcal{L}$–M–S filter architecture.

by the library of selectable processing masks $\mathcal{L}$ and the number of median filters N) can be designed using a recursive procedure, which combines the selection of different masks with the optimal stack filter design. At each iteration one more processing mask for the median filtering is selected from the library. The outputs of the previously selected and newly added median filers are fed to the stack filter. The procedure in [70] is used for optimal stack filter design. The added processing mask is always the one among all masks of the library $\mathcal{L}$ which minimizes the sum of absolute errors in the output image with respect to the ideal uncorrupted image. The recursive procedure has the main advantage of avoiding the computational complexity involved by an exhaustive search for the global optimum. Only $N \cdot P$ stack filter designs are performed, where the number of input variables for the stack increases from 2 to N. P denotes the number of processing masks in $\mathcal{L}$. The performance achieved in image restoration by this filtering architecture was experimentally shown to approach the filtering performance achieved when optimal stack filters with larger masks are used.

The $\mathcal{L}$–M–S filters include the previously proposed multilevel median filtering architectures, when the processing masks for the median filters are fixed, and the stack filter is also a median filter. However, the $\mathcal{L}$–M–S filters have a more flexible architecture, allowing more degrees of freedom to fit particular image features.

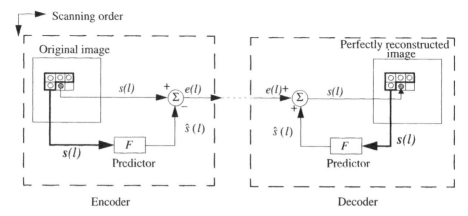

Fig. 2.5   Sequential prediction method.

2.4   PREDICTION METHODS BASED ON BOOLEAN FILTERS FOR LOSSLESS IMAGE COMPRESSION

Image compression[1] techniques are essential tools for efficient transmission and archiving of image data. Two main classes of compression methods exist, the discriminative feature being their ability to preserve the complete information contained in the original image: reversible (lossless), for which perfect reconstruction of the image is possible and irreversible (lossy), which focuses on obtaining higher compression rates but allows only approximate reconstruction. The bitrates reached in lossless compression are far from the spectacular compression bitrates attained by lossy methods (tenths to hundredths of the uncompressed image bitrate). However, there are several applications where the complete recovery of the original image is strictly required (e.g., medical or satellite imaging [3, 44, 60, 62]) and thus efficient lossless compression is a highly desired objective.

The "spatial redundancy" is the key property of image data that is exploited differently by various compression methods in order to achieve good compression rates. One way to take advantage of this redundancy is to figure out a *prediction model* very accurate for the image and then to encode and transmit only the unpredictable information, namely the difference between the real (true) and the predicted pixel values, as illustrated in Fig. 2.5.

In this section a brief presentation of prediction methods used in lossless image compression is given and it is shown that Boolean and stack filters, optimally designed using the sum of absolute errors criterion, can be used, with good compression performance, for prediction. A good explanation for this behavior can be found in

---

[1]Parts of this section have been published by Kluwer Academic Publishers in *Multidimensional Systems and Signal Processing*, vol.10, pp. 161-187, April 1999. They are included in this chapter by copyright permission from Kluwer Academic Publishers.

the connection between the sum of absolute prediction errors and the entropy of these errors, which is a lower bound of the bitrate obtained using an independent source model for encoding the errors. An Error-Entropy optimal design procedure can be derived based on the fast procedures presented in Section 2.2.2.

2.4.1   Prediction Methods for Lossless Image Compression

For lossless image compression, two of the major types of prediction methods are *the sequential prediction methods* and *the multiresolution prediction methods* [44, 60, 62].

2.4.1.1 Sequential prediction methods   The prototype of *sequential prediction method* consists in using for prediction the pixels inside a causal mask (in a raster scan order), and obtaining the prediction value as a combination of these pixels, as illustrated in Fig. 2.5.

The number of pixels inside the prediction mask is usually selected as a trade-off between the prediction accuracy and computational cost. Most of the currently used prediction schemes employ **linear predictors**. The predicted value for each position, denoted by $\hat{s}(l)$ is computed as

$$\hat{s}(l) = \sum_{i=1}^{N} a_i s_i(l), \qquad (2.26)$$

where $\mathbf{s}(l) = \{s_1(l), \ldots, s_N(l)\}$ denotes the vector containing the values of the pixels inside the prediction mask, and $a_1, \ldots, a_N$ are the coefficients of the linear predictor. Linear predictors have been used in different prediction schemes, with either fixed or adaptive coefficients, under different adaptive algorithms.

Fixed linear predictors use fixed coefficients which must be known at the receiver (they do not vary during transmission). Seven simple and very efficient linear predictors are used in JPEG still image compression standard [72]. These predictors sometimes use a prediction mask with only one or two pixels. Although these predictors are very simple and do not adapt to the image content, they are standardized and the most commonly used.

Optimal linear predictors differ from fixed predictors in that they use optimal coefficients for weighting the values of the pixels inside the prediction mask. The optimal coefficients are obtained by minimizing the mean square of prediction errors for the data to be transmitted. The latter does not necessarily minimize the bitrate. Moreover they require the transmission of prediction coefficients to the receiver, along with the prediction errors. In [44] it is emphasized that optimal linear predictors do not generally outperform fixed prediction schemes.

Adaptive linear predictors are more efficient prediction methods than fixed or optimal prediction methods, as the coefficients are adapted to local changes in image characteristics. Several proposals for a new lossless image compression standard presented in [45] include such adaptive predictors. For the adaptation of the coefficients they use either vertical and horizontal gradients in the neighborhood of the current pixel or the prediction errors corresponding to the pixels inside the prediction mask. Some of these schemes perform coefficients adaptation based on the local statistics. The drawback of such adaptive schemes is that they are computationally more expensive than nonadaptive schemes. More efficient adaptive schemes are obtained by hard adaptation, that is, switching the predictor between several fixed linear predictors based on the computation of local statistics at each position. An example of such a switched linear prediction scheme is the gradient adjusted predictor [75], where the selection is performed according to the amplitude of local vertical and horizontal gradients. This type of predictor achieves the best performance in terms of bitrate of prediction errors among all other linear predictors presented in the literature [45, 75].

Block adaptive prediction schemes were proposed to overcome the difficulty incurred by the computation of statistics for each pixel [46]. The image is subdivided into small blocks, and for each block a linear predictor is selected from a given set of linear predictors.

Nonlinear predictors based on order-statistics filters were proposed and shown to outperform linear prediction methods in [33, 43, 47]. In [43] the median of a set of three linear predictors (selected from JPEG standard) is used to predict the current pixel value. According to the comparative study in [45] this nonlinear prediction scheme achieves the lowest bitrates for prediction errors among several types of linear switched and adaptive predictors.

2.4.1.2 Multiresolution prediction methods The sequential prediction methods are restricted by causal constraints to use only partially the two-dimensional spatial information. One way to overcome this disadvantage is to use *multiresolution*, or *hierarchical*, or *pyramidal* methods where each level of the hierarchy represents the entire image, at different spatial resolutions [11]. Hierarchical prediction comprises a number of different stages of sequential prediction where the original image is subsampled at different resolutions. The lowest resolution level is encoded using a sequential prediction method. For the higher resolution levels the prediction, also referred to as *interpolation*, makes use of the 360 degree information available at immediately lower resolution. More accurate prediction is obtained for the higher-resolution levels.

Various multiresolution prediction structures using linear predictors have been proposed [11, 29, 63]. A multiresolution prediction structure starting with the original image subsampled horizontally and vertically by a factor of 2 is presented in Fig. 2.6. The principle is similar when starting with lower resolutions (subsampling by factors of 8, 4, or 2 have been commonly used for various predictors). Each stage has its own scanning order, prediction mask, and predictor.

In the first stage, the pixels obtained by subsampling the original image by 2 horizontally and vertically (circles in Fig. 2.6) are sequentially predicted, using a causal prediction mask. This stage achieves the transmission of the subsampled version of the original image to the decoder. Stage II will continue with the sequential

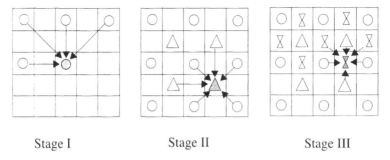

Stage I   Stage II   Stage III

Fig. 2.6 Predicting at different resolution levels.

prediction of the pixels at locations represented by triangles in Fig. 2.6. The prediction mask may now contain any combination of pixels (not restricted by causality) from the image transmitted in Stage I and from the pixels already available in the sequential Stage II. The prediction accuracy is significantly increased at this stage, due to the availability of pixels in the subsampled image (the term interpolation is often used instead of prediction, due to the apparent noncausality of the prediction mask). Half of the image pixels are available at the end of the second stage of sequential prediction. For the remaining pixels, predicted in Stage III, the information available is even richer than in the second stage. Six pixels out of eight in 8-connectivity are available and therefore the prediction will have a very high accuracy.

Different interpolation techniques based on median filters and FIR median hybrid filters have been proposed and shown to perform better than linear interpolative techniques [32, 83]. We analyze the performances of MAE-optimal Boolean and FIR-Boolean hybrid filters for multiresolution prediction and compare them with other reported results.

2.4.2 Optimal Boolean and Stack Filters Under the Entropy of Errors (EE) Criterion

2.4.2.1 Modeling the entropy of prediction errors Consider an integer-valued image with L pixels, the current pixel being denoted $s(l)$, where the lexicographic ordering is used to transform the 2-D image domain to 1-D vectors:

$$s(l) \in \{0, 1, \ldots, M\}, \ l = 1, \ldots, L.$$

The processing window $\mathcal{N}_{s(l)}$ selects N pixels forming the input vector $\mathbf{s}(l)$ for predicting $s(l)$. $\mathcal{N}_{s(l)}$ is to be causal with respect to the given scanning order. We propose to use a Boolean filter for predicting the value of $s(l)$ using the information from $\mathbf{s}(l)$ [30, 70]:

$$\hat{s}(l) = \sum_{m=1}^{M} f(\mathcal{T}^{(m)}(\mathbf{s}(l))), \ l = 1, \ldots, L. \tag{2.27}$$

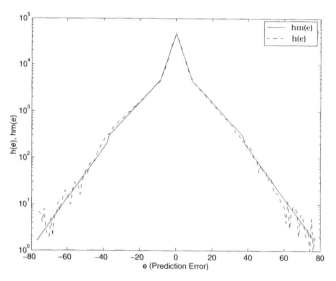

Fig. 2.7 Modeling the histogram of prediction errors using piecewise exponential functions. (- -) $h(e)$ is the experimental histogram (image Barb1, Boolean predictor, with 10-pixel mask); (–) $h_m(e)$ is a fitted model to the histogram, with $k = 3$ exponentials: $\lambda_1 = 0.2680$, $\lambda_2 = 0.0968$, $\lambda_3 = 0.1245$.

A good evaluation of the lower bound on the average number of bits needed to encode the prediction error:

$$e_f(l) = s(l) - \sum_{m=1}^{M} f(T_m(s(l))) \tag{2.28}$$

is the entropy

$$H(p_{e_f}) = - \sum_{e=-(M)}^{M} p_{e_f}(e) \log_2 p_{e_f}(e), \tag{2.29}$$

where $p_{e_f}(e)$ is the relative frequency of occurrence of the prediction error value e, $e \in \{-M, -(M-1), \ldots, 0, \ldots, (M-1), M\}$. Denoting by $h_f(e)$ the histogram of the prediction errors of an image

$$h_f(e) = Card\{l | e_f(l) = e\}, \tag{2.30}$$

the relative frequency of error occurrence will be $p_{e_f}(e) = h_f(e)/L$, and hence

$$H(h_f) = \log_2 L - \frac{1}{L} \sum_{e=-M}^{M} h_f(e) \log_2 h_f(e) \tag{2.31}$$

$$= \log_2 L - \frac{1}{L} \sum_{l=1}^{L} \log_2 h_f(e_f(l)), \tag{2.32}$$

where Eq. (2.32) is simply a rephrasing of Eq. (2.31), but with a more useful interpretation, as the sample mean over the image of the quantity $\log_2 h_f(e_f(l))$. In order to introduce the modeling environment for computing the entropy the following aspects regarding the histogram $h_f(e_f(l))$ are considered.

The histograms of natural images manifest a high variability (various number of local maxima, types of monotonicity between local minima and maxima: ramp, exponential, etc.) [59]. Totally different patterns are encountered with the prediction error histograms. They are highly regular for the range of medium and small errors; they present a symmetrical lobe, centered at $e = 0$ and have monotonic sides, the monotonicity being piecewise exponential. The very large errors occur only seldom, and no pattern can be fit on this range of the histogram (Fig. 2.7). Therefore we will model the histogram as a piecewise symmetrical doubly exponential function:

$$\hat{h}_\Theta(e) = \begin{cases} \mu_1 2^{-\lambda_1 |e|} & \text{for } \theta_0 = 0 \le |e| < \theta_1 \\ \mu_2 2^{-\lambda_2 |e|} & \text{for } \theta_1 \le |e| < \theta_2 \\ \dots & \dots \\ \mu_p 2^{-\lambda_p |e|} & \text{for } \theta_{p-1} \le |e| < \theta_p = M, \end{cases} \tag{2.33}$$

where Θ groups all parameters of Eq. (2.33), $\Theta = \{\theta_1, \dots, \theta_{p-1}, \mu_1, \dots, \mu_p, \lambda_1, \dots, \lambda_p\}$. The case $p = 1$ corresponds to the Laplacian probability distribution of the errors and has been long advocated when analyzing the errors in differential pulse code modulation (DPCM) structures [22]. The fitting of $\hat{h}_\Theta(e)$ to an experimental histogram, $h_f(e)$ can be accomplished by a least square (LS) procedure minimizing the criterion:

$$
\begin{aligned}
J_{h_f}(\Theta) &= \sum_{|e|=0}^{M} \left(\log_2(h_f(e)) - \log_2(\hat{h}_\Theta(e)) \right)^2 \tag{2.34} \\
&= \sum_{|e|=0}^{\theta_1 - 1} (\log_2(h_f(e)) - \log_2 \mu_1 + \lambda_1 |e|)^2 + \dots \\
&\quad + \sum_{|e|=\theta_{p-1}}^{\theta_p - 1} (\log_2(h_f(e)) - \log_2 \mu_p + \lambda_p |e|)^2, \tag{2.35}
\end{aligned}
$$

where the number of segments, p, and the segment boundaries, $\{\theta_1, \dots, \theta_{p-1}\}$, are selected by the user (e.g., by inspection of the curve h_f), while the parameters $\{\mu_1, \dots, \mu_p\}$ and $\{\lambda_1, \dots, \lambda_p\}$ will immediately result by solving the least square estimation problem in Eq. (2.34) with $2p$ parameters, which in turn, can be decoupled to minimizing p LS criteria of Eq. (2.35), each depending on two parameters.

2.4.2.2   *The EE-optimal Boolean prediction problem*   We denote by Θ_f the optimal parameters minimizing Eq. (2.34) for the experimental histogram h_f. Let us denote

$$\mathcal{E}_k(f) = \{l | \theta_{k-1} \le |e_f(l)| < \theta_k\}, \tag{2.36}$$

the set of pixel locations resulting when modeling the histogram h_f with the model in Eq. (2.33) for $p = 2$. Now the entropy of the prediction error in Eq. (2.32) can be written as

$$
\begin{aligned}
H(h_f) &= \log_2 L - \frac{1}{L} \sum_{l=1}^{L} \log_2 h_f(e_f(l)) && (2.37) \\[2mm]
&\approx \log_2 L - \frac{1}{L} \sum_{l=1}^{L} \log_2 \hat{h}_{\Theta_f}(e_f(l)) \\[2mm]
&= \log_2 L - \frac{1}{L} \left\{ \sum_{l \in \mathcal{E}_1(f)} (\log_2 \mu_{1f} - \lambda_{1f} |e_f(l)|) \right. \\[2mm]
&\qquad\qquad \left. + \sum_{l \in \mathcal{E}_2(f)} (\log_2 \mu_{2f} - \lambda_{2f} |e_f(l)|) \right\} && (2.38) \\[2mm]
&= \mu_{0f} + \frac{1}{L} \left\{ \sum_{l \in \mathcal{E}_1(f)} \lambda_{1f} |e_f(l)| + \sum_{l \in \mathcal{E}_2(f)} \lambda_{2f} |e_f(l)| \right\}, && (2.39)
\end{aligned}
$$

where

$$\mu_{0f} = \log_2 L - \frac{1}{L} Card(\mathcal{E}_1(f)) \log_2 \mu_{1f} - \frac{1}{L} Card(\mathcal{E}_2(f)) \log_2 \mu_{2f}. \tag{2.40}$$

The connection between MAE criterion and EE criterion   Observe that in the case $p = 1$ of modeling the histogram using only one exponential function (or, equivalent, using p exponentials with the same rates $\lambda_1 = \ldots = \lambda_p$), the entropy reduces to

$$H(h_f) \approx \mu_{0f} + \frac{\lambda_{1f}}{L} \sum_l |e_f(l)|. \tag{2.41}$$

In our experiments we found that the parameters μ_{0f} and λ_{1f} do not depend significantly on f in a neighborhood around the filter minimizing $\sum_l |e_f(l)|$, that is, the entropy will depend almost linearly on the sum of absolute errors criterion. This explains why generally a low entropy of error is obtained with MAE-optimal designed Boolean predictors, optimal when fitting a $p = 1$ model to the experimental histogram.

The connection between weighted MAE criterion and EE criterion   When using the histogram model with $p = 2$, improvements over the case $p = 1$ are expected. The problem of designing a predictor optimal with respect to the entropy of the errors is equivalent to finding the Boolean function f^* that minimizes the criterion

$$H(h_f) \approx \mu_{0f} + \frac{1}{L} \left\{ \sum_{l \in \mathcal{E}_1(f)} \lambda_{1f} |e_f(l)| + \sum_{l \in \mathcal{E}_2(f)} \lambda_{2f} |e_f(l)| \right\}. \quad (2.42)$$

Note that the criterion

$$H_{WMAE}(f) = \frac{1}{L} \left\{ \sum_{l \in \mathcal{E}_1} \lambda_1 |e_f(l)| + \sum_{l \in \mathcal{E}_2} \lambda_2 |e_f(l)| \right\} \quad (2.43)$$

is a weighted MAE criterion for some given sets $\mathcal{E}_1, \mathcal{E}_2$ and constants λ_1, λ_2 and obviously is much simpler to minimize than the criterion in Eq. (2.42).

Exact solution of the minimization problem   We introduce the EE-optimal Boolean predictor as the predictor minimizing the highly nonlinear criterion in Eq. (2.37). However, the exact solution to the minimization problem involves extensive search over the set of all 2^{2^N} predictors. The evaluation of the criterion in Eq. (2.37) for each filter is clearly nontractable for common values of the prediction mask size ($N \approx 10$).

2.4.2.3   EE-suboptimal design procedure

An effective technique for minimizing the criterion in Eq. (2.37) is to freeze the sets $\mathcal{E}_1(f)$ and $\mathcal{E}_2(f)$ to the sets $\mathcal{E}_1(f^0)$ and $\mathcal{E}_2(f^0)$ corresponding to the MAE-optimal solution, f^0, and, instead of solving the modeling problems that find $\lambda_1(f^{k-1})$ and $\lambda_2(f^{k-1})$, to only check the filters minimizing the criterion in Eq. (2.43) corresponding to any values of the pairs (λ_1, λ_2). The problem is thus reduced to carefully planing the search over a reduced set of functions f, which minimize the criterion in Eq. (2.43) for different values of λ_1 and λ_2. Since only a small number of such optimal f exists, this technique will prove more efficient than exhaustive search.

Moreover, since the criterion in Eq. (2.43) was obtained by a series of approximations of the criterion in Eq. (2.37), the final solution will be selected from the limited search space of functions f, and it will be the one minimizing the criterion in Eq. (2.37).

First we concentrate on finding the set of functions f, which are minimum points of the criteria in Eq. (2.43), for all values of λ_1 and λ_2. The criterion in Eq. (2.43) has the same mathematical form as the criterion found in the optimization of nonlinear filters with structural constraints [69], and we customize in the following that procedure for the prediction setting. The criterion to be minimized in the predictor design is the following upper bound of H_{WMAE} in Eq. (2.43):

$$J^b(f) \;=\; \frac{\lambda_1}{L} \sum_{l \in \mathcal{E}_1} \sum_{m=1}^{M} |\mathcal{T}^{(m)}(s(l)) - f(\mathcal{T}^{(m)}(\mathbf{s}(l)))|$$

$$+ \frac{\lambda_2}{L} \sum_{l \in \mathcal{E}_2} \sum_{m=1}^{M} |\mathcal{T}^{(m)}(s(l)) - f(\mathcal{T}^{(m)}(\mathbf{s}(l)))|. \tag{2.44}$$

The cost coefficients for the binary contexts $\mathbf{b}_i \in \{0,1\}^N$ for the two sets of pixels are

$$c^{[\mathcal{E}_k]}(\mathbf{b}_i) = N_0^{[\mathcal{E}_k]}(\mathbf{b}_i) - N_1^{[\mathcal{E}_k]}(\mathbf{b}_i), \quad k = 1,2, \tag{2.45}$$

where $N_0^{[\mathcal{E}_k]}(\mathbf{b}_i)$ and $N_1^{[\mathcal{E}_k]}(\mathbf{b}_i)$ are defined by Eqs. (2.18) and (2.19). We additionally restrict $l \in \mathcal{E}_k$. Using the mixing cost coefficients for the binary vectors $\mathbf{b}_i \in \{0,1\}^N$, we have

$$c^{[mix]}(\mathbf{b}_i) \;=\; \lambda_1 c^{[\mathcal{E}_1]}(\mathbf{b}_i) + \lambda_2 c^{[\mathcal{E}_2]}(\mathbf{b}_i), \tag{2.46}$$

and the criterion in Eq. (2.44) becomes

$$J^b(f) = C_0^{[mix]} + \frac{1}{L} \sum_{i=1}^{2^N} c^{[mix]}(\mathbf{b}_i) f(\mathbf{b}_i), \tag{2.47}$$

where $C_0^{[mix]}$ does not depend on f. The predictor f^{λ_1,λ_2} minimizing J^b in Eq. (2.47), for fixed λ_1 and λ_2, is obtained by assigning the values of $f^{\lambda_1,\lambda_2}(\mathbf{b}_i)$ according to the sign of $c^{[mix]}(\mathbf{b}_i)$ as in Eq. (2.20).

There are $K \leq 2^N$ positive numbers, $\{\nu_1, \ldots, \nu_K\}$ with $\nu_0 = 0 < \nu_1 < \nu_2$ $\ldots < \nu_K < \nu_{K+1} = \infty$, such that to any $\lambda_1, \lambda_2 > 0$ with $\lambda_1/\lambda_2 \in (\nu_k, \nu_{k+1})$, there corresponds the same Boolean function f^{λ_1,λ_2} minimizing J^b in Eq. (2.47). The cost coefficients in Eq. (2.47) are obtained from Eq. (2.46) with fixed pixel sets $\mathcal{E}_1$ and $\mathcal{E}_2$.

This result is proved by the fact that the Boolean function minimizing the criterion in Eq. (2.47), $f^{\lambda_1,\lambda_2}(\mathbf{b}_i)$, is assigned according to the sign of $c^{[mix]}(\mathbf{b}_i) = \lambda_1 c^{[\mathcal{E}_1]}(\mathbf{b}_i) + \lambda_2 c^{[\mathcal{E}_2]}(\mathbf{b}_i) = \lambda_2(\lambda_1/\lambda_2 c^{[\mathcal{E}_1]}(\mathbf{b}_i) + c^{[\mathcal{E}_2]}(\mathbf{b}_i))$, which in turn depends only on the ratio $\nu = \lambda_1/\lambda_2$. If the value

$$\nu_{change} = -\frac{c^{[\mathcal{E}_2]}(\mathbf{b}_i)}{c^{[\mathcal{E}_1]}(\mathbf{b}_i)} \tag{2.48}$$

is positive, then there will be a change in $f^{\lambda_1,\lambda_2}(\mathbf{b}_i)$ only when $\nu = \lambda_1/\lambda_2$ crosses the value ν_{change}. Now we collect all the positive ν_{change} corresponding to all $\mathbf{b}_i$, (at most 2^N numbers) and then order them, obtaining the string $\nu_0 = 0, \nu_1, \ldots, \nu_K$, $\nu_{K+1} = \infty$. For any pair λ_1, λ_2 with $\lambda_1/\lambda_2 \in (\nu_k, \nu_{k+1})$, $k = 0, 1, \ldots, K$, the predictor f^{λ_1,λ_2} minimizing the criterion in Eq. (2.47) will be the same.

The possible solutions f^{λ_1, λ_2} will change from the MAE-optimal Boolean predictor for the pixels in the set $\mathcal{E}_1$ (easily predictable), to the MAE-optimal Boolean predictor for the pixels $\mathcal{E}_2$ (difficult to predict). The number $K + 1$ of different solutions f^{λ_1, λ_2} is in practice much lower than 2^N, since ν_{change} in Eq. (2.48) is usually negative for most of the binary contexts $\mathbf{b}_i$ (in our present prediction experiments we found on average 4 different predictors for $N = 3$ and about 200 different predictors for $N = 10$). Table 2.1 summarizes the main steps of the EE-suboptimal design procedure presented in this paragraph. We will denote in the sequel, for short, this procedure by *EE-optimal design*, even if the result of the procedure may be only suboptimal, due to all the approximations in the derivation of the algorithm.

However, since the partition of the image into the sets $\mathcal{E}_1(f^0)$ and $\mathcal{E}_2(f^0)$ corresponds to the optimal MAE solution, the resulting solution of our procedure will have entropy less than or equal to that of the MAE-optimal filter, as was experimentally confirmed in all our experiments. We can easily check the entropy in Eq. (2.31) for the different filters to find the best filter to use as predictor. This search can be made more efficient by observing that the scalar λ parameterizes the filters to be tested, and therefore efficient line search methods can be used in Step 5 (e.g., the golden section method).

2.4.3 Optimal Boolean, Stack, and FIR-Boolean Hybrid Filters in Different Prediction Schemes

Certain features of optimal Boolean and stack predictors for small prediction masks ($N < 8$) make them attractive in DPCM compression schemes : (1) the procedure for the optimal design is simpler than the procedure for linear filter design; (2) the code-length necessary for transmitting the filter parameters is smaller for Boolean and stack filters than for linear filters. For these reasons the optimal stack and Boolean filters used for sequential prediction are well tuned to the image to be transmitted, while the cost incurred by the transmission of predictor information is not so high, globally resulting in an efficient lossless compression.

The two-level filtering architecture including linear and Boolean filters, the FIR-Boolean hybrid filters, can be successfully used for prediction. In the first level, several linear predictors are used in parallel. The outputs of the linear combinations of the values inside the prediction mask are then fed into a Boolean filter, which yields the predicted value. The Boolean filter is optimally designed for each image, while the linear predictors are fixed.

We performed several experiments in order to analyze the alternatives in setting predictor types (Boolean, stack, FIR-Boolean hybrid filters), structural parameters (size of prediction mask), and prediction type (sequential or multiresolution). We experimented with images from the (standard) JPEG test image set to obtain a meaningful comparison of our procedure with other previously reported prediction methods. We take as a measure of performance the entropy H of the errors (given by (2.29), (2.31), or (2.32)), which is a lower bound on the required code length per pixel obtained with independent source coding techniques (Huffman, arithmetic

Table 2.1   EE-(sub)optimal design procedure

Step 1 Find the MAE-optimal Boolean predictor f^0 by using Eqs. (2.18)-(2.20).

Step 2 Find the sets $\mathcal{E}_1$ (easily predictable pixels) and $\mathcal{E}_2$ (difficult to predict pixels), for the predictor f^0 by using Eq. (2.36).

Step 3 Compute the cost coefficients $\{c^{[\mathcal{E}_1]}(\mathbf{b}_i)\}_{i=1}^{2^N}$, $\{c^{[\mathcal{E}_2]}(\mathbf{b}_i)\}_{i=1}^{2^N}$ by using Eq. (2.45).

Step 4 Find the values of $\nu = \lambda_1/\lambda_2 \in [0, \infty)$ where f^{λ_1, λ_2} may change:

For each binary context $\mathbf{b}_i$: if $c^{[\mathcal{E}_1]}(\mathbf{b}_i)$ and $c^{[\mathcal{E}_2]}(\mathbf{b}_i)$ have different signs, then compute

$$\nu_{change} = -\frac{c^{[\mathcal{E}_2]}(\mathbf{b}_i)}{c^{[\mathcal{E}_1]}(\mathbf{b}_i)}.$$

Order increasingly the values ν_{change} obtaining the sequence $\nu_1, \ldots, \nu_K$. Assign $\nu_0 = 0, \nu_{K+1} = \nu_K + 1$.

Step 5 For each $k = 1, \ldots, K+1$

- compute the mixing factors $\lambda_2 = 1$ and $\lambda_1 = (\nu_{k-1} + \nu_k)/2$
- compute the set of mixed cost coefficients $\{c^{[mix]}(\mathbf{b}_i)\}_{i=1}^{2^N}$ by using Eq. (2.46)
- assign the Boolean function f^{λ_1, λ_2} as in Eq. (2.20)

$$f^{\lambda_1, \lambda_2}(\mathbf{b}_i) = \begin{cases} 0 & c^{[mix]}(\mathbf{b}_i) > 0 \\ 1 & c^{[mix]}(\mathbf{b}_i) \leq 0 \end{cases}$$

- compute the entropy of errors for the predictor f^{λ_1, λ_2} by using Eqs. (2.30) and (2.31).

Step 6 Select the best, with respect to Eq. (2.31), predictor obtained in the search at Step 5.

---

coding). However, the code length needed to encode the predictors themselves has to be taken into account, when this code length becomes significant (note that when using an optimal predictor, say of size $N = 10$, for the overall image, the code length for encoding the predictor will be only 0.0025 bits per pixel for the 576×720 images used).

Table 2.2 Summary of the best performances: the entropy of prediction error for different predictors.

Image	JPEG	MedAP	GAP MAE Optimal	Boolean (10) EE Optimal	Boolean (10) Sequential	FBH(6,8) Multi-resolution	FBH(6,8)
Balloon	3.172	3.122	3.046	3.100	3.018	3.024	2.986
Barb1	5.302	5.208	5.137	5.007	4.973	5.061	4.934
Barb2	5.236	5.182	5.190	5.056	5.036	5.047	4.991
Boats	4.469	4.313	4.286	4.293	4.265	4.256	4.221
Girl	4.225	4.210	4.103	4.108	4.055	4.075	4.015
Goldhill	4.875	4.717	4.676	4.708	4.696	4.690	4.678
Hotel	4.943	4.735	4.661	4.637	4.625	4.649	4.639
Zelda	4.179	4.116	3.951	3.964	3.918	3.968	3.844
Average	4.550	4.450	4.363	4.359	4.323	4.346	4.288

Fig. 2.8 Ten-pixels and six-pixels prediction masks used in Boolean and FIR-Boolean hybrid predictors; the black circle represents the predicted pixel.

We summarize in Table 2.2 the best prediction results obtained in our global prediction approach and compare them with the performance of standard algorithms (the best global and block JPEG [44]), the median adaptive predictor (MedAP) proposed in [43], and the best prediction results we found in the literature, the gradient adjusted predictor (GAP) used by CALIC system [75]. The median adaptive predictor uses 3-pixels prediction masks, and it is the median of three values: the nearest two pixels and a linear combination of the three pixels. This is a form of fixed FIR-Boolean hybrid predictor. The gradient adjusted predictor uses a 6-pixels prediction mask, and it is an adaptive linear predictor, selecting one from a set of fixed linear combinations of pixels. The selected combination is decided based on the value of a local gradient containing information about the presence, the amplitude, and direction of local edges.

In Table 2.2, FBH(6,8) denotes the FIR-Boolean hybrid filter with a 6-pixels prediction mask and 8 inputs to the Boolean filter. The inputs in the Boolean filter stage consist of the 6 values of the pixels in the mask, and two additional linear combinations of these values, involving usually the pixels that are spatially nearest to the predicted one. Two of the JPEG linear predictors were used in the sequential prediction case. The 10-pixels and 6-pixels prediction masks used in our experiments are presented in Fig. 2.8.

In multiresolution prediction the prediction/interpolation masks for the second and third stages are those presented in Fig. 2.6. In the first stage, the sequential prediction is achieved for the subsampled image using the same type of predictors as in sequential methods.

From our experiments we conclude that Boolean filters, and mainly FIR-Boolean hybrid filters optimally designed achieve good prediction performance and are an interesting alternative for optimal prediction methods, which deserve further investigation on how to combine the prediction stage with a most suitable error coding method. In addition multiresolution methods have similar performance with sequential prediction methods and they provide useful progressive decoding methods, which are especially convenient in image transmission.

In the following subsections we discuss block-optimal prediction as an alternative method to tune optimal predictors to local characteristics and reduce the errors.

2.4.3.1 *Fixed size block-optimal prediction*   Locally adaptive (block-optimal) Boolean filters are introduced in [53] for image restoration applications, where they are shown to outperform the one-block optimal Boolean filters. Similar improvement can be obtained in the present prediction application, but at the cost of increasing the predictor complexity.

The image is subdivided into small blocks, and one MAE-optimal Boolean predictor is fitted to each block. All the predictors must be transmitted altogether with the prediction errors, in order to recover the image at the decoder.

The prediction efficiency is measured using a cumulative global criterion, $\mathcal{R}$, which includes both the entropy of the residual image and the additional bitrate needed to encode each local predictor. For the regular partition of the image into n_b equally sized blocks, the cumulative criterion must be computed as

$$\mathcal{R} = H + 2^N \cdot \frac{n_b}{T}, \tag{2.49}$$

where H is the entropy of the errors (computed using the histogram of the *whole error image*), N denotes the prediction window size, and 2^N is the number of bits necessary for coding each predictor.

From our experiments [56] we concluded that block-optimal Boolean prediction with equally sized blocks is a good solution for small prediction masks (3-5 pixels), where the cost for transmitting the optimal predictor for each block stays small. Block-optimal Boolean prediction does not improve the coding performance for large prediction masks. Use of many block-optimal Boolean filters with small prediction masks (on many blocks) results in values for the performance criterion $\mathcal{R}$ close to those obtained when only one optimal Boolean filter with a large prediction mask (for the whole image) is used.

From both a conceptual and an algorithmic point of view, the regular block decomposition of the image is simple, but it may be inefficient. For example, in flat areas different filters for each block may not improve much the prediction

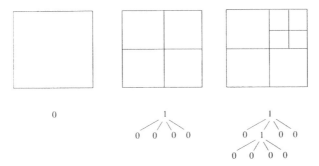

Fig. 2.9   Quadtree codes for three types of blocks.

performance, despite the high price paid for transmitting each filter. In areas with a high density of details, the splitting into very low size blocks might improve significantly the overall criterion $\mathcal{R}$. A tree procedure is used adaptively setting the size in the image partition is presented in the following paragraph.

2.4.3.2   Adaptive size block-optimal prediction   Quadtrees are widely used combined with image compression techniques [71]. We use a top-down strategy to perform an adaptive block size partition. We start with one large block, and we check whether subdividing it into four equally sized subblocks would yield improvement in performance (by reducing the global criterion $\mathcal{R}$). The highest weight in the criterion $\mathcal{R}$ is given by the entropy of prediction errors for the whole image, H. As a result the decision on further subdividing each block becomes highly dependent on how the rest of the image is already partitioned, and hence this decision may be proved only suboptimal.

Reiterating the top-down splitting decisions starting from the final partition obtained in the first top-down pass may provide better decisions, but with an increase in computational effort. We start the procedure using an initial partition of the image into blocks of equal (medium) size, $L \times L$ (32×32 in our experiments), for which the global criterion $\mathcal{R}$ is computed from Eq. (2.49). The image will be scanned, block by block. A recursive procedure decides whether or not to split the current block and one bit, b, is allocated to encode the decision for each subblock. A quadtree is built and coded for each $L \times L$ block, as shown in Fig. 2.9.

Quadtree procedure for subdividing images for prediction   Below we outline the recursive splitting procedure. We denote by $\mathcal{B}$ the block that is currently being checked whether to split it or not. We denote by $h_{out}(e)$ the histogram of the errors at pixel locations outside block $\mathcal{B}$.

1. Compute the optimal predictor $f_{\mathcal{B}}$ for the overall block and $f_{\mathcal{B}_1}, \ldots, f_{\mathcal{B}_4}$ the optimal predictors for the subblocks $\mathcal{B}_1, \ldots, \mathcal{B}_4$.

2. Compute the prediction errors and the histograms $h_B(e)$, $h_{B_1}(e), \ldots, h_{B_4}(e)$ and the entropy of these errors for one block, $H_{\text{no-split}} = H(h_{out}(e) + h_B(e))$, and for the four blocks $H_{\text{split}} = H(h_{out}(e) + h_{B_1}(e) + \ldots + h_{B_4}(e))$.

3. Decide to split the block if

$$H_{\text{no-split}} > H_{\text{split}} + \frac{1}{L}(3 \times 2^N + 4), \tag{2.50}$$

where 3×2^N is the cost for extra-coding the predictors for the four subblocks and that for transmitting the code for the quadtree branching is 4 bits.

4. If the block is split, and if it is not at the lowest level accepted for splitting (4×4 in our experiments), iterate the present procedure for each of the resulting subblocks.

We experimented using this procedure for adaptive size block-optimal Boolean and FIR-Boolean hybrid predictors. Prediction masks with 3 to 5 pixels were used. Images were first divided into regular blocks of size $L \times L = 32 \times 32$ and the deepest level where block-splitting was allowed was set to 4×4 pixel blocks. The quadtree block dividing procedure can be performed for all the $L \times L$ blocks, but since the search for adaptive splitting is only suboptimal, the procedure was repeated several times, starting from the partition found after a complete pass through the image. The iterations were stopped when no additional improvement in the cumulative criterion $\mathcal{R}$ was obtained (2 to 4 iterations are usually required). The results showed that the average values of the cumulative criterion are smaller than those when fixed block sizes are used. Fig. 2.10 shows the final partition of image Zelda obtained using the quadtree procedure.

The best prediction performance in our experiments was obtained with block-optimal FIR-Boolean hybrid filters for small prediction masks of 3 to 6 pixels (the adaptive size block-optimal prediction slightly outperforming regular block-optimal prediction).

2.4.4   Gradient Adaptive Boolean Prediction

Usually prediction schemes have poor accuracy near edges in images. The amplitude and direction of edges can thus be used in the procedure for adaptive design of different predictors for different types of edges, as presented in [45].

The optimal Boolean predictor can be adapted to achieve better performance by using the local gradient computed in a causal neighborhood of the current pixel. First the overall optimal Boolean predictor is designed for an image on a medium processing mask (10 pixels in our experiments) in a first pass through the image and transmitted to the decoder. This predictor is used for flat regions and if no adaptation has been already performed. During the encoding (decoding) process, the presence of an edge is detected using the gradient obtained as the difference between the local maximum and minimum within a small mask, which is included

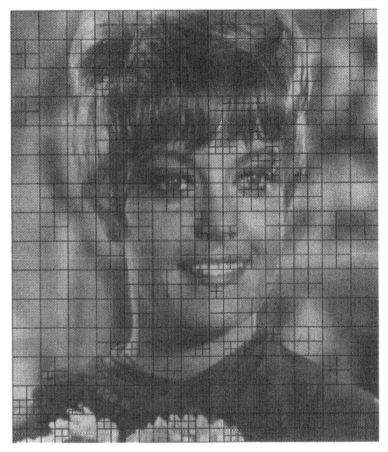

Fig. 2.10 Final partition of image Zelda resulted using the adaptive size block-optimal Boolean prediction with 4-pixels prediction mask.

in the initial processing mask. The amplitude of the gradient is compared to a threshold, which is decided from the histogram of the gradient values inside the image. If the presence of an edge is detected, the Boolean predictor is selected from several predictors, adaptively designed during the encoding (decoding) process. Two switching schemes are proposed. The first scheme uses for switching the positions of maximum and minimum values inside the small processing mask. The second scheme switches between different Boolean predictors based on the binary pattern obtained by thresholding the small prediction mask at its midrange (the mean between the maximum and minimum values).

In the experiments, a 6-pixels mask was used for edge detection and 4 additional pixels were used for the optimal Boolean predictor design. The detection and prediction masks are those presented in Fig. 2.8.

This scheme has the advantage that no additional information is transmitted besides the optimal Boolean predictor itself. In addition the prediction performance is improved while only a low increase in computational complexity is required. When the small prediction mask contains 6 pixels, 30 adaptive predictors are used for the case of the local gradient switching scheme, and 64 predictors for the binary pattern switching scheme.

2.4.5   Lossless Compression Scheme Based on Boolean Predictors and Contextual Modeling of Prediction Errors

The compression capabilities of adaptive Boolean predictors can be assessed by including them into a complete compression scheme. It was shown previously that adaptive prediction alone cannot completely remove the redundancy of the pixel values inside the processing mask [45]. The contextual modeling of prediction errors can therefore further reduce the bitrate. In order to efficiently encode prediction errors, a technique for *error feedback and modeling* was proposed in [45] and [75]. A similar technique of modeling prediction errors with compound *energy* and *texture* contexts can be used in conjunction with the adaptive Boolean predictors. The *energy* contexts are obtained by quantizing to 16 levels the amplitude of the gradient (max-min) over the 6-pixels processing mask. The quantization levels were selected such that the errors are evenly distributed. The *texture* contexts are obtained as binary patterns by thresholding at the midrange value the pixels inside the 6-pixels processing mask (like in local gradient adaptive prediction), resulting in a total number of 1024 contexts. In the *error feedback*, the average error value for each context is computed and subtracted from the real error, resulting in the residual to be transmitted. The average error value for each context is updated during the encoding/decoding process. Finally, the prediction residuals are encoded by an m-ary arithmetic coder using the 16 energy contexts. The arithmetic coder is a public domain package[2] based on the work developed in [48].

The entropy of prediction errors after error feedback and the bitrate after arithmetic encoding for the images in the test set are presented in Table 2.3.

The proposed scheme was used to compress a medical image database, containing colored anatomical images in a 24-bits/pixels raw format. Each image takes more than 7 Mbytes in the uncompressed representation. Experiments performed over four such images have shown that an average compression ration of 50% is achieved with the proposed scheme, compared to a 44% compression ratio achieved by JPEG lossless compressor and 31% compression ratio obtained with Lempel-Ziv compressors.

---

[2]http://www.cs.mu.oz.au/~alistair/arith_coder/

Table 2.3   Lossless coder using gradient adaptive Boolean prediction: entropy of errors after contextual modeling and feedback, bitrate after arithmetic encoding

Image	Entropy of Errors	Bitrate
Balloon	2.86	2.95
Barb1	4.44	4.54
Barb2	4.55	4.66
Boats	3.87	3.96
Girl	3.82	3.91
Goldhill	4.44	4.54
Hotel	4.28	4.39
Zelda	3.77	3.85
Average	4.00	4.10

2.5   EDGE DETECTION BASED ON BOOLEAN FILTERS

Edges denote regions of rapid changes in local image properties, such as object boundaries, shadow boundaries, and changes in material properties. The edge detection problem was addressed under different instances, and many solutions were proposed. In a broad sense "edge detection" refers to the process of detecting discontinuities, edge extraction and edge linking. In a narrow sense the term "edge detection" refers only to the process of detecting the discontinuities in the image plane. Edges in images are usually modeled as step functions, and differential operators seem to be well suited to enhance them; the difference image is thresholded at a suitably selected value to yield the edge image. Real images are usually corrupted by noise, and a filtering stage is necessary in order to suppress noise before applying the differential operators. Edges are usually modeled by fixed-height step immersed in noise, which is generally assumed to have a Gaussian distribution. Different criteria have been proposed for achieving good detection and localization of edges. When the noise contains impulses (outliers), such as noise produced in the acquisition or transmission of images, the edge detection solution may utilize nonlinear impulse removing filters. The following two types of solutions use nonlinear filters for detecting edges from images corrupted with impulsive noise:

1. Prefiltering the image for impulsive noise cancellation. The use of median filters may provide acceptable solutions for filtering images when there are no fine details. If detail preservation is desired, stack filters can provide acceptable performance. Several authors have proposed order statistics based filters for edge enhancing and noise smoothing. LUM (lower-upper-middle) filters and compare and select filters can switch the output between several ranked values of the pixels inside a neighborhood of the current pixel. The rank of the value of the current pixel, e.g., in [19] and [20] or the average value of the pixels in the neighborhood are used for switching.

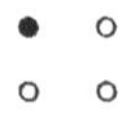

Fig. 2.11   Four-pixels processing mask for detecting thin edges using the morphological gradient. The reference pixel is the black pixel.

2. Using nonlinear operators with low-noise sensitivity for the gradient computation. This solution was proposed in [57, 79]. In [57] the use of quasi-range operator is proposed for the computed local gradient. The quasi-range k, defined as the difference between the kth largest and kth smallest pixel values inside the processing mask, exhibits good robustness under noise conditions when the value of k is large. In [79] a filtering structure, denoting *the difference of estimates*, was proposed to reduce the noise sensitivity of the nonlinear gradient. In this approach the local minimum and local maximum are replaced by optimally designed stack filters. The edge detector is the difference between the outputs of the two stack filters. These filters are able to achieve a better detail preservation than order statistics filters. For the noise-free case the filters reduce to the local maximum and local minimum. The optimal design with a reference stack filter was performed in [79] using the training approach under the MAE criterion and the procedure in [36].

2.5.1   Stack Filters and Morphological Edge Operators

The dilation and erosion residues were introduced in [31] as edge operators whose outputs can be thresholded to obtain image edge maps. The dilation residue is formed by computing the difference between the dilation of an image and the image itself. The erosion residue is defined as the difference between the image and its erosion. The stack filters that are equivalents to erosion and dilation are the minimum, and the maximum values inside the processing mask, respectively. These operators lead to spatially biased estimates of the edge locations, as the dilation residue enhances the outer boundary of bright objects on a dark background, and the erosion residue enhances the inner boundary of bright objects on a dark background. In order to eliminate this bias, the morphological gradient, defined as the sum of dilation and erosion residues, is used:

$$g(\mathbf{x}) = \max(\mathbf{x}) - \min(\mathbf{x}). \tag{2.51}$$

The output of the morphological gradient depends on the size and shape of the processing mask. For 3×3 masks, as was used in [31], thick edges are generated. In [79] the same differential operator was used for a 2×2 processing mask, shown in Fig. 2.11. The reference pixel is arbitrarily considered at position as the black pixel. The output of the operator should be ideally assigned to the center of the 2×2 mask. Unfortunately, that is not a position in the grid of the image, so we move the output

Fig. 2.12 Twelve-pixels processing mask for edge detection using stack filters in noisy images. The reference pixel is the black pixel.

to the position corresponding to the black pixel. The edges obtained are thin, but the edge operator defined using maximum and minimum is very sensitive to noise. To reduce the effect of noise, in [79] a new difference operator, denoted **the difference of estimates**, was proposed. The operator is defined as

$$DoE(X) = S_d(\mathbf{x}) - S_e(\mathbf{x}). \tag{2.52}$$

This is the difference between two stack filters, S_d and S_e. These filters can be designed as optimal stack filters with a reference filter. The reference filter consists of the local maximum and the local minimum. The objective is that the difference of estimates achieves a good approximation of the difference between the 2×2 maximum and the 2×2 minimum in the noiseless image. Large processing masks are needed to design S_d and S_e. In [52] a processing mask containing 12 pixels, as shown in Fig. 2.12 is used.

In Fig. 2.13 edge detection is applied to a noisy synthetic image. The quasi-range edge detector for a 3×3 processing mask, $x_{(8)} - x_{(2)}$, and the optimal Boolean filter with reference Boolean function have been used and the results have been thresholded. Figure 2.14 presents the edges detected in noisy Lenna and Peppers images shown in (a), by using the quasi-range (b), and the optimal Boolean filters (c) designed for the synthetic image in Fig. 2.13.

2.6 CONCLUSIONS

This chapter treats the optimal Boolean and stack filter design for simple and combined architectures focusing novel applications in image processing: lossless compression and edge detection.

The optimal design of Boolean and stack filters is addressed for the case where a training set containing a target image and its noise-corrupted version is available. The optimal design with a reference Boolean function is introduced and used in applications where filters are designed to simultaneously achieve noise rejection and a desired processing effect, specified by the reference function.

The optimal design procedure proposed is based on two processing stages. In the first stage, the threshold level statistics are computed over the whole data set

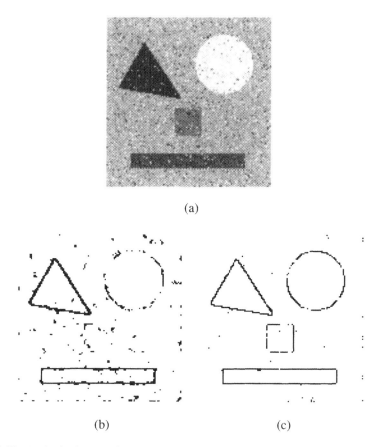

(a)

(b)   (c)

Fig. 2.13 (a) Noisy image. (b) Edges detected using quasi-range edge detector. (c) Edges detected using optimal Boolean edge detector.

while in the second stage, the Boolean filter minimizing the sum of absolute errors at all threshold levels is found. If the solution is searched inside stack filters class, a fast algorithm projects the Boolean filter onto the class of positive Boolean functions. The projection algorithm is faster than two well-known stack filter design algorithms: the nonadaptive linear programming procedure [8] and the fast adaptive algorithm described in [35].

A new solution to the optimal Boolean and stack filter design under structural constraints is proposed. This solution, using a mixed criterion that linearly combines the two parts of the problem, namely the noise rejection and the detail preservation, allows more flexibility than the traditional solutions, where the constraints were imposed directly in the binary domain [9, 14, 25, 37].

The design of the optimal Boolean and stack filters in image restoration becomes unfeasible for large processing masks, due to the exponential growth of the filter parameters with the mask size. For this reason different architectures have been

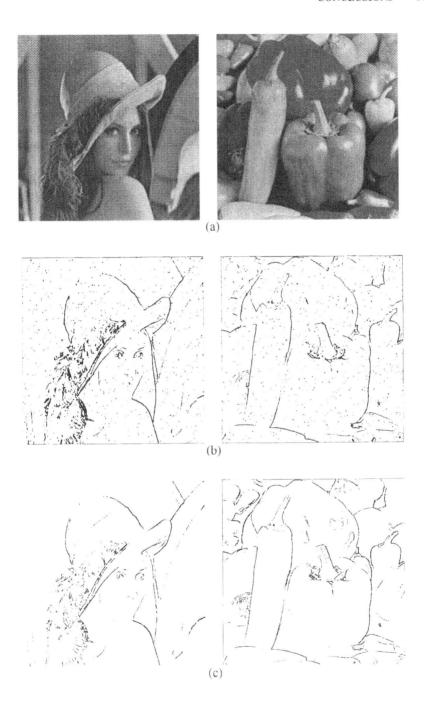

Fig. 2.14   (a) Noisy images. (b) Edges detected using quasi-range edge detector. (c) Edges detected using optimal Boolean edge detector.

proposed in the past that involve several filtering levels of optimal stack filters [66, 82]. A new and flexible multilevel filtering architecture aiming at overcoming the drawback of small processing masks, namely the $\mathcal{L}$–M–S filtering structure was presented. This structure consists of one level of several selectable median filters and an optimal stack filter in the second level, generalizing traditional multilevel median filtering architectures [2, 49]. The flexibility of this architecture comes from the possibility to select different filters for the first processing level from a library of median filters. A tractable procedure can be used for selecting the median filters from the library.

Optimal Boolean and FIR-Boolean hybrid filters are used in another area of image processing: prediction for lossless compression. The connections between the sum of absolute errors and the entropy of prediction errors, when the distribution of prediction errors is piecewise exponential, are revealed. A procedure to design Boolean filters under the Error-Entropy criterion is proposed based on these connections. The performance of optimal Boolean and FIR-Boolean hybrid predictors in sequential and hierarchical structures, using nonadaptive and block-adaptive predictors is investigated. Simulation results show that the proposed predictors outperform classical fixed linear and nonlinear predictors [45, 72]. Adaptive Boolean prediction based on local gradient reduces even more the entropy of prediction errors. The compression results obtained using the proposed predictors for a medical image database encourages the use in lossless compression schemes of this type of predictors. However, the redundancy is not completely removed from the prediction errors, and other methods should be investigated for further reducing the entropy of encoded data.

The optimal design with a reference Boolean function is used for edge detection in noisy images. In this application noise rejection is a supplementary requirement besides the desired feature extraction task. The edge detectors obtained using optimal Boolean filters achieve a more flexible trade-off between noise rejection and detail preservation than range edge detectors [57], the latter being well-known edge detectors based on nonlinear filters.

The major contribution of this chapter is in the exploitation of the efficient design procedures for optimal Boolean and stack filters to fundamental image processing tasks. For other applications, such as image segmentation, motion estimation, processing of color images or video sequences, solutions based on these optimal filters deserve further investigation.

REFERENCES

1. S. Agaian, J. Astola, and K. Egiazarian. *Binary Polynomial Transforms and Nonlinear Digital Filters*. Dekker, NY, 1995.

2. G.R. Arce and R.E. Foster. Detail preserving ranked-order based filters for image processing. *IEEE Trans. Acoustics, Speech, and Signal Processing*, 37(1): 83-98, January 1989.

3. R.B. Arps and Th.K. Truong. Comparison of international standards for lossless still image compression. *Proc. IEEE*, 82(6): 889-899, June 1994.

4. J. Astola and P. Kuosmanen. *Fundamentals of Nonlinear Digital Filtering*. CRC Press, Boca Raton, FL, 1997.

5. J.T. Astola, K.O. Egiazarian, and D.Z. Gevorkian. Spectral methods for threshold Boolean filtering. In *Proc. SPIE Visual Communications and Image Processing*, Vol. 2501, Taipei, May 1995, pp. 221-230.

6. J.A. Bangham and R.V. Aldridge. Multiscale decomposition using median and morphological filters. In *Proc. IEEE Winter Workshop Nonlinear Digital Signal Processing*, Tampere, Finland, January, 1993, pp. 6.1-1–6.1-4.

7. E.J. Coyle. Rank order operators and the mean absolute error criterion. *IEEE Trans. Acoustics, Speech, and Signal Processing*, 36(1): 63-76, January 1988.

8. E.J. Coyle and J.-H. Lin. Stack filters and the mean absolute error criterion. *IEEE Trans. Acoustics, Speech, and Signal Processing*, 36(8): 1244-1254, August 1988.

9. E.J. Coyle, J.-H. Lin, and M. Gabbouj. Optimal stack filtering and the estimation and structural approaches to image processing. *IEEE Trans. Acoustics, Speech, and Signal Processing*, 37(12): 2037-2066, December 1989.

10. E.R. Dougherty and J. Astola. *An Introduction to Nonlinear Image Processing*, Vol. TT16. SPIE Optical Engineering Press, Bellingham, WA, 1994.

11. T. Endoh and Y. Yamazaki. Progressive coding scheme for multilevel images. In *Proc. Picture Coding Symp.*, 1986, pp. 21-22.

12. J.P. Fitch, E.J. Coyle, and N.C. Gallagher Jr. Threshold decomposition of multi-dimensional ranked order operations. *IEEE Trans. Circuits and Systems*, 32(5): 445-450, May 1985.

13. M. Gabbouj and E.J. Coyle. Minimax optimization over the class of stack filters. In *Proc. SPIE Visual Communications and Image Processing*, Vol. 1360, Lausanne, Switzerland, October 1-4 1990, pp. 143-154.

14. M. Gabbouj and E.J. Coyle. Minimum mean absolute error stack filtering with structural constraints and goals. *IEEE Trans. Acoustics, Speech, and Signal Processing*, 38(6): 955-968, June 1990.

15. M. Gabbouj, E.J. Coyle, and N.C. Gallagher Jr. An overview of median and stack filtering. *Circuits, Systems and Signal Processing*, 11(1): 7-45, 1992.

16. M. Gabbouj, P.T. Yu, and E.J. Coyle. Convergence behavior and root signal sets of stack filters. *Circuits, Systems and Signal Processing*, 11(1): 171-194, 1992.

17. C. Giardina and E.R. Dougherty. *Morphological Methods in Signal and Image Processing*. Prentice-Hall, Englewood Cliffs, NJ, 1988.

18. R. Haralick, S. Sternberg, and X. Zhuang. Image analysis using mathematical morphology. *IEEE Trans. Pattern Analysis and Machine Intelligence*, 9(4): 532-550, July 1987.

19. R. Hardie and C.G. Boncelet. Gradient-based edge detection using nonlinear edge enhancing prefilters. *IEEE Trans. Image Processing*, 4(11): 1572-1577, November 1995.

20. R.C. Hardie and C.G. Boncelet. LUM filters: A class of rank order based filters for smoothing and sharpening. *IEEE Trans. Signal Processing*, 41(3): 1061-1076, March 1993.

21. P. Heinonen and Y. Neuvo. FIR- median hybrid filters. *IEEE Trans. Acoustics, Speech, and Signal Processing*, 35(7): 832-838, June 1987.

22. A.K. Jain. *Fundamentals of Digital Image Processing*. Englewood Cliffs, Prentice-Hall, NJ 1989.

23. D.H. Kang, J.H. Choi, Y.H. Lee, and C. Lee. Applications of a DPCM system with median predictors for image coding. *IEEE Trans. Consumer Electronics*, 38(3): 429-435, August 1991.

24. R. Kresch and D. Malah. Multi-parameter skeleton decomposition. In J. Serra and P. Soille, eds., *Mathematical Morphology and Its Applications to Image Processing*. Kluwer Academic, Dordrecht, 1994, pp. 141-148.

25. P. Kuosmanen and J. Astola. Optimal stack filters under rank selection and structural constraint. *Signal Processing*, 41(3): 309-338, February 1995.

26. P. Kuosmanen and J. Astola. Soft morphological filtering. *J. Math. Imaging and Vision*, 5: 231-262, September 1995.

27. P. Kuosmanen and J. Astola. Breakdown points, breakdown probabilities, mid-point sensitivity curves, and optimization of stack filters. *Circuits, Systems and Signal Processing*, 15(2): 165-211, 1996.

28. P. Kuosmanen, L. Koskinen, and J. Astola. The connection between generalized soft morphological filters and stack filters. In *Proc. IEEE Winter Workshop Nonlinear Digital Signal Processing*, Tampere, Finland, January 1993, pp. 236-239.

29. H. Lee, Y. Kim, and S. Oh. Lossless compression of medical images by prediction and classification. *Optical Engineering*, 33(1): 160-166, January 1994.

30. K.D. Lee and Y.H. Lee. Threshold Boolean filters. *IEEE Trans. Signal Processing*, 42(8): 2022-2036, August 1994.

31. S.J. Lee, R.M. Haralick, and L.G. Shapiro. Morphologic edge detection. *IEEE J. Robotics and Automation*, 3(2): 142-156, April 1987.

32. A. Lehtonen and M. Renfors. Nonlinear quincunx interpolation filtering. In *Proc. SPIE Visual Communications and Image Processing*, Vol. 1360, Lausanne, 1990, pp. 131-141.

33. C-C. Lien, C-L. Huang, and I-C. Chang. A new ADPCM image coder using frequency weighted directional filter. In *Proc. SPIE Visual Communications and Image Processing*, Vol. 2501, Taipei, Taiwan, 1995, pp. 647-657.

34. J.-H. Lin and E.J. Coyle. Generalized stack filters and mean absolute error nonlinear filtering. *IEEE Trans. Acoustics, Speech, and Signal Processing*, 38(4): 663-678, April 1990.

35. J.-H. Lin and Y.-T. Kim. Fast algorithms for training stack filters. *IEEE Trans. Signal Processing*, 42(4): 772-781, April 1994.

36. J.-H. Lin, T.M. Sellke, and E.J. Coyle. Adaptive stack filtering under the mean absolute error criterion. *IEEE Trans. Acoustics, Speech, and Signal Processing*, 38(6): 938-954, June 1990.

37. Y. Lin, R. Yang, M. Gabbouj, and Y. Neuvo. Weighted median filters: A tutorial. *IEEE Trans. Circuits and Systems*, 43(3): 157-192, March 1996.

38. H.G. Longbotham and A.C. Bovik. Theory of order statistics filters and their relantionship to linear FIR filters. *IEEE Trans. Acoustics, Speech, and Signal Processing*, 37(2): 275-287, February 1989.

39. H.G. Longbotham and D. Eberly. The WMMR filters: A class of robust edge enhancers. *IEEE Trans. Signal Processing*, 41(4): 1680-1685, April 1993.

40. P. Maragos. Pattern spectrum and multiscale shape representation. *IEEE Trans. Pattern Analysis and Machine Intelligence*, 11(7): 1228-1244, July 1989.

41. P. Maragos and R.W. Schafer. Morphological filters - Part II: their relation to median, order statistics and stack filters. *IEEE Trans. Acoustics, Speech, and Signal Processing*, 35(8): 1170-1184, Augusr 1987.

42. P. Maragos and R.W. Schafer. A representation theory for morphological signal and image processing. *IEEE Trans. Pattern Analysis and Machine Intelligence*, 11(6): 586-599, June 1989.

43. St. Martucci. Reversible compression of HDTV images using median adaptive prediction and arithmetic coding. In *Proc. 1990 IEEE Int. Symp. Circuits and Systems*, New Orleans, LA, 1990, pp. 1310-1313.

44. N.D. Memon and K. Sayood. Lossless image compression: A comparative study. In *Proc. SPIE Still Image Compression*, Vol. 2418, San Jose, CA, 1995, pp. 8-20.

45. N.D. Memon and K. Sayood. A comparison of prediction schemes proposed for a new lossless image compression standard. In *Proc. 1996 IEEE Int. Symp. Circuits and Systems*, Vol. II, Atlanta, GA, 1996, pp. 309-312.

46. N.D. Memon, V. Sippy, and X. Wu. An asymmetric lossless image compression technique. In *Proc. 1995 IEEE Int. Conf. Image Processing*, Vol. III, Washington, DC, 1995, pp. 97-100.

47. R. Mickos, X. Song, T. Sun, T.G. Campbell, and Y. Neuvo. Median structures for image sequence prediction. *IEEE Trans. Consumer Electronics*, 38(4): 795-804, November 1992.

48. A. Moffat, R.M. Neal, and I.H. Witten. Arithmetic coding revisited. Presented of 1995 Data Compression Conf., June 1995.

49. A. Nieminen, P. Heinonen, and Y. Neuvo. A new class of detail-preserving filters for image processing. *IEEE Trans. Pattern Analysis and Machine Intelligence*, 9(1): 74-91, January 1987.

50. D. Petrescu. *Optimal Design and Image Processing Applications of Boolean, Stack and Morphological*. Ph.D. dissertation. Tampere University of Technology, 1997.

51. D. Petrescu, I. Tăbuş, and M. Gabbouj. Adaptive skeletonization using multistage Boolean and stack filtering. In *Proc. VII European Signal Processing Conf.*, Edinburgh, UK, September 1994, pp. 951-954.

52. D. Petrescu, I. Tăbuş, and M. Gabbouj. Edge detection based on optimal stack filtering under given noise distribution. In *Proc. 1995 European Conf. Circuit Theory and Design*, Istanbul, Turkey, August 1995, pp. 1023-1026.

53. D. Petrescu, I. Tăbuş, and M. Gabbouj. Locally adaptive techniques for stack filtering. In *Proc. VIII European Signal Processing Conf.*, Trieste, Italy, September 1996, pp. 587-590.

54. D. Petrescu, I. Tăbuş, and M. Gabbouj. FIR-Boolean hybrid filtering architecture and applications to image processing. In *Proc. 1997 Finnish Signal Processing Symp.*, Pori, Finland, May 1997, pp. 124-128.

55. D. Petrescu, I. Tăbuş, and M. Gabbouj. $\mathcal{L}$–M–S filters for image restoration applications. *IEEE Trans. Image Processing*, 8(9): 1299-1305, September 1999.

56. D. Petrescu, I. Tăbuş, and M. Gabbouj. Prediction capabilities of Boolean and stack filters for lossless image compression. *Multidimensional Systems and Signal Processing*, 10: 161-187, April 1999.

57. I. Pitas and A.N. Venetsanopoulos. Edge detectors based on nonlinear filters. *IEEE Trans. Pattern Analysis and Machine Intelligence*, 8(7): 538-550, July 1986.

58. I. Pitas and A.N. Venetsanopoulos. *Nonlinear Digital Filters. Principles and Applications.* Kluwer Academic, Boston, 1990.

59. W.K. Pratt. *Digital Image Processing.* Wiley, NY, 1990.

60. T.V. Ramabadran and K. Chen. The use of contextual information in the reversible compression of medical images. *IEEE Trans. Medical Imaging,* 11(2): 185-195, 1992.

61. S.J. Reeves. On the selection of median structure for image filtering. *IEEE Trans. Circuits and Systems II-Analog and Digital Signal Processing,* 42(8): 556-558, August 1995.

62. P. Roos, M.A. Viergever, M.C.A. Van Dijke, and J.H. Peters. Reversible intraframe compression of medical images. *IEEE Trans. Medical Imaging,* 7(4): 328-336, December 1988.

63. A. Said and W.A. Pearlman. An image multiresolution representation for lossless and lossy compression. *IEEE Trans. Image Processing,* 5(9): 1303-1311, September 1996.

64. J. Serra. *Image Analysis and Mathematical Morphology. Theoretical Advances,* Vol. 2. Academic Press, London, 1988.

65. I. Shmulevich, T.M. Sellke, M. Gabbouj, and E.J. Coyle. Stack filters and free distributive lattices. In *Proc. IEEE Workshop Nonlinear Signal and Image Processing,* Vol. II, Greece, 1995, pp. 927-930.

66. T. Sun, B. Zeng, and Y. Neuvo. Multilevel stack filtering for image processing applications. In *Proc. SPIE Nonlinear Image Processing III,* Vol. 1658, November 1992, pp. 46-55.

67. I. Tăbuş. *Training and Model Based Approaches for Optimal Stack and Boolean Filtering with Applications in Image Processing.* Ph.D. dissertation. Tampere University of Technology, 1995.

68. I. Tăbuş, D. Petrescu, and M. Gabbouj. Multilayer Boolean and stack filter design. In *Proc. SPIE Nonlinear Image Processing V,* Vol. 2180, San Jose, CA, February 1994, pp. 2-10.

69. I. Tăbuş, D. Petrescu, and M. Gabbouj. Training based optimal stack filter design under structural constraints. In *Proc. 1996 IEEE Int. Conf. Image Processing,* Lausanne, Switzwerland, September 1996.

70. I. Tăbuş, D. Petrescu, and M. Gabbouj. A training framework for stack and Boolean filtering-Fast optimal design procedures and robustness case study. *IEEE Trans. Image Processing Special Issue Nonlinear Image Processing,* 5(6): 809-826, June 1996.

71. J. Vaisey and A. Gersho. Image compression with variable block size segmentation. *IEEE Trans. Signal Processing*, 40(8): 2040-2060, August 1992.

72. G.K. Wallace. The JPEG still picture compression standard. *Communications of the ACM*, 34(4): 31-44, 1991.

73. X. Wang. Generalized multistage median filter. *IEEE Trans. Image Processing*, 1(4): 543-545, October 1992.

74. P.D. Wendt, E.J. Coyle, and N.C. Gallagher Jr. Stack filters. *IEEE Trans. Acoustics, Speech, and Signal Processing*, 34(8): 898-911, August 1986.

75. X. Wu and N.D. Memon. CALIC–A context based adaptive lossless image codec. In *Proc. 1996 IEEE Int. Conf. Acoustics, Speech, and Signal Processing*, Vol. IV, Atlanta, GA, May 1996, pp. 1890-1893.

76. R. Yang, Y. Lin, M. Gabbouj, J. Astola, and Y. Neuvo. Optimal weighted median filtering under structural constraints. *IEEE Trans. Signal Processing*, 43(3): 591-604, March 1995.

77. L. Yin, J. Astola, and Y. Neuvo. Adaptive stack filtering with application to image processing. *IEEE Trans. Signal Processing*, 41(1): 162-184, January 1993.

78. O. Yli-Harja, J. Astola, and Y. Neuvo. Analysis of the properties of median and weighted median filters using the threshold logic and stack filter representations. *IEEE Trans. Signal Processing*, 39(2): 395-410, February 1991.

79. J. Yoo, C.A. Bouman, E.J. Delp, and E.J. Coyle. The nonlinear prefiltering and difference of estimates approaches to edge detection: applications of stack filters. *Computer Vision, Graphics, Image Processing: Graphical Models and Image Processing*, 55: 140-159, March 1993.

80. B. Zeng. Optimal median-type filtering under structural constraints. *IEEE Trans. Image Processing*, 4(7): 921-931, July 1995.

81. B. Zeng, M. Gabbouj, and Y. Neuvo. A unified design method for rank order, stack and generalized stack filters based on classical Bayes decision. *IEEE Trans. Circuits and Systems*, 38(9): 1003-1020, September 1991.

82. B. Zeng, Y. Neuvo, and A.N. Venetsanopoulos. Optimal multistage stack filtering for image restoration. In *Proc. 1992 IEEE Int. Symp. Circuits and Systems*, November 1992, pp. 117-120.

83. B. Zeng and A.N. Venetsanopoulos. A comparative study of several nonlinear image interpolation schemes. In *Proc. SPIE Visual Communications and Image Processing*, Vol. 1818, Boston, 1992, pp. 21-29.

84. B. Zeng, H. Zhou, and Y. Neuvo. FIR stack hybrid filters. *Optical Engineering*, 30(7): 965-975, July 1991.

3 Image Processing Using Rational Functions

G. RAMPONI, S. MARSI, and S. CARRATO

Department of Electronics
University of Trieste, Italy

3.1 INTRODUCTION

This chapter overviews the state of the art of the *rational filter* (RF) approach to image processing, and suggests possible uses for operators of this family in smoothing different types of noise, in image interpolation, and in contrast enhancement.

Rational functions, namely the ratio of two polynomials, have been recently proposed to represent the input/output relation in a nonlinear signal processing system. The motivation for their introduction basically was to overcome the limitations which are typical of the more conventional polynomial approach, the discrete Volterra series, such as its need for a large amount of parameters when the order of the nonlinearity is high.

The RF can be very effective in removing both short-tailed and medium-tailed noise corrupting an image. It can act as an edge-preserving smoother, conjugating the noise attenuation capability of a linear lowpass filter and the sensitivity to high-frequency details of its embedded edge sensor. The RF can also be extended to a 3-D operator for sequences of images, even without using explicit motion detection.

The same approach can also be used to design effective interpolation operators for highly compressed data. Conventional linear techniques yield blurred results, because they tend to smooth details. The principle of the RF technique is to obtain each interpolated value as an average of its neighborhood, using weights that privilege neighbors that are more similar to each other. The rational filter paradigm can be exploited for devising novel enhancement techniques too. Operators of this type may be interpreted as Laplacian filters modulated by an edge-sensing term that confers higher emphasis to fine details but avoids noise amplification.

The implementation of this family of operators is also studied in this chapter, aiming at exploiting its characteristics of simplicity in order to obtain fast and economical real-time structures. Both very large scale integration (VLSI) realization, in application-specific integrated circuits (ASIC) or field-programmable gate arrays (FPGA), and digital signal processor (DSP)-based implementations will be considered.

The chapter is organized as follows: In Section 3.2 a survey of the theoretical bases of the rational operators is provided. In Section 3.3 several applications are presented, in particular in the areas of noise reduction, image interpolation, and contrast enhancement. The realization of these operators, both using DSP's and via hardware implementations, is discussed in Section 3.4. Finally, conclusions are drawn and possible directions of future work are analyzed in Section 3.5.

3.2 RATIONAL FILTERS: A THEORETICAL SURVEY

In this section we introduce the few mathematical concepts that constitute the foundations for the rational operators. These concepts trace back to the last century, when the mathematician Vito Volterra devised the series that took his name to formally represent continuous nonlinear time-invariant systems with memory [1]. However, their first use in the digital signal processing field is, of course, by far more recent. For this purpose a discrete version of the Volterra series is required [2, 3, 4], that takes the following expression:

$$
\begin{aligned}
y(n) \; = \; & h_0 \\
+ \; & \sum_{i_1=-\infty}^{\infty} h_1(i_1) x(n-i_1) \\
+ \; & \sum_{i_1=-\infty}^{\infty} \sum_{i_2=-\infty}^{\infty} h_2(i_1, i_2) x(n-i_1) x(n-i_2) \\
+ \; & \cdots \\
+ \; & \sum_{i_1=-\infty}^{\infty} \cdots \sum_{i_p=-\infty}^{\infty} h_p(i_1, \ldots, i_p) x(n-i_1) \cdots x(n-i_p) \\
+ \; & \cdots ,
\end{aligned}
\tag{3.1}
$$

where $y(n)$ and $x(n)$ are the output and the input signals, respectively. In a more compact form, using the Volterra operators $\bar{h}_p[x(n)]$, we have

$$
y(n) = h_0 + \sum_{p=1}^{\infty} \bar{h}_p[x(n)] .
\tag{3.2}
$$

The terms $\bar{h}_p = h_p(i_1, \ldots, i_p)$ are called Volterra kernels and are symmetric functions of their arguments. h_0 represents an offset component, $h_1[i_1]$ is the impulse response of a digital noncausal IIR filter, and $h_p[i_1, \ldots, i_p]$ is a generalized pth order impulse response. If the lower limit of the summations in Eq. (3.1) is zero, a causal system results, while if the upper limits are changed to a finite value N, the system has a finite memory. In this case $h_1(i_1)$ is an FIR filter. For the higher-order kernels the effect of the nonlinearity on the output depends only on the present and a finite set of past input values. The truncated Volterra series of order P is obtained by replacing the infinity in the summation of Eq. (3.2) with the finite integer P. Finite-support, finite-memory polynomial filters result.

The truncated discrete Volterra series is a powerful mathematical tool but may show some limitations when used as a polynomial operator acting on an input sequence. It may be observed in fact that the complexity in each component of Eq. (3.1) is N^p coefficients. By taking into account the symmetry property, the number of independent coefficients can be expressed as

$$N_p = \frac{(N + p - 1)!}{(N - 1)! \; p!}. \tag{3.3}$$

This means that if we are using, for example, a simple quadratic filter operating on a 3×3 support, we have to deal with $N_p = 45$ coefficients; the coefficients become $N_p = 325$ if the support is 5×5. In general, when the degree of the nonlinearity and the memory of the system grow, the number of the required parameters increases at a very fast pace. The same happens to the computational load when the operator is realized. Rational operators, whose input/output relation can be expressed as the ratio of two polynomials, have been devised exactly to eschew this problem.

Credit is due to Leung and Haykin for their seminal paper on the usage of rational functions, formulated as the ratio of two polynomials, in a nonlinear signal processing system [5]. Similarly to a polynomial function, a rational function is a universal approximator; that is, with enough parameters and enough data to optimize them, it can approximate any function arbitrarily well. Moreover the rational polynomial function can achieve the desired level of accuracy with a lower complexity, and it possesses better extrapolation capabilities than a polynomial function. It has also been demonstrated that a linear adaptive algorithm can be devised for determining the parameters of this structure. By this approach the two problems of estimating the direction of arrival of plane waves on an array of sensors and of detecting radar targets in clutter were tackled. In [5], a rational function with a linear numerator and a linear denominator has been used. A major obstacle for these functions potentially is the complexity: if a high order is used, many parameters are required and this can cause slow convergence.

For one-dimensional (1-D) signals, a rational filter can take the general expression

$$y(n) = \frac{a_0 + \sum_{i=0}^{N-1} a_{1i} x(n - i) + \sum_{i=0}^{N-1} \sum_{j=0}^{N-1} a_{2ij} x(n - i) x(n - j) + \cdots}{b_0 + \sum_{i=0}^{N-1} b_{1i} x(n - i) + \sum_{i=0}^{N-1} \sum_{j=0}^{N-1} b_{2ij} x(n - i) x(n - j) + \cdots}. \tag{3.4}$$

To exploit it for still image and image sequence processing, a multidimensional operator will be required. The complexity of the extension of rational operators to the multidimensional case is in general extremely large. The independent variables should be arranged in suitable vectors to derive a compact notation, conveniently ordering the input data and the filter coefficients. For example, a quadratic two-dimensional (2-D) polynomial filter can be expressed as

$$
y(n_1, n_2) = h_0 + \sum_{i_1=0}^{N_1-1} \sum_{i_2=0}^{N_2-1} h_1(i_1, i_2) x(n_1 - i_1, n_2 - i_2) \tag{3.5}
$$

$$
+ \sum_{i_{11}=0}^{N_1-1} \sum_{i_{12}=0}^{N_2-1} \sum_{i_{21}=0}^{N_1-1} \sum_{i_{22}=0}^{N_2-1} h_2(i_{11}, i_{12}, i_{21}, i_{22})
$$

$$
x(n_1 - i_{11}, n_2 - i_{12}) x(n_1 - i_{21}, n_2 - i_{22}),
$$

where $y(n_1, n_2)$ and $x(n_1, n_2)$ are the output and the input images, respectively. It should be observed that symmetries exist in the nonlinear kernels. For example, in the quadratic kernel $h_2(i_{11}, i_{12}, i_{21}, i_{22}) = h_2(i_{21}, i_{22}, i_{11}, i_{12})$. This helps to control the complexity of the operator. Moreover many constraints can be derived if isotropic conditions are imposed on the filter [6]. Analogous expressions can be derived for rational operators, by separately manipulating the numerator and the denominator of their input/output equation.

It is fortunate that filters defined on very small supports often yield remarkable results in practical applications. Moreover, a low-order filter is often sufficient to obtain significant improvements with respect to the use of conventional linear filters.

3.3   APPLICATIONS

This section overviews the applications of rational operators in different fields, such as noise smoothing in images and image sequences, image interpolation, and contrast enhancement. As it will be seen, in all cases very simple operators that are designed using obvious heuristic criteria can produce good results.

3.3.1   Noise Smoothing in Images and Image Sequences

All the numerous algorithms for noise smoothing in images that can be found in the recent literature make specific provisions to avoid blurring the details of the images. Often this is achieved at the expense of a reduced smoothing ability or at high computational costs. A good compromise between the two contrasting effects of noise smoothing and detail preservation can be achieved by resorting to the rational filter scheme. The underpinning in this case is to use the numerator of the rational function to introduce the required smoothing action, and the denominator to control this action according to the presence of significant luminance transitions. Formally, a possible expression for one such filter in the 1-D case is obtained starting from a simple linear lowpass filter of the type

$$y(n) = \frac{x(n-1) + x(n+1)}{A} + x(n)\left(1 - \frac{2}{A}\right), \tag{3.6}$$

where $A \geq 3$, which is used at the numerator. A suitable denominator is needed, that may be simply based on the squared difference of the two samples that are adjacent to the reference one. This term will act as a sensor of the luminance transitions which represent important details and must therefore be preserved. The overall operator becomes [7]

$$y(n) = \frac{x(n-1) + x(n+1)}{k[x(n-1) - x(n+1)]^2 + A} +$$
$$x(n)\left[1 - \frac{2}{k[x(n-1) - x(n+1)]^2 + A}\right]. \tag{3.7}$$

In this equation k is a parameter that should take positive values, and it permits the balance between the smoothing and edge preservation effects. More precisely, if k is very large, $y[n] \simeq x[n]$ and the filter has no effect; if $k \simeq 0$, the rational filter becomes the simple linear lowpass filter of Eq. (3.6); for intermediate values of k, the output of the sensor $[x(n-1) - x(n+1)]^2$ modulates the response of the filter.

This operator can be extended in various ways to obtain a multidimensional (M-D) filter. For image processing, the simplest approach is to sum the effects of different 1-D filters operating in various directions [7]. Another more effective operator results taking the expression

$$y(m,n) = x(m,n) + \sum_{i=1}^{7} \sum_{j=i+1}^{8} \frac{x_i + x_j - 2x(m,n)}{k(x_i - x_j)^2 + A}, \tag{3.8}$$

where the set $\{x_i\}$, $i = 1, \ldots, 8$, is formed by the arbitrarily ordered eight pixels that are adjacent to the reference one, $x(m,n)$.

It should be observed that this and the other rational filters can be applied more than once on the input data to obtain a stronger smoothing action. For repeated applications, in uniform areas where the $(x_i - x_j)^2$ terms are negligible, the filter is equivalent to a linear lowpass filter of a larger size; on the contrary, in the vicinity of a detail the mask of the equivalent filter takes an asymmetric shape that tends to cover pixels similar to the reference one. Hence the required edge preservation is obtained.

An example of the performances of this operator is presented in Fig. 3.1. In particular, Fig. 3.1(a) shows an original test image which has been corrupted with additive noise (signal-to-noise ratio, SNR = 6.0 dB). Figure 3.1(b) shows the same image after processing with the operator of Eq. (3.8). The quality improvement is apparent, and the details have been preserved. The processed image is characterized by a significantly increased SNR (13.2 dB). The noise in Fig. 3.1(a) has Gaussian

(a)   (b)

Fig. 3.1   (a) Corrupted test image, and (b) its filtered version.

distribution, but it should be mentioned that the rational filter successfully deals also with other types of noise, such as the additive noise with contaminated Gaussian distribution or the multiplicative noise with Laplacian distribution. A modified filter has also been designed for the processing of SAR images contaminated by speckle noise[8]. Moreover a new class of hybrid rational filters has been recently introduced [9]. For multichannel image processing, a vector median-rational hybrid filter has been used, where a rational operator is applied to the outputs of three subfilters, one of which is a central weighted vector median operator. The resulting filter also preserves well the chromaticity of the original image. The situation is slightly more complicated when processing a sequence of images. In fact the presence of motion in the scene has to be taken into account if motion artifacts are to be avoided.

A nonlinear spatiotemporal filter capable of significantly attenuating noise in image sequences without corrupting image details is presented in [10]. The operator is based on the extension of the 2-D filter (Eq. (3.8)) to the 3-D case:

$$y_0^t = x_0^t + f_{spatial} + f_{temp}, \tag{3.9}$$

with

$$f_{spatial} = \sum_{i,j \in I} \frac{x_i^t + x_j^t - 2x_0^t}{k_S(x_i^t - x_j^t)^2 + A_S},$$

$$f_{temp} = \sum_{i,j \in J} \frac{x_i^{t-1} + x_j^{t+1} - 2x_0^t}{k_T(x_i^{t-1} - x_j^{t+1})^2 + A_T},$$

where the superscripts $t-1$ and $t+1$ refer to pixels in the previous and in the following frame, respectively, I is a set of indices defining the pixels used in the spatial operator, and J is another set defining the nine spatiotemporal filtering directions. These definitions are given pictorially in Fig. 3.2. The parameters k_S, A_S, k_T and A_T may

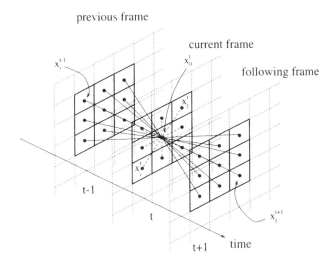

Fig. 3.2 Pixels and directions considered by the spatial and spatiotemporal parts of the operator proposed in [10]. Dashed lines: spatial directions; solid lines: spatiotemporal directions (reprinted by copyright permission from IEEE).

allow for a different tuning of the spatiotemporal part of the filter with respect to the purely spatial one. Robustness to motion artifacts is provided by a simple movement detector, which suppresses the temporal part in presence of fast motion. In particular, no motion may be assumed if

$$|x_0^t - \frac{x_o^{t-1} + x_0^{t+1}}{2}| < \theta, \qquad (3.10)$$

where θ is a suitable threshold. It may be shown that this simple control is sufficient to avoid "image flow" artifacts.

The operator is shown to be able to effectively filter different types of noise, namely the Gaussian, the contaminated Gaussian, and the impulsive noise, if a suitable choice of the parameters A and k which appear at the denominators of both the spatial and the temporal parts is done.

In the just mentioned approach, some a priori information on the noise characteristics has to be available. If this is not the case, it is necessary to be able to estimate the noise type, in order to suitably adjust the parameters. A possible approach is presented in [12], where a simple noise type discriminator, based on the evaluation of the kurtosis of an estimate of the present noise, is used to control a set of three smoothing filters based on the rational operator. The kurtosis in fact is well known to be related to the length of the tails of a distribution, being $k = 3$ for a Gaussian distribution, $k > 3$ for contaminated Gaussian, and $k \gg 3$ for impulsive noise [13]. In particular, a small part of each frame is analyzed, supposing that the noise is spatially uniform; within this window an estimate of the noise is extracted by computing the difference between these data $y = x + n$, where x is the ideal noise-free image and n is noise, and the same data are filtered using a median filter

$$z = y - \text{median}(y). \tag{3.11}$$

Because of the noise reduction and edge-preserving properties of the median filter [14], the resulting signal, z, is composed approximately of noise only [15], that is $z \simeq n$; k is then estimated on z to provide an indication on the type of noise.

The general form of the three filters is the same as in Eq. (3.9); however, the filters support is restricted to two frames only, in order to reduce the frame memory cost. Consequently the temporal filters structure is slightly changed:

$$f_{temp}^{(gauss)} = \sum_{i \in J} \frac{-x_i^{t-1} + x_0^t}{k_{T1}(x_i^{t-1} - x_0^t)^2 + A_{T1}} \tag{3.12}$$

$$f_{temp}^{(contgauss)} = \sum_{i \in J} \frac{-x_i^{t-1} + x_0^t}{[k_{T2}(x_i^{t-1} - x_0^t)^2 + k_{T3}(x_i^{t-1} - x_i^t)^2]/2 + A_{T2}} \tag{3.13}$$

$$f_{temp}^{(imp)} = \sum_{i \in J} \frac{-x_i^{t-1} + x_0^t}{k_{T4}(x_i^{t-1} - x_0^t)^2 \cdot (x_i^{t-1} - x_i^t)^2 + A_{T3}} \tag{3.14}$$

where $i, j \in J$ describe sets of temporal filtering directions, as shown in Fig. 3.3, and k_{Tl} and $A_{Tl}, l = 1, \ldots, 4$, are suitable filter parameters which, like for the spatial part, are chosen by minimizing the mean square error (MSE) between the filter output and the original, noise-free image. It has to be noted that for the contaminated Gaussian and the impulsive noises it is not recommended to use the same control strategy as for the Gaussian noise: the difference $(x_i^{t-1} - x_0^t)$ may be large due to a noise peak, instead of an edge, with consequent loss of the noise filtering action. In turn the same difference can be "corrected" by averaging with or multiplying by another difference, i.e., $(x_i^{t-1} - x_i^t)$, so that the denominator remains low also in presence of isolated noisy pixels and the desired lowpass behavior is obtained. In particular, it is advisable to average the two differences in the contaminated Gaussian case, due to the simultaneous presence of long- and short-tailed noise. On the other hand, multiplication is preferred in the impulsive noise case to obtain a stronger correction. This is what is done in Eqs. (3.13) and (3.14), respectively. For example, in Fig. 3.4 it may be seen that the operator is able to effectively reduce the contaminated Gaussian noise.

3.3.2   Image Interpolation

The problem of image interpolation by large scaling factors can be addressed using the rational filter approach. As a representative example, we examine the problem of interpolating the dc components of JPEG or MPEG-coded images and image se-

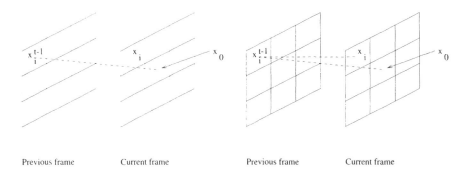

Previous frame   Current frame   Previous frame   Current frame

Fig. 3.3 Example of directions used by the temporal part of the rational filters, for (left) Gaussian noise, and (right) contaminated Gaussian noise. There is a total of 9 possible directions, according to the possible position of x_i and x_i^{t-1} on the respective masks; only one direction has been drawn for the sake of clarity.

Fig. 3.4 Performance of the 2-frame spatiotemporal rational filter (right) on an image corrupted by contaminated Gaussian noise (left) [reprinted by copyright permission from IEEE].

quences[1]. A nonlinear interpolator has been proposed in [16] that is able to accurately reconstruct the scene yielding both a sharp image and low blocking artifacts. The proposed technique accurately reconstructs edges of any orientation and areas rich in details as well. Following this approach, it is possible to have images of good quality

---

[1]The dc (direct current) component is the zero-order Fourier (or any other transform) coefficient.

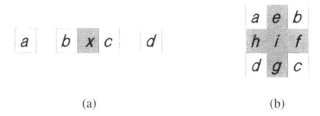

(a)   (b)

Fig. 3.5 Samples considered for (a) one- and (b) two-dimensional interpolation. Data in the white boxes are available, the others have to be interpolated.

at low bit rates using standard coding techniques; of course, the ac components may be transmitted later, possibly on demand, if the actual image is eventually needed[2].

In this example we want to interpolate a compressed image obtained by considering only the dc terms of the cosine transform of the 8×8 blocks of a JPEG image or an MPEG I-frame, or by approximating the dc terms of an MPEG P- or B-frame. The image can be recovered at its original size after interpolating for three times the data by a factor of two. Conventional linear techniques, for example, [17], generally yield blurred results or introduce ringing effects. Moreover they tend to smooth the outliers. However, in such dc images the outliers are always due to significant details rather than to noise. It is convenient consequently to use nonlinear interpolators that are able to preserve steep edges and small details. A further problem is that because of the large magnification factor, any artifact that may be introduced by the interpolator is generally amplified and becomes quite visible in the final image. In particular, separable operators often create jagged diagonal edges.

Consider first a one-dimensional interpolator which operates on four consecutive samples of decimated data, a, b, c, and d, in order to reconstruct x, the central one; see Fig. 3.5(a). A possible nonlinear interpolator is

$$x = \frac{w_b b + w_c c}{w_b + w_c}, \tag{3.15}$$

with

$$w_b = 1 + k\left[(b - d)^2 + (c - d)^2\right] \tag{3.16}$$

$$w_c = 1 + k\left[(c - a)^2 + (b - a)^2\right], \tag{3.17}$$

which accurately preserves the relative sizes of the objects after the various interpolation steps. With the parameter k, the user can control the amount of nonlinearity in the proposed scheme. The value $k = 0$ yields the bilinear interpolator.

---

[2]The ac (alternate current) components are the first, second, etc., order Fourier (or any other transform) coefficients.

An analogous operator has been introduced in [18], where a rational function is used in order to weigh the contributions to x of its two adjacent samples b and c according to the differences $(a - b)$ and $(c - d)$. The resulting operator is able to reconstruct sharp edges more accurately than a linear filter, because it is able to roughly estimate their position with sub-pixel accuracy. The approach used in [18] is more suitable for small interpolation factors.

The same concept can be extended to two dimensions to define an image interpolation technique. The above-mentioned function can be used row- or columnwise to evaluate the pixels, as e, f, g, and h in Fig. 3.5(b), which lie close to two already known pixels. It is easily seen that this solution is quite accurate for horizontal and vertical edges.

The interpolation of the pixels in the central position, i in Fig. 3.5, is more complicated. Both standard linear operators and the operator described above can be directly used, along the vertical or the horizontal position, using already interpolated pixels. However, with this approach applied for three times oblique edges tend to become jagged. Better results may be obtained using a nonseparable method. In particular, a weighted average of couples of pixels that are adjacent to i along four directions may be computed: couples of similar pixels are privileged. Hence

$$i = \frac{w_{ac}(a + c) + w_{bd}(b + d) + w_{eg}(e + g) + w_{fh}(f + h)}{2(w_{ac} + w_{bd} + w_{eg} + w_{fh})}, \tag{3.18}$$

where

$$w_{ac} = 1/(1 + k(a - c)^2) \tag{3.19}$$

$$w_{bd} = 1/(1 + k(b - d)^2) \tag{3.20}$$

$$w_{eg} = 1/(1 + k(e - g)^2) \tag{3.21}$$

$$w_{fh} = 1/(1 + k(f - h)^2). \tag{3.22}$$

It may be noted that although the algorithm is nonseparable, there is no difference in performing the interpolation first along the rows and then along the columns and vice versa, and that the operator is isotropic with respect to the horizontal and the vertical directions.

It is interesting to observe the behavior of the filter on diagonal edges, by testing it on a synthetic image. In Fig. 3.6 a uniform disk of diameter 160 pixels has been considered. The image obtained with the rational interpolator starting from the dc components of 8×8 blocks is shown. It may be noted that edges are accurately reconstructed in all directions, and no artifacts are introduced. As a comparison, the corresponding image obtained with a standard interpolation algorithm, namely the Keys interpolator [17], is significantly more blurred and has a higher MSE: 146.4 instead of 54.8.

Another possible application is in the interpolation of color images [19]: any rational filter can be used to process multichannel signals by applying it separately to each component, but this in general is not desirable when correlation exists among

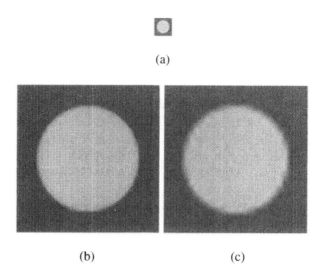

(a)

(b)   (c)

Fig. 3.6   (a) Interpolation of a synthetic image from its 32×32 dc components to its original 256×256 size using (b) the nonlinear interpolator with $k = 0.01$, and (c) a classical interpolation algorithm [17].

components because false color may be introduced. Better results are obtained using vector extensions of the operator, such as by extending it to vector form:

$$\mathbf{y} = \frac{w_2 \mathbf{x_2} + w_3 \mathbf{x_3}}{w_2 + w_3}, \tag{3.23}$$

with

$$w_2 = 1 + k\|\mathbf{x_2} - \mathbf{x_4}\|_p, \quad w_3 = 1 + k\|\mathbf{x_3} - \mathbf{x_1}\|_p. \tag{3.24}$$

Here $\mathbf{y}$, $\mathbf{x_1}$, $\mathbf{x_2}$, $\mathbf{x_3}$, and $\mathbf{x_4}$ are vector samples, for example, RGB triplets, and $\|.\|_p$ indicates an l_p norm.

A somehow different approach is proposed in [20], where a high quality $2\times$ interpolator for both synthetic and natural images is proposed. In this case, the interpolator is equivalent to a pixel replication followed by a suitable nonlinear correction of all the samples, i.e., possibly changing also the original values. The latter is needed to avoid blocking artifacts both for sharp details, in particular for diagonal edges and lines, and when smooth transitions between different levels of gray are present, as typically happens in natural images.

With reference to Fig. 3.7, the pixel to be interpolated is first split into four pixels. Then the value of each of these pixels is corrected using the information in a suitable subset $\mathcal{S}_k$ of neighboring pixels. This permits to correctly operate in presence of both sharp and smooth edges.

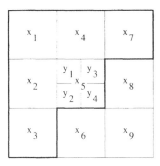

Fig. 3.7 Pixels used in the interpolation process. The bold lines identify the mask subset used to interpolate pixel y_1.

Let us consider an example. To interpolate y_1 the pixels belonging to $S_1 \triangleq \{x_1, x_2, x_3, x_4, x_5, x_7\}$ are used, and taking into consideration the obvious symmetry, the following set of pixel differences is considered:

$$
\begin{aligned}
d_1 &= x_1 - x_5, \\
d_2 &= (x_2 + x_4)/2 - x_5, \\
d_3 &= (x_3 + x_7)/2 - x_5.
\end{aligned}
$$

In order to correctly treat the edges, the operator must be able to "recognize" them, to avoid blurring. Nonlinear interpolation can yield steep edges [18], typically by emphasizing abrupt large transitions in the signal. However, this approach can generate under- or overshoots at the proximity of the edges; this behavior may be desirable in natural images, but it is very annoying in synthetic ones. In order to avoid this drawback, it is possible to impose to the operator a height-independent response. This may be easily obtained by weighting the above-mentioned differences using ratios of polynomials of the same order, such as of order two. Consequently the interpolator may have the form

$$
y_1 = x_5 + \frac{d_1 w_1 + d_2 w_2 + d_3 w_3}{w_4 + \epsilon}, \tag{3.25}
$$

where the term ϵ is needed to avoid a null denominator, and the weights w_j ($j = 1, \ldots, 4$) are computed as

$$
\begin{aligned}
w_j = \; & p_{1j}(x_1 - x_5)^2 + \\
& p_{2j}(x_2 - x_4)^2 + \\
& p_{3j}(x_3 - x_7)^2 + \\
& p_{4j}((x_1 - x_2)^2 + (x_1 - x_4)^2) \\
& p_{5j}((x_1 - x_3)^2 + (x_1 - x_7)^2)
\end{aligned}
$$

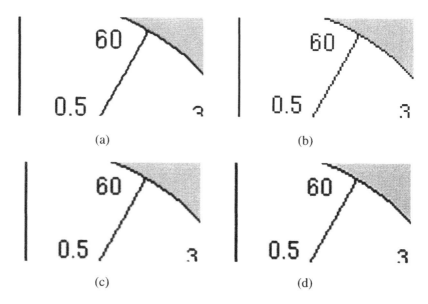

Fig. 3.8 (a) Performance of the proposed rational operator compared with those obtained using (b) pixel repetition, (c) bilinear convolution and (d) bicubic convolution.

$$p_{6j}((x_2 - x_5)^2 + (x_4 - x_5)^2)$$
$$p_{7j}((x_3 - x_5)^2 + (x_7 - x_5)^2)$$
$$p_{8j}((x_3 - x_2)^2 + (x_7 - x_4)^2)$$
$$p_{9j}((x_3 - x_4)^2 + (x_7 - x_2)^2)$$

for $j = 1, \ldots, 4$. Optimal values for the parameters p_{ij}, $i = 1, \ldots, 9$, may be easily found using a suitable optimization technique.

In Fig. 3.8 results of $2\times$ interpolation on a synthetic image are shown using the rational interpolator described, and several standard techniques. It may be seen that a significantly better image is obtained using the rational operator, in terms of both the sharpness and the absence of blocking artifacts.

Finally, it is worth mentioning that rational interpolators have also been successfully used for the restoration of old movies [21] corrupted by scratches and dirt spots. After localization of the stationary and random defects, a spatial interpolation scheme is used to reconstruct the missing data. The algorithm first checks the existence of edges in order to take them into consideration. The edge orientation is estimated and the most convenient data to be used in the reconstruction of the missing pixels are selected. In this way the obtained edges are free from blockiness and jaggedness.

3.3.3 Contrast Enhancement

In this subsection a rational filtering approach is suggested that may improve the results obtainable from the classic unsharp masking (UM) technique. The UM technique can be realized by processing the image with a highpass filter, usually a Laplacian, multiplying the result by a scaling factor, and by adding it to the original data. This technique is very popular but suffers from two drawbacks which can significantly reduce its benefits: noise sensitivity and excessive overshoot on sharp details. The former problem comes from the fact that the UM method assigns an emphasis to the high-frequency components of the input, amplifying a part of the spectrum in which the SNR is usually low. On the opposite, wide and abrupt luminance transitions in the input image can produce overshoot effects that become visually unpleasant, such as black or white streaking along the border of the objects.

Hence the basic technique has been modified in various ways, in particular, with the purpose of reducing the noise amplification problem. A quite trivial approach consists in substituting a bandpass filter for the highpass one. This limits the noise effects but also precludes effective detail enhancement in most images. In more sophisticated approaches, nonlinear operators such as order statistics, polynomial, and logarithmic are used to generate the correction signal that is added to the image. Another possibility is to resort to linear but adaptive techniques.

The rational-based method presented in this subsection permits one to select details having low and medium sharpness and to enhance them. On the other hand, noise amplification is very limited, and steep edges, which do not need further emphasis, remain almost unaffected.

For simplicity we describe the 1-D operator. Its extension to the 2-D case is straightforward [22]. The enhanced signal $y(n)$ obtained from the linear UM scheme is

$$y(n) = x(n) + \lambda z(n), \tag{3.26}$$

where $z(n)$ is the correction term computed as the output of a suitable enhancing filter and λ is a positive scaling factor. A common choice in the linear case is to employ the Laplacian of $x(n)$ as the signal $z(n)$:

$$z(n) = L(n) = 2x(n) - x(n-1) - x(n+1). \tag{3.27}$$

A modified $z(n)$ signal obtained as the product of the Laplacian and a rational control term $c(n)$ can give better results. The output of the filter becomes

$$y(n) = x(n) + \lambda c(n)L(n), \tag{3.28}$$

with

$$c(n) = \frac{g(n)}{kg^2(n) + h} \tag{3.29}$$

where $g(n)$ is a measure of the local activity of the signal which will be discussed in the following. Parameters k and h are proper positive factors. This rational

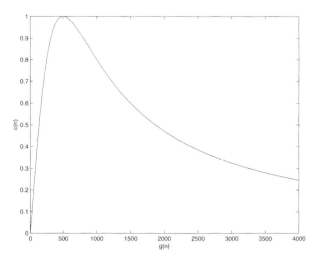

Fig. 3.9   Plot of the control function $c(n)$ for $h = 250$ and $k = 0.001$ (reprinted by copyright permission from SPIE).

control term is plotted as a function of $g(n)$ and for $h = 250$ and $k = 0.001$ in Fig. 3.9. Since this control term is multiplied by the output of the Laplacian filter, it will permit the sharpening effect only for low- and medium-amplitude luminance transitions (portions of the plot around the peak). At the same time the amplification of noise is avoided where the activity is deemed small ($g(n)$ is close to 0). Finally, undesirable overshoot effects are avoided too, thanks to the fact that the control function tends to zero if the activity is large. The position g_0 of the peak can be selected according to the luminance distribution of the data by choosing the values of k and h. For a normalized control function whose maximum amplitude should be 1, it can be easily shown that $h = g_0/2$ and $k = 1/(2g_0)$. Then, to achieve the desired level of sharpness, the value of λ can be set as in the conventional UM method.

The activity measure $g(n)$ should be capable of representing important image features. For this purpose, a squared bandpass filter of the type

$$g(n) = [x(n+1) - x(n-1)]^2 \tag{3.30}$$

can be adopted. Results that can be obtained with the rational UM method are depicted in Fig. 3.10. Figure 3.10(a) shows a 256×256 original test image, while Fig. 3.10(b) shows the output of a linear UM operator ($\lambda = 1$). In Fig. 3.10(c) the output of the rational UM method ($\lambda = 1.2, g_0 = 20$) is presented. It is clear that noise amplification is much reduced in the image processed with the rational filter, while the detail enhancement is approximately the same. Moreover some streaking effects recognizable in Fig. 3.10(b) along the sharpest edges are no longer present.

(a)

(b)   (c)

Fig. 3.10   (a) Original test image; (b) result of the linear UM operator; (c) result of the rational UM algorithm.

From a computational viewpoint, it must be stressed that the proposed solution maintains almost the same simplicity as the original linear UM method, and this is by far simpler than many nonlinear or adaptive methods that can be found in the literature.

3.4   REALIZATIONS

In this section we discuss the realization of rational operators both in software, using DSP's and in hardware, and adopting various solutions from ASIC design to FPGA implementation. It will be seen that the best choice is strictly dependent on both the complexity of the specific operator and the constraints of the implementation itself, typically in terms of speed and cost.

3.4.1   Overview on Hardware Implementation Issues

Even if rational operators often present a simple structure from an analytic point of view, their implementation on a physical circuit is not always trivial. In particular, if the circuit designer attempts to realize the various functions, step by step, as described in the algorithm equations, presumably he or she will have to manage a large amount of complex operators (e.g., adders, multipliers, dividers, and square powers). This approach typically leads to a very large size, slow, and high-cost circuit. To obtain an effective realization of the algorithm on a circuit, there are mainly two ways, particularly suitable for these kinds of filters, for the designer to accomplish this:

1. Share the various elementary operators like adders, multipliers, dividers, and schedule them in a pipeline sequential structure.

2. Simplify any elementary operators using a suitable approximation.

Of course, these basic steps can lead to different implementation strategies depending on the physical support on which the circuit will be realized. There are mainly two different ways to realize the circuit and both can be split into at least two more subcases: the circuit can be realized in a dedicated integrated circuit, a so-called ASIC, or it can be implemented on programmable devices, like gate arrays, or FPGA. Moreover, in the ASIC category, the project can be designed as a full custom circuit, or it can be developed using a cell-based technique. On the other hand, the project can vary considerably among different FPGA architectures and granularities. In particular, fine grain FPGA's are comparable to other cell-based technologies such as standard cells, while the use of coarse grain FPGA's implies the usage of suitable macros for the implementation of some operators, like adders or look-up tables (LUT), rather than the more universal approach of description at the gate level and implementation.

3.4.2   Pipeline Structure

Since the rational filter is a structure composed by several elementary operators, a significant benefit in terms of performance and area reduction can be obtained through a pipeline architecture. The rational function can be considered to work as a loop that repeatedly executes the operations in its body until an exit condition becomes true. Two are the ways to implement this function:

1. As an asynchronous structure. In this case typically the input values cannot change until all the computations have been done and the result is stable at the outputs. In this way the delay time of the rational structure limits the computational frequency of the entire structure.

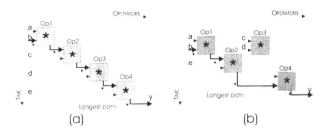

Fig. 3.11 Examples of asynchronous structures: (left) serial, and (right) tree.

2. Using a synchronous pipeline structure. In this case the process can schedule consecutive loop iterations to partially overlap in time. A new loop iteration is initiated before the current iteration has finished, so the frequency can be significantly improved and the design throughput increases too.

The primary reason for using overlap loop pipelining is to increase the throughput of the design, but this kind of structure obviously uses a large amount of registers, which leads to an increase of the design area. Consequently a trade-off is often necessary.

The two timing-related aspects of a pipeline loop that affect throughput are the *initiation interval*, which is the number of clock cycles between the start of two consecutive loop iterations, and the *latency*, which is the number of clock cycles required to execute all the operations in a single loop iteration. The initiation interval is related to the throughput rate; while the latency is related to the input-to-output delay.

Example

Let us implement a system to evaluate the following equation:

$$y = a \cdot b \cdot c \cdot d \cdot e. \tag{3.31}$$

If we want to realize the system in an asynchronous architecture, we need to realize four multipliers and to connect them according to one of the structures depicted in Fig. 3.11. The system delay is obviously proportional to the delay of the single multipliers and to the number of multipliers present in the longest path that connects the inputs to the output.

On the other hand, if we are going to use a pipeline structure as depicted in Fig. 3.12, the advantages with respect to the asynchronous structure are twofold: the design area can be reduced and the evaluation frequency can be improved. If a single multiplier, realized as an asynchronous circuit, computes its output within a clock cycle, in the remaining clock cycles it is possible to reuse the same operator for the evaluation of different data. These data can consist either in new input samples or in the output of a previous multiplication, included the output of the present operator itself. In this phase a constraint on the initiation interval is useful to define a trade-off between the frequency of

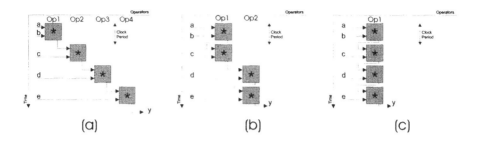

Fig. 3.12   Examples of pipeline structures.

the input samples and the throughput degree of the design, that is, between the computation delay and the design area.

In Fig. 3.12 three different pipeline structures with the same four clock cycle latencies are depicted. In Fig. 3.12(a) the initiation interval is one clock cycle. The structure uses four different multipliers but it is able to afford one output sample in every clock cycle. In Fig. 3.12(c) the initiation interval is equal to the latency. There is no overlap between the loop cycles, while the structure is composed by only one multiplier that is reused four times in the same loop cycle. The design area is significantly reduced, while the circuit affords one output sample only every four clock cycles. In Fig. 3.12(b) an intermediate case between the previous two is represented. The initiation interval is two clock cycles. In this way a sample is output every two clock cycles, while the circuit is realized using two different multiplier structures. Of course, many other different pipeline structures can be defined that present different properties and a different trade-off between circuit complexity and time delay.

The application of these concepts to rational filters is very useful: the number of the elementary operators like adders or multipliers can be reduced, with respect to a complete asynchronous realization, by a factor equal to the number of initiation interval cycles. The latency interval has to be set in agreement with the entire group of operations to evaluate. A short latency improves the throughput degree but requires a high number of elementary operators. A longer latency can permit a small reduction in the number of registers used by the circuit to store the partial results. A too long latency increases the input to output delay without any substantial advantage.

Example

The algorithm proposed in Eq. (3.25) presents a complex expression, and several operators are involved in the evaluation of the final result. A significant simplification can be obtained in realizing the circuit by using a pipeline structure and scheduling the various operations. Considering that some operators can be realized with a delay lower than the clock period, it is evident that the related elementary operations can be performed in a single clock cycle. Using

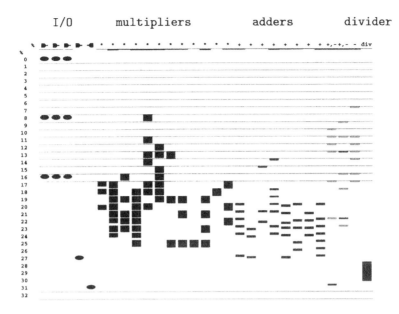

Fig. 3.13 Pipeline scheduling for the proposed algorithm. The diagram shows the use of the various units versus time.

a memory cell to store the result of these operations, the single operator can then be re-employed during other clock cycles to perform other similar tasks. In such a case it would be convenient to realize all the elementary operators without simplifications or approximations. A general-purpose structure, in fact, may be easily re-used in different parts of the algorithm.

The entire circuit has been implemented in a sequential form with an extensive use of flip-flops. The proposed architecture presents a pipeline with a latency of 32 clock cycles and an initiation interval of 8 cycles. The clock cycle has been kept under 19 ns. In such a way it is possible to use the proposed chip to process in real time a 360×288 CIF digital sequence at 60 frames per second (fps). In particular, in this kind of sequence the samples come with a cadence of 160 ns; thus the entire 152 ns initial interval of the pipeline can be consumed before a new input sample appears. A complete scheme of the pipeline is represented in Fig. 3.13. However, in the figure all the transitions have been neglected for the sake of clarity.

In the physical realization of the circuit all the various elementary operators (e.g., adders, multipliers, dividers) have been developed in a combinatorial form using carry look ahead and Wallace tree architectures [23]. The final circuit, realized with a standard cell 0.6μ CMOS technology, is depicted in Fig. 3.14. The size of the final circuit is 6.4×6 mm with a total area of 38 mm^2.

Fig. 3.14 Chip layout.

The core covers an area of 25 mm², and it is composed by 19.000 standard cell instances. The simulated power consumption is approximately 390 mW.

3.4.3 Realization of the Elementary Operators

When considering each operator that has to be included in a pipeline structure as described in the previous section, it is important to note that in rational filters not all these operators play the same role with respect to the final result. Indeed, it is essential to identify the behavior of the entire rational operator and to distinguish its various components. In particular, there will be some parts of the algorithm that must be computed with high precision, while in other components the precision may be significantly reduced. Subsequently, since the precision of the operator is proportional to its complexity and consequently to its cost. An appropriate approach may be arrange the precision of these parts to obtain a good trade-off between performance and cost.

In particular, in rational algorithms there are two recurring operators that are quite expensive if realized with canonical functions but can be simplified, or sometimes substituted, by simpler techniques without significant degeneration of the final result. These operations are the division and the square power computations.

3.4.3.1 The division operator Division is, of course, always present in a rational algorithm. This operation is absolutely fundamental in the equations that describe the algorithm itself. Consequently, before realizing the divider, it is important to analyze this operation with reference to the behavior of the algorithm.

Usually the final result of the algorithm is computed through a weighted sum of two or more terms. These terms are commonly simple functions of the input samples, while the weights are evaluated through a suitable strategy defined ad hoc by the designer. The divider generally plays a role in the evaluation of these weights. It aims at normalizing the weights, that is, either to maintain their sum equal to unity or to keep them within a predefined range. A great benefit in the physical realization of the circuit can be reached by exploiting the possibility of a system being able to evaluate these weights, thus avoiding the use of a canonical divider operator. The normalization of the weights can be realized, indeed, through other strategies like shift registers or LUTs. Moreover often the weights do not need to be evaluated with high precision, while their relationship constraints must be granted. Then, if a coarse quantization is acceptable, a further simplification can be included both in the system that evaluates the weights and in the part dedicated to the evaluation of the weighted sum. For example, let us consider an equation that is typical of a rational algorithm:

$$y = \frac{w_1 x_1 + w_2 x_2}{w_1 + w_2}, \tag{3.32}$$

where x_1 and x_2 are two generic inputs. The equation can be rewritten as

$$y = \hat{w}_1 x_1 + \hat{w}_2 x_2. \tag{3.33}$$

It is obvious that the result will fall between x_1 and x_2 and that the weights

$$\hat{w}_1 = \frac{w_1}{w_1 + w_2}, \tag{3.34}$$

$$\hat{w}_2 = \frac{w_2}{w_1 + w_2}, \tag{3.35}$$

will define whether y is closer to x_1 or to x_2. In such a case, the precision used in the weights $(\hat{w}_1, \hat{w}_2)$ is not critical for the final result, while it is essential that the sum of the weights equals one. Now $\hat{w}_1$ can be evaluated using, for example, a LUT. $\hat{w}_2$, being its complement, can be very easily obtained, in that a normalized fixed point notation can be adopted for their representation since both the weights magnitudes are between $\{0\,0\,0 \cdots 0\,0\} = 0$ and $\{1\,1\,1 \cdots 1\,1\} = 1$. Note that the complement of a number can be simply computed through the inverse of all its bits.

As was mentioned earlier, if an approximation for the weights is acceptable, the input for the LUT can be composed by just the more significant bits of w_1 and w_2 and thus a significant reduction in the depth of the LUT can be achieved. A further simplification can be realized if the weights are chosen to be a subset of all the possible values. This subset, for example, can be composed of a certain numbers of values obtained as powers of two and all their complementary quantities, as for example the subset

$$\{0; 1/8; 1/4; 1/2; 3/4; 7/8; 1\}. \tag{3.36}$$

In this way the dimension of the LUT can be further reduced, and also the multipliers used in the evaluation of the weighted sum represented in Eq. (3.33) can be conse-

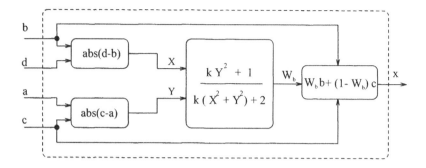

Fig. 3.15   Implementation of Eq. (3.15).

quently simplified. Several simulations have shown that this sort of approximation is acceptable in several rational algorithms.

Example

The algorithm for image interpolation proposed in Eq. (3.15) is numerically quite simple, so it can be easily implemented on an ASIC circuit or on an FPGA. The possible hardware implementation to obtain a high-speed performance is shown in Fig. 3.15, where the main algorithm has been subdivided into three different subblocks.

Due to the complexity of the realization of some operations like multiplication, division and power elevation, and also due to the fact that these operators cannot reach high-speed performance with standard sequential methods, it is advisable to use a look-up table to group most of these operators. On the other hand, to limit memory requirements, it may be useful to compute $|a - c|$ and $|b - d|$ first, and subsequently to use these 8-bit values as addresses for the LUT memory.

In this way a simplification is attained by avoiding the use of the two's complement representation: all the internal signals of the circuit are positive integer numbers, and thus they can be realized using unsigned representation. The block that computes the absolute value of the difference between two numbers is simple to realize, so it may be convenient to use two operators working in parallel in order to increase the speed performance.

As was mentioned earlier, a considerable simplification can be obtained in noting that the values of the coefficient w_b in Fig. 3.15 are between 0 and 1. These values are weights that keep the interpolated pixel similar to its left or its right neighbor, so a coarse quantization of these values does not much affect the result, while it can produce an important simplification in the hardware implementation. In particular, very good results (a peak SNR, PSNR, just 0.04 dB less than that of the floating point simulation) are obtained using a sigmoid-like quantization composed of seven steps as in Eq.(3.36).

Table 3.1   Codes for the w_b and w_c quantized coefficients

w	Code	w	Code
0	000	1	111
1/8	001	7/8	110
1/4	010	3/4	101
1/2	011	1/2	100

The advantage of this approach is twofold.

- The cost to compute the approximated coefficient can be reduced.
- The final weighted sum is significantly simplified, since a shift-register and an adder can replace the multipliers in the block that evaluates the weighted sum.

A further simplification can be obtained using a suitable 3-bit code to represent the w_b and w_c coefficients as in Table 3.1. In such a way a symmetrical code is realized in which w_b and w_c are put in symmetrical positions in the code table. Consequently, the evaluation of $w_c = (1 - w_b)$ can be computed directly from the code of w_b using a simple 3-bit inverter.

The proposed code for w_b and w_c weights is suitable for a further simplification. While one of the two multipliers can be replaced by a simple shift register and one adder, the second can be realized using only a shift register. To explore this possibility, we need, of course, a circuit to exchange the input pixels values b and c between the two different "simplified multipliers" according to the weights values. This information can be easily obtained from the codes listed in Table 3.1, and in particular from their most significant bit. In this way a couple of multiplexers can be used to exchange the input buses between the two simplified multipliers, as can be seen Fig. 3.16. Consequently, the information about w_b and w_c, that is fed to the two simplified multipliers can be reduced to only 2 bits.

Good performance can be reached using a LUT as core of the circuit. Unfortunately, this solution limits the possible implementation of the circuit on an FPGA. In fact, even if we can limit the memory size to 16 bit of address × 3 bits of word, the result is too large to be implemented inside an FPGA, at least using the present technology. Of course, we could attain this memory with an external chip, but in this way the performance speed will be compromised.

3.4.3.2   The square power operator   This kind of operator is often used in a rational filter structure and usually in conjunction with a subtraction operator. Typically it allows the evaluation of the distance between two data, weighted through a nonlinear function, that emphasizes large differences compared to small ones. Even though the realization of such an operator is not expensive, it is useful to simplify it by

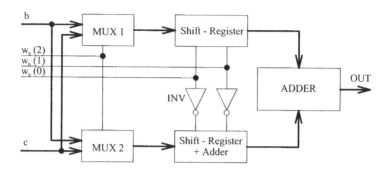

Fig. 3.16   Proposed hardware implementation for the weighted-sum block.

examining its behavior. In particular, since this operation is used to realize a sensor for the difference between two samples, it is not essential to evaluate it through a full-precision square power unit, and some approximated solutions can be exploited. A coarse approximation for such operator may be reached by substituting the square power with an absolute value. This solution, however, is typically not acceptable because it significantly compromises the actual performance of the algorithm. In this case, indeed, the emphasis on large differences is absent, and the entire algorithm behavior may be prejudiced. A different, more appropriate solution may be achieved using a piecewise linear approximation. Let us consider a fixed point unsigned arithmetic. The input data are composed of N bits

$$\{x_N, x_{N-1}, \ldots, x_1, x_0\}, \tag{3.37}$$

while the output data contain $2N$ bits:

$$\{y_{2N}, y_{2N-1}, \ldots, y_1, y_0\}. \tag{3.38}$$

Calling m the position of the most significant bit in the input data, which is the position of the leftmost bit set to one, we can create an intermediate bit stream composed of the bits

$$\{x_m, x_{m-2}, x_{m-3}, \ldots, x_1, x_0\}. \tag{3.39}$$

Applying to this variable a left shift by $m + x_{m-1}$, a good approximation of the square power can be obtained. In Fig. 3.17 an example of the application of this algorithm is depicted, while in Fig. 3.18 a comparison diagram between the proposed algorithm and a full precision square power operator is represented.

A peculiarity of the proposed algorithm is its capability to obtain an approximation error proportional to the input magnitude. In other words, the smaller the inputs, the higher is the precision. The fundamental advantage of such an approximation is its simplicity. Indeed, its realization requires only a shifting structure, a recursive unit to evaluate the most significant bit position and a few other simple and low-cost structures that are easy to synthesize. Notwithstanding its good performances and

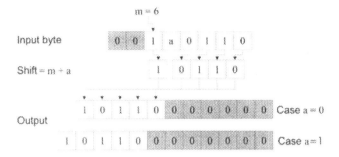

Fig. 3.17   An example of the piecewise approximation described in the text: The 8 bits input sample presents the 2 most significant bits equal to zero; in the first step, an intermediate bit stream composed of 5 bits is realized. This stream is composed of the 4 least significant bits of the input sample and a "1" in the fifth position. In the second step, the intermediate bit stream is shifted to the left by a suitable number of positions defined by the sum of two conditions: (1) the position of the first "1" in the input sample, and (2) the value of the bit in position $m\text{-}1$.

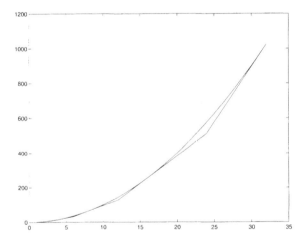

Fig. 3.18   A comparison between the ideal square power function and the approximate proposed operator.

its evident simplicity, this simplified operator does not have to be a direct substitute for the square power operator in the data path structure, but this is suitable when an approximated evaluation is sufficient, as it often happens in rational filters.

3.4.4 Realizations of Rational Operators on DSP's

Several rational operators have also been implemented on commercial DSPs. The advantages of a DSP implementation are of course the low-cost development and the ease of modification or upgrade, so this is a good choice for demonstrators, for example. From the other side, for large-scale production a VLSI implementation can lead to lower overall costs. Moreover a DSP has an inherent maximum computational capability that sets an upper limit to the algorithm complexity. If more computing power is needed, it is necessary to resort to multiple DSP systems, which may be difficult to program efficiently.

3.4.4.1 Realization of a spatiotemporal filter on Texas Instruments TMS320C80 DSP In [10] a spatiotemporal filter is presented that operates on three frames. The filter is implemented on the Texas Instruments 'C80 DSP. Processing speeds of 10 fps are reported on CIF format video. A quadruple buffer scheme has been used: each frame is divided into four vertical portions, and each portion is assigned to one of the four parallel processors (PP) available in the 'C80 DSP. Meanwhile the master processor coordinates their activities and controls data acquisition and display. The code has been optimized using the three basic types of parallelism of the PP's:

- *Multiprocessor parallelism.* Each PP can be used as a separate independent unit. In particular, in this system each image is divided into 4 horizontal, 72-pixel-high strips, and each PP takes care of a different strip.

- *Internal instruction parallelism.* Both the master and the parallel processors can execute simultaneously up to 4 RISC-like instructions if these refer to different subunits such as global address unit, local address unit, arithmetic and logical unit (ALU), and so on.

- *Internal ALU parallelism.* Each 32-bit ALU can split into four 8-bit sub-ALUs.

In order to reduce the computation time per pixel, fixed point arithmetic and LUT to compute the divisions are used. A simplified algorithm flowchart is depicted in Fig. 3.19.

For what concerns this implementation, the major problem consists in the small amount of memory available on the chip. Indeed, each image strip has to be further subdivided into small segments. Each of these segments is loaded in the internal memory, processed, and the corresponding output is saved in the external memory, before moving to the following segment. This approach implies a heavy data traffic with the external memory. This traffic may overload the transfer controller, which is the unit which takes care of both *direct memory access* (DMA) and memory interface, with consequent significant decrease in system overall performance.

If a frame rate higher than 10 fps is needed, it is possible to simplify the operator by considering only one direction in the spatiotemporal part of the RF. In this case the natural choice is the straight temporal direction obtained with x_0^t, x_0^{t-1} and x_0^{t+1}, that

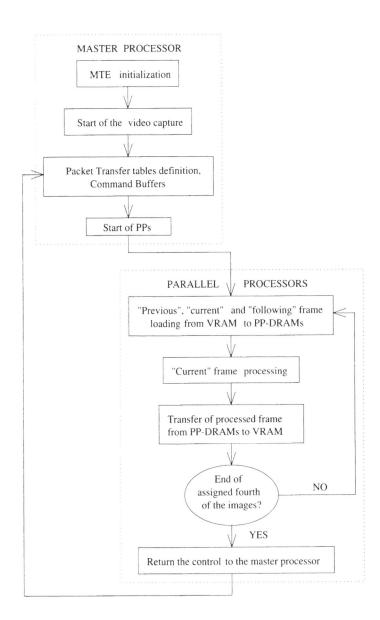

Fig. 3.19   Video processing flow for the implementation of the rational filter on the Texas 'C80 processor.

is, with the pixels in the three input frames that have the same spatial position as x_0^t. Then the temporal part of the filter can operate only in case of absence of motion, such as, with static background. By this simplification, a video-processing frequency of 17 fps has been achieved with the CIF format. It is worth noting that according to the cycles count provided by the DSP simulator, 25 fps would be possible. However, this does not happen in practice due to the already mentioned memory access conflicts.

3.4.4.2 Realization of a spatiotemporal filter on the Philips Trimedia TM1000 DSP A similar approach is proposed in [12]. There, however, the filter operates on two temporally adjacent frames only, and the target architecture is the Philips Trimedia TM1000 DSP, which is a DSP mainly addressed to multimedia consumer electronics applications. In this case, 12.5 fps are obtained for CIF format video. Note that approximately the same frame rate is obtained as in the previous case, notwithstanding the greater simplicity and lower cost of this DSP with respect to the Texas 'C80. This is due both to the architecture of the Trimedia, which is a very long instruction word (VLIW) processor and to the very high efficiency of its C compiler. It is worth noting that a consequence of the much simpler internal architecture is the strong positive impact on the ease of central processor unit (CPU) programming, since no problems relating to interprocessor communication and synchronization have to be considered.

Also in this case attention is given to the computation of the divisions, and look-up tables are used. Moreover quantization of the entries for these tables are used to reduce their size. It is well known that in modern DSPs speed is often limited by the cache, so if a very large look-up table occupies most of it, data and/or code loading speed from the memory are significantly reduced. Finally, for what concerns the implementation on the TM1000, the set of custom operations provided by DSP manufacturer is used in order to optimize the speed of data processing, together with loop optimization techniques, restricted pointers and profiling procedures.

3.4.4.3 Realization of an image interpolator on the Philips Trimedia TM1000 DSP
Similar considerations of course apply to the implementation of rational based interpolators on DSPs. A rational $2\times$ interpolator is implemented on the Philips Trimedia DSP. The algorithm operates at 10 fps, providing standard definition television format images from CIF format ones. In order to keep the computational cost low, the nonlinear operator of Eq. (3.25) was simplified by removing one of the three rational parts and neglecting some of the terms of the polynomials. To optimize the DSP performance, loop unrolling and #define commands with several custom operations are used.

The loop unrolling procedure can be explained with an example. Let us consider the following C source where a convolution problem is proposed:

```
for(k=0;k<NMAX;k++){
        c[k]=0;
        for(j=0;j<8;j++){
        c[k]+=b[j]*a[k-j];
```

```
        }
}
```

The following C source is the previously described code unrolled four times: every time the k variable is incremented, four assignments are used instead of performing the loop.

```
for(k=0;k<NMAX;k+=4){
c[0]=b[0]*a[0]+b[1]*a[-1]+b[2]*a[-2]+b[3]*a[-3]+
     b[4]*a[-4]+b[5]*a[-5]+b[6]*a[-6]+b[7]*a[-7];
c[1]=b[0]*a[1]+b[1]*a[0]+b[2]*a[-1]+b[3]*a[-2]+
     b[4]*a[-3]+b[5]*a[-4]+b[6]*a[-5]+b[7]*a[-6];
c[2]=b[0]*a[2]+b[1]*a[1]+b[2]*a[0]+b[3]*a[-1]+
     b[4]*a[-2]+b[5]*a[-3]+b[6]*a[-4]+b[7]*a[-5];
c[3]=b[0]*a[3]+b[1]*a[2]+b[2]*a[1]+b[3]*a[0]+
     b[4]*a[-1]+b[5]*a[-2]+b[6]*a[-3]+b[7]*a[-4];
a+=4; c+=4; } }
```

The unrolled code is obviously larger but works faster. Similarly macro definitions that use the #define command instead of procedures significantly improve the execution speed.

3.5   CONCLUSIONS AND OPEN ISSUES

In this chapter the possibilities offered by algorithms based on rational functions for the processing of images are examined. Methods are presented for the reduction of different types of noise in images and image sequences, for the enhancement of the contrast, and for interpolation.

An important problem that is unsolved is the design of the type of operator, that is often based on heuristic criteria. This problem is the subject of future research.

It should also be seen that several aspects have to be taken into account when deciding the type of implementation. The various possible solutions (ASIC, FPGA, DSP) present different levels of development cost, device cost, and speed, so the choice depends on the performance constraints and on the production volume needed. For what concerns the hardware realizations, efficient implementations can be obtained if care is given to the realization of critical operators, such as the square operator or the divider, where trade-offs are sought between data resolution and overall precision. When considering DSPs, in turn, memory management turns out to be the main issue, because of possible bottlenecks in the DMA and/or in the internal cache.

REFERENCES

1. V. Volterra. Sopra le funzioni che dipendono da altre funzioni. *Rendiconti Regia Accademia dei Lincei*, 1887.

2. V.J. Mathews. Adaptive polynomial filters. *IEEE Signal Processing Magazine*, 8(3): 10-26, July 1991.

3. G.L. Sicuranza. Quadratic filters for signal processing. *IEEE Proc.*, 80(8): 1263-1285, August 1992.

4. V.J. Mathews and G.L. Sicuranza. *Polynomial Signal Processing*. Wiley, New York, 2000.

5. H. Leung and S. Haykin. Detection and estimation using an adaptive rational function filter. *IEEE Trans. Signal Processing*, 42(12): 3366-3376, December 1994.

6. G. Ramponi. Bi-impulse response design of isotropic quadratic filters. *IEEE Proc.*, 78(4): 665-677, April 1990.

7. G. Ramponi. The rational filter for image smoothing. *IEEE Signal Processing Letters*, 3(3): 63-65, March 1996.

8. G. Ramponi and C. Moloney. Smoothing speckled images using an adaptive rational operator. *IEEE Signal Processing Letters*, 4(3): 68-71, March 1997.

9. L. Khriji and M. Gabbouj. Vector median-rational hybrid filters for multichannel image processing. *IEEE Signal Processing Letters*, 6(7): 186-190, July 1999.

10. F. Cocchia, S. Carrato, and G. Ramponi. Design and real-time implementation of a 3-D rational filter for edge preserving smoothing. *IEEE Trans. Consumer Electronics*, 43(4): 1291-1300, November 1997.

11. S. Marsi, R. Castagno, and G. Ramponi. A simple algorithm for the reduction of blocking artifacts in images and its implementation. *IEEE Trans. Consumer Electronics*, 44(3): 1062-1070, August 1998.

12. L. Tenze, S. Carrato, C. Alessandretti, and S. Olivieri. Design and real-time implementation of a low-cost noise reduction video system. In *Proc. COST 254 Workshop on Intelligent Communitation Technologies and applications, with Emphasis on Mobile Communications*, Neuchatel, Switzerland, May 1999, pp. 36-40.

13. E. Lloyd. *Handbook of Applicable Mathematics*. Wiley, Chichester, 1980.

14. I. Pitas and A. N. Venetsanopoulos. *Nonlinear Digital Filters*. Kluwer Academic, Boston, 1990.

15. S. I. Olsen. Estimation of noise in images: An evaluation. *CVGIP*, 55(4): 319-323, July 1993.

16. G. Ramponi and S. Carrato. Interpolation of the dc component of coded images using a rational filter. In *Proc. 1997 IEEE Int. Conf. Image Processing*, Vol. I, Santa Barbara, CA, October 1997, pp. 389-392.

17. R.G. Keys. Cubic convolution interpolation for digital image processing. *IEEE Trans. Acoustics, Speech, Signal Processing*, 29(6): 1153-1160, December 1981.

18. S. Carrato, G. Ramponi, and S. Marsi. A simple edge-sensitive image interpolation filter. In *Proc. 1996 IEEE Int. Conf. Image Processing*, Vol. III, Lausanne, Switzerland, September 1996, pp. 711-714.

19. L. Khriji, F.A. Cheikh, M. Gabbouj, and G. Ramponi. Color image interpolation using vector rational filters. In *Proc. SPIE Conf. Nonlinear Image Processing IX*, Vol. 3304, San Jose, CA, January 1998, pp. 26-29.

20. S. Carrato and L. Tenze. A high quality 2× image interpolator. *IEEE Signal Processing Letters*, 7(5): 132-134, June 2000.

21. L. Khriji, M. Gabbouj, G. Ramponi, and E.D. Ferrandiere. Old movie restoration using rational spatial interpolators. In *Proc. 6th IEEE Int. Conf. Electronics, Circuits and Systems*, Vol. 2, Cyprus, September 1999, pp. 1151-1154.

22. G. Ramponi and A. Polesel. A rational unsharp masking technique. *Electronic Imaging*, 7(2): 333-338, April 1998.

23. M.R. Zargham. *Computer Architecture: Single and Parallel Systems*. Prentice-Hall, Englewood Cliffs, NJ, 1998.

4 Mathematical Morphology and Motion Picture Restoration

E. DECENCIERE FERRANDIERE

Center of Mathematical Morphology
Ecole des Mines de Paris
Fontainebleau, France

4.1 INTRODUCTION

The field of digital motion picture restoration has experienced rapid technological growth during the last few years. There are several reasons for this development. First, there is the problem that most old motion pictures and video material are going through inexorable deterioration. The common figure stated is that half of the motion pictures shot before 1950 has disappeared [10]. Films have to be preserved and many of them have to be restored. Classical physicochemical restoration methods can correct some defects, but not all. In addition the development of the audiovisual market has put a high demand on high-quality broadcasting material. Old motion pictures could contribute to fill this demand, provided that they are restored. Besides, restoration tools are also of interest for the postproduction stage of the modern film industry. They can be used to correct defects caused by accidents during the shooting process. Finally, the development of digital restoration tools has become possible thanks to the increase of the ratio "computing power" over "computing price." It is now possible to imagine, for instance, digital restoration systems based on PCs.

Most existing digital restoration systems are essentially manual: the user has to detect and correct the defects mostly by hand. For example, the restoration of *Snow White* in 1993 was done this way. The achieved quality was excellent, but the cost was very high. In order to make digital restoration affordable for more films, the cost factor has to be reduced. One way of achieving this is to increase the degree of automation in the restoration process. This work goes in that direction. Given the high-quality results required from restoration work, complete automation is extremely difficult to achieve.

Several restoration algorithms have been developed at the Center of Mathematical Morphology. They treat jitter, flicker, scratches, and blotches. A prototype restoration system named *SARSA*, which stands in French for *Système Automatique de Restauration de Séquences Animées* (Automatic Restoration System for Animated Sequences), was developed based on these tools. This work was partly founded by the European ESPRIT-LTR NOBLESSE project.

Gray level and color images are considered in this chapter, but it is supposed that the input image sequences do not contain any scene cuts. This hypothesis is realistic given that researchers working on video indexing have developed cut detection algorithms.

Most of the restoration methods hereafter described are based on mathematical morphology operators. The reader interested by mathematical morphology is referred to the books on the subject [42, 43, 41, 46].

4.2 DIGITAL IMAGE SEQUENCE RESTORATION: A HISTORICAL OVERVIEW

In 1989, in an article about motion picture restoration, this author talked about digital restoration methods only in the conclusion, and as a long term possibility [37]. Four years later, in 1993, the motion picture *Snow White*, from Walt Disney studios, was the first film to be restored using only digital techniques, thanks to the Cineon system that then belonged to Kodak. However, the work was done practically image by image and was therefore very expensive.

The same year the Fox Movietone News preservation project started. The aim of the project was to scan over 13 million meters of 35mm film, and to store them on digital tapes. The transfer was finished three years later [45].

The European Union has financed several projects for the development of digital tools for film and video restoration. The EUREKA LIMELIGHT project (07/1994-01/1997) aimed at developing high-resolution scanning tools and restoration algorithms. As far as we know they did not publish any results. The objective of the ACTS AURORA project, started in September 1995, was to develop fast digital restoration tools for television archives. Researchers working on that project have published interesting articles on digital restoration. We can cite for example J.H. Chenot [2], A.C. Kokaram [22], D. Lyon [28], and P.M.B. van Roosmalen [49, 50]. As said previously, the work presented in this chapter was financed by the ESPRIT LTR NOBLESSE project (01/1996-12/1998). Motion picture restoration was one of the main application fields for the nonlinear tools developed within the project.

Finally, more and more private companies are offering digital restoration services. Their restoration techniques remain confidential, but as far as we know there is room for improvement.

This historical overview shows that the field of digital motion picture and video restoration has experienced extremely fast development. Its dynamic growth should continue for the years to come.

4.3   PHILOSOPHY OF THE APPROACH

Originally the objective of this work was to develop automatic tools for digital motion picture restoration. By automatic it is meant that the user of the restoration algorithm should only have to choose some parameters, and then the algorithm should treat a whole sequence of images or several sequences on the fly. Afterward, the user would check that the restoration worked correctly and be able to cancel or undo some corrections.

Another good characteristic for a restoration algorithm is to be able to evaluate the confidence of its own results, that is, how well the detected feature, which is supposed to be a defect, corresponds to the defect model. Such a *restoration quality measure* would greatly help the user during the verification phase.

At the beginning of this work we were not concerned about the speed of the algorithms. Since, an automatic algorithm is theoretically able to work alone, the delays are not critical. However, given the huge amount of information contained in a film, the algorithms cannot afford to be too slow. For example, a full resolution motion picture frame contains as much as 12 million pixels. At 24 frames per second this means that an hour of film contains more than 10^{12} pixels! For video sequences, these figures should be divided approximately by 25. If the complete restoration of a single frame took one minute (which, given the above figures, could be considered fast for a motion picture frame), the total processing time would be of 1500 hours. A network of 20 PCs would then need three days to restore one hour of film, provided that networking and parallelization problems are solved. So in the end, we had to also develop efficient algorithms, even if speed was not the main priority.

In the following sections some of the most common defects in old motion pictures are treated:

- *Flicker.* This defect corresponds to an abnormal variation of the luminance in the scene. It is found essentially in very old motion pictures.

- *Jitter.* Many old motion pictures present parasite vibrations, very annoying for the spectator.

- *Scratches.* Vertical scratches are very common in old motion pictures, as well as in more recent films. They appear as vertical light or dark lines and stay approximately at the same position through many frames.

- *Local random defects.* All defects that cover a relatively small region of each frame and that most of the time do not appear in the same position between consecutive frames that belong to this category.

The order of application of the algorithms was chosen according to the following rule: "Apply first the algorithms that simplify the most the work of the following algorithms." Different restoration algorithms for these defects are fully described in the Ph.D. thesis of Etienne Decencière [6]. However, at that time we only had little experience about their possible applications to real industrial problems. Now we are able to keep only the best ones. Despite these achievements automatic film restoration remains an extremely difficult task and far from been completely solved. Some of the algorithms described in this chapter have shortcomings. It will be explained why they do not always work as expected, and possible solutions will be proposed.

The initial work was only done with black and white images. It will be explained here, when necessary, how to extend the algorithms to color images.

4.4   FLICKER CORRECTION

Flicker appears as an abnormal luminance variation in the film. It is typical of old black and white motion pictures. Therefore flicker correction is applied to gray level images.

Let us have a look at the evolution of the mean, the minimum, and the maximum gray level of two image sequences from old motion pictures (see Figs. 4.1 and 4.2). In order to reduce the effects of noise, we considered either the 1% of pixels with the smallest gray level value or the 1% pixels with the largest gray level value. Both sequences are in very bad shape and present defects other than flicker. The first sequence comes from a film by the Lumière brothers, *Arrivée d' un train en gare de Sidney*[1]. It shows moderate flicker (see Fig. 4.1) and is severely damaged by other defects, like dark crackles and local random defects. The second sequence comes from a film by Max Linder. Flicker is in this case very severe (see Fig. 4.2). It can be seen that the mean, maximum, and minimum gray level values for these two sequences vary significantly with time. These variations are caused by flicker but also by changes in the scene.

When flicker is very severe, information can be lost. This happens when the image is saturated. It is then very difficult to recover the initial information. Some frames of the Max Linder sequence present this problem. Moreover flicker does not always homogeneously affect each image. Often the defect is more severe near the borders of the image. However, the presented algorithm is based on the assumptions that there is no saturation problem and that the effect of flicker is homogeneous on each image.

Flicker correction methods have not been found in the literature. However, P. Richardson and D. Suter propose a method for histogram equalization intended for improving the results of a motion compensation algorithm [38]. For each frame I_k independently from the others, they compute the minimal gray level value $m(I_k)$ and

---

[1]Circa 1896 © Association frères Lumière.

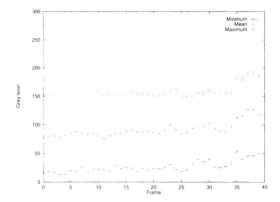

Fig. 4.1 Minimum, mean, and maximum gray levels for the images of the Lumière brothers sequence.

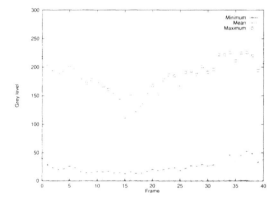

Fig. 4.2 Minimum, mean, and maximum gray levels for the images of the Max Linder sequence.

the maximal gray level value $M(\mathbf{I}_k)$. In order to reduce the effect of noise, when calculating $m(\mathbf{I}_k)$ and $M(\mathbf{I}_k)$ the authors remove from the histogram the bN pixels that have the smallest gray level value, as well as the bN pixels which have the largest gray level value, where N is the total number of pixels of image $\mathbf{I}_k$. A practical value for parameter b is 0.01. Afterward, they use $m(\mathbf{I}_k)$ and $M(\mathbf{I}_k)$ to normalize $\mathbf{I}_k$ and to obtain in this way a new image $\mathbf{I}_k^*$ defined as (P is a pixel from $\mathbf{I}_k$):

$$\mathbf{I}_k(P) \leq m(\mathbf{I}_k) \quad \Rightarrow \quad \mathbf{I}_k^*(P) = 0,$$
$$m(\mathbf{I}_k) < \mathbf{I}_k(P) < M(\mathbf{I}_k) \quad \Rightarrow \quad \mathbf{I}_k^*(P) = \frac{\mathbf{I}_k(P) - m(\mathbf{I}_k)}{M(\mathbf{I}_k) - m(\mathbf{I}_k)} G,$$
$$M(\mathbf{I}_k) \leq \mathbf{I}_k(P) \quad \Rightarrow \quad \mathbf{I}_k^*(P) = G,$$

where G is the maximal gray level value of the images. The flicker correction method is inspired by this approach.

4.4.1   Algorithm Description

The flicker correction algorithm is based on two hypotheses. The first is that the degradation can be modeled as an affine transformation of the original image $\mathbf{I}_k^o$:

$$\mathbf{I}_k = \alpha_k \mathbf{I}_k^o + \beta_k. \tag{4.1}$$

This relation is true in particular for the mean gray level $\bar{\mathbf{I}}_k$, minimum gray level $m(\mathbf{I}_k)$, and maximum gray level $M(\mathbf{I}_k)$ of the images:

$$\bar{\mathbf{I}}_k = \alpha_k \bar{\mathbf{I}}_k^o + \beta_k, \tag{4.2}$$

$$M(\mathbf{I}_k) = \alpha_k M(\mathbf{I}_k^o) + \beta_k, \tag{4.3}$$

$$m(\mathbf{I}_k) = \alpha_k m(\mathbf{I}_k^o) + \beta_k. \tag{4.4}$$

In order to calculate the parameters of the affine transformation, a second hypothesis is adopted. It is supposed that the mean gray level value of the original image $\bar{\mathbf{I}}_k^o$, as well as the difference between the maximum and minimum gray level values $(M(\mathbf{I}_k^o) - m(\mathbf{I}_k^o))$, are constant through time.

Based on these assumptions, an iterative algorithm is built: we suppose that the previous image $\mathbf{I}_{k-1}$ has already been restored. Thus $\bar{\mathbf{I}}_{k-1}^o$ and $M(\mathbf{I}_{k-1}^o) - m(\mathbf{I}_{k-1}^o)$ are known. As a consequence, given the second hypothesis, $\bar{\mathbf{I}}_k^o$ and $M(\mathbf{I}_k^o) - m(\mathbf{I}_k^o)$ are deduced. Therefore we obtain:

$$\alpha_k = \frac{M(\mathbf{I}_k) - m(\mathbf{I}_k)}{M(\mathbf{I}_k^o) - m(\mathbf{I}_k^o)}, \tag{4.5}$$

$$\beta_k = \bar{\mathbf{I}}_k - \frac{M(\mathbf{I}_k) - m(\mathbf{I}_k)}{M(\mathbf{I}_k^o) - m(\mathbf{I}_k^o)} \bar{\mathbf{I}}_k^o. \tag{4.6}$$

Finally the restored frame is obtained:

$$\mathbf{I}_k^o = \frac{\mathbf{I}_k - \beta_k}{\alpha_k}. \tag{4.7}$$

However, this method has two drawbacks. First, it is very sensitive to noise because of the use of the minimum and maximum gray level values. In order to reduce this problem, neither the bN pixels with smallest gray level value are taken into account for calculating $M(\mathbf{I}_k)$ and $m(\mathbf{I}_k)$ nor the bN pixels with largest gray level value. Second, the basic hypothesis about the constancy of the reference gray level values is too strict. For instance, there can be illumination changes in the scene, or new bright or dark objects. In order to solve this inconvenience, a variation through time of the mean and extreme values is allowed. This variation between consecutive

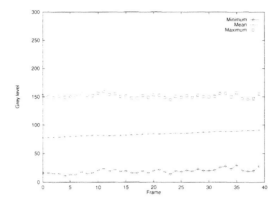

Fig. 4.3   Minimum, mean, and maximum of the restored Lumière brothers sequence.

frames is equal to δ_1 for the mean gray level value, and to δ_2 for the extreme values. For example, $M(\mathbf{I}_k^o)$ is computed in the following way:

- If $M(\mathbf{I}_{k-1}^o) - \delta_2 \leq M(\mathbf{I}_k) \leq M(\mathbf{I}_{k-1}^o) + \delta_2$ then $M(\mathbf{I}_k^o) = M(\mathbf{I}_k)$.

- If $M(\mathbf{I}_k) < M(\mathbf{I}_{k-1}^o) - \delta_2$ then $M(\mathbf{I}_k^o) = M(\mathbf{I}_{k-1}^o) - \delta_2$.

- If $M(\mathbf{I}_{k-1}^o) + \delta_2 < M(\mathbf{I}_k)$ then $M(\mathbf{I}_k^o) = M(\mathbf{I}_{k-1}^o) + \delta_2$.

The same method is used to compute $m(\mathbf{I}_k^o)$ and $\bar{\mathbf{I}}_k^o$. Practical values for δ_1 and δ_2 are respectively 0.5 and 1.

4.4.2   Restoration Quality Measure

Based on these hypotheses (linear deformation of the histogram, slow variation for the minimum, mean, and maximum gray levels) we have computed a set of restoration parameters α_k and β_k. If these hypotheses hold, then the histogram of the restored image $\mathbf{I}^{o_k}$ should be very similar to the histogram of the precedent, restored image $\mathbf{I}^{o_{k-1}}$. This is easy to check by computing a histogram distance (e.g., using the mean absolute difference between histogram classes). This value is a good restoration quality measure.

4.4.3   Results and Comments

For moderate flicker let us come back to the Lumière brothers sequence. The application of the above algorithm to this sequence improves the overall quality. Flicker completely disappears. Figure 4.4.3 shows the new values of the mean, maximum and minimum gray levels along time. The improvement appears even more clearly when looking at the animated sequence.

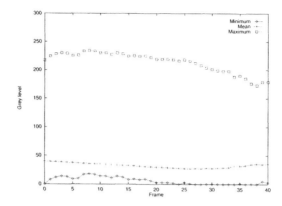

Fig. 4.4   Minimum, maximum, and mean gray level values for the restored Max Linder sequence.

In Fig. 4.4 we show the minimum, mean, and maximum gray level values for the images of the restored Max Linder sequence, using the same parameters as for the Lumière brothers sequence. The mean is now stable and the extremal values of the histogram vary smoothly. When the sequence is visualized, it is effectively observed that its global quality has improved, but some problems still remain, especially for the frames where flicker caused saturation.

This method gives good results as long as the basic assumptions of the algorithms are respected. It is also very fast. However, as we said above, in many cases flicker does not affect the whole image in the same way; the degradation is often worse near the borders, for example. As noted in [6], this approach can be generalized by dividing each frame into blocks and by applying this method inside each block. A convenient interpolation of the correction parameters between blocks should then be used to avoid block artifacts. The deformation caused by flicker is more complex than an affine transformation. A more complex model could be used, but we think that for most cases the affine transformation is a good approximation to the real deformation. Finally, when flicker causes saturation, the problem becomes more complex, because missing gray levels have to be interpolated.

Once that flicker has been corrected, motion estimation algorithms, which are often based on a constant illumination assumption, perform better. Therefore the flicker correction algorithm is applied before the jitter correction algorithm, which is described hereafter.

4.5 JITTER CORRECTION

Jitter appears in the image sequence as parasite vibrations, similar to those produced by an amateur with a hand-held camera. This kind of defect can be caused, for example, by mechanical problems during scanning or during transfer with a telecine.

A method of jitter correction has been proposed by T. Vlachos and G. Thomas [51]. In order to estimate the global translation between two frames, the authors compute the motion of several blocks and then, by means of a voting procedure, keep the best translation vector. However, it was not clear how the authors differentiated between the natural global motion of the scene and jitter. Our approach is similar, but more importance has been given to the separation between the jitter and the natural motion of the scene.

4.5.1 Description of the Correction Method

The jitter for each image has been modeled as a global translation, and the series of these translations as a high-frequency signal. The jitter correction algorithm consists of two steps.

First, a global translation between consecutive frames is calculated, based on the luminance of the original images. Different methods can be used to compute this translation. A computationally intensive one has been used: all possible translations within a given range are tested, and for each of them the global mean absolute error between $\mathbf{I}_{k-1}$ translated, and $\mathbf{I}_k$ is computed. This method is accurate and robust, but slow. Other faster global translation estimation techniques can be used. The resulting translation vector is equal to the addition of two components: the translation component of the natural background motion of the scene, which varies slowly with time, and the jitter effect, which is a high-frequency signal.

Second, the sequence of translations is filtered in order to separate jitter from the natural global movement of the scene. This is simply done by considering that slow variations of the sequence of translations correspond to the background movement, whereas high-frequency oscillations correspond to jitter. This means that the acceleration in the scene is supposed to be almost constant, which is most of the time true. A mean filter is used to separate both components.

4.5.2 Comments

The jitter correction method has been applied to the Lumière brothers sequence, where there is forward camera motion and a moderate jitter. The results were very good [6]. However, each step of the algorithm can be improved. On the one hand, the global translation estimation, which has a precision of one pixel, can be improved to obtain a half-pixel resolution, and can be accelerated. On the other hand, the filtering used to separate the high-frequency components from the slow varying ones can be improved. A more adequate linear filter could be used, or even a nonlinear filter like a median filter.

After application of the jitter correction algorithm, the image sequence is steadier. This facilitates the work of algorithms that use temporal information, like the local random defects detection algorithm. However, before moving on to the description of that algorithm, the scratch restoration algorithm is described.

4.6   RESTORATION OF VERTICAL SCRATCHES

A very common kind of defect in old and even modern motion pictures is vertical scratches. They appear as light or dark vertical lines that run across many frames of the film. They tend to appear at the same position in consecutive frames. They are caused by friction against the film. Not only are these defects common, they are also very annoying for the spectator. The restoration of these defects seems easy at first view, but in fact they have proved to be very difficult to correct in a satisfactory way.

The literature on this subject is small. Morris describes in his Ph.D. thesis a scratch correction method based on Bayesian techniques [34]. The resulting algorithm is apparently very slow. In [21] Kokaram proposes an improvement of this method: he first computes the lines that are candidates for scratches; then, through a Bayesian refinement, he keeps the best candidates. The interpolation is made by means of a 2-D autoregressive filter. In both publications the restoration phase is divided into two steps: a detection phase and an interpolation phase. Our algorithm will be divided into the same two steps, but morphological tools will be used for the detection and interpolation. Only light scratches will be considered. The proposed method can be applied to dark scratches simply by inverting the original image.

4.6.1   Scratch Detection

Given that scratches appear approximately in the same position in consecutive frames, a temporal approach for their detection is not convenient. A frame by frame approach will be used instead.

Two images have been chosen to illustrate and test the algorithm. They both come from the Polish film *Manewry Milosne*, by J. Nowina-Przybylski and K. Tom. The first, Fig. 4.5(a), shows an example of a scratch that does not cover the whole height of the image. In the second image, Fig. 4.5(b), structures that are almost vertical, besides the scratch, are seen. Both images are of size 768×576.

The description of the detection algorithm follows:

1. A top-hat transformation, using as structuring element a horizontal segment of length D_x, is applied to the original image. This transformation detects vertical structures whose width is smaller than D_x. The result is shown in Fig. 4.6(a) and (b).

(a)   (b)

Fig. 4.5   Test images.

2. The mean of each column of the image is computed; a one-dimensional image is obtained. This allows us to improve the contrast of perfectly vertical structures. Vertical scratches will produce a high value in the one-dimensional image, see Fig. 4.6(c) and (d).

3. Given that scratches are vertical and thin, the same top-hat as previously is applied to the one-dimensional image. Thus other vertical structures are erased, Fig. 4.6(e) and (f).

4. By thresholding the resulting one-dimensional image with a value of g_1, markers are obtained for the columns of the original image that contains vertical scratches. However, some scratches do not cover the whole height of the image, so it will be considered that a pixel belongs to a scratch if the corresponding column has been marked and if its value in the first top-hat image, Fig. 4.6(a) and (b), is larger than a constant g_2. Finally, masks for the scratches are obtained, Fig. 4.6(g) and (h).

For both examples the parameters were $D_x = 5$, $g_1 = 30$, and $g_2 = 4$.

4.6.2   Scratch Interpolation

Once the scratches have been correctly detected, the missing data have to be interpolated. The gray level values of the pixels close to the scratches are often disrupted by the scratch. This is the reason why the detection mask is dilated before the interpolation. A structuring element of size 3×3 is used (see Fig. 4.7).

Since scratches are thin, a simple interpolation method is used: a morphological opening with the horizontal structuring element used during the detection phase. Of course, the result of the interpolation is only applied within the detection mask; the rest of the image remains unmodified. In Fig. 4.8 a detail of the result of this interpolation is shown.

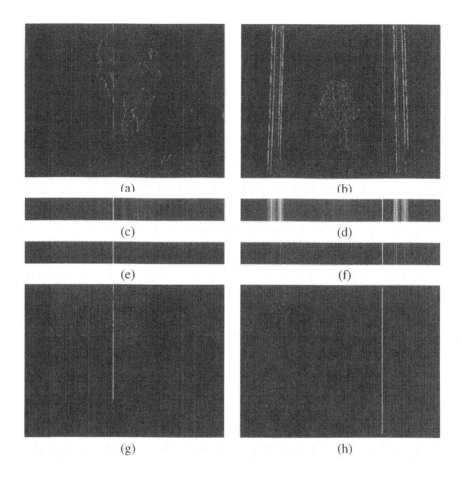

Fig. 4.6 (a) and (b): Result of the application of the top-hat with a horizontal structuring element. (c) and (d): Mean of the columns of the top-hat image. (e) and (f): Result of the application of the same top-hat to the one-dimensional image. (g) and (h): Resulting detection masks. All images presented in this figure have been normalized.

4.6.3 Conclusion

The proposed restoration method is very fast. It successfully detects perfectly vertical scratches. The use of the detection mask allows an easy verification of the results, and allows some corrections to be canceled easily.

The quality of the restored image seems correct, but when the animated sequence is visualized, the interpolated regions remain visible. The interpolation method needs to be improved. Interpolation tests have been undertaken using rational filters (see Chapter 3). The resulting textures are more realistic [18, 19].

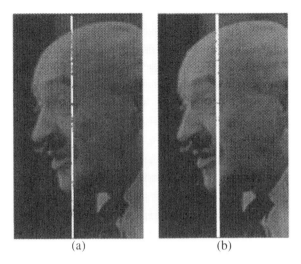

Fig. 4.7   Detection mask detail. (a) Detection mask, superimposed on the original image. (b) Dilatation of the detection mask.

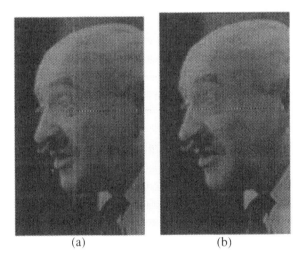

Fig. 4.8   Zoom on the original image (a) and result of the interpolation (b).

The detection method can produce false detections when other perfectly vertical structures appear in the scene. In that sense the detection model is not precise enough. Moreover we have seen since then that some scratches are not always perfectly vertical, so they are not detected by the algorithm. In that sense the scratch model is too strict.

Finally a restoration quality measure has not been defined for this kind of defect. This algorithm was originally developed for black and white films. Since then the detection phase has been applied to color images. This was achieved by applying the algorithm to the luminance of the images. The interpolation phase is more difficult to generalize, because it can be dangerous to interpolate each component of the color image independently from the others.

The next step in the restoration chain is the correction of local random defects.

4.7   RESTORATION OF LOCAL RANDOM DEFECTS

Among motion picture defects, scratches, blotches, and crackles can be treated in a similar way because they affect a small region of each frame and most of the time they do not appear in the same position in consecutive frames. In this section they will be addressed as *local random defects*.

The simplest sort of local random defects is probably noise. Noise reduction in image sequences is often based on a first motion compensation step, followed by filtering along the computed trajectories [15, 9]. Research continues in this domain with more modern tools, but they are still based on a preliminary motion compensation step (e.g., [35, 20, 48]). G. Arce has shown in [1] that satisfactory spatiotemporal filtering can be obtained without motion compensation. To that end he uses multistage order statistic filters. The results are interesting, but in the presence of motion the resulting filtering is only spatial.

Specific publications on image sequence restoration are scarcer. As far as we know the first publication on digital restoration of local random defects was written by R. Storey; he used simple temporal connectivity criteria to detect film dirt [47]. The result was satisfactory provided motion was small.

S. Geman, D.E. McClure, and D. Geman were among the first to apply Bayesian restoration techniques to image sequences [11]. Researchers from Cambridge University have been working on motion picture restoration since then. They have continued the development of Bayesian techniques (e.g., [34]), and they have proposed other restoration methods, like the spike detection index [23, 24] and the 3-D autoregressive model [27]. Some of the resulting publications are [23, 34, 31, 32, 33, 25, 26, 24, 13, 12]. The main restoration algorithms are described in [22]. The 3-D autoregressive model has also been studied by researchers from the Singapour University [14, 17, 3].

All these local random defects restoration algorithms follow the same structure: a motion compensation phase, often based on hierarchical block matching, is followed by a detection phase that gives the position of the defects, and finally an interpolation step reconstructs the missing data.

The same structure will be adopted, except that we will try to avoid the motion compensation phase, or at least to simplify it, for example by considering only camera motion. It will be shown that the use of the morphological opening by reconstruction can build an efficient local random defects detection algorithm that will work even in presence of moderate motion.

Temporal closings and openings by reconstruction were for the first time applied to image sequences to remove impulsive noise by Pardàs and Serra [36]. In [4, 5] a detection method was proposed for local random defects in motion pictures based on the morphological opening by reconstruction. The method was defect oriented in the sense that it had to be adapted to each kind of local defect. In [7] it was shown that the morphological opening by reconstruction was a very useful tool for image sequence analysis, and in [6] different detection methods based on this tool were described. We will see that the reconstruction step results into a motion compensation as long as objects remain connected along time. Only individual object motion will be considered during the interpolation phase, which will not only interpolate missing pixels but will also correct the eventual false detections produced by the displacement of small objects.

We begin with the modeling of local defects and with some basic definitions. Then we describe the detection method, based on morphological openings and closings by reconstruction, and present some results. Next we describe two interpolation methods.

4.7.1   Local Random Defects Model

Local random defects are those motion picture defects that have the following two characteristics:

1. They are local, meaning that they do not affect the whole frame.

2. They are thin along the time axis, meaning that they do not appear in the same position in consecutive frames.

Somehow local random defects are local not only from a spatial point of view but also from a temporal point of view. Note also that local random defects are either darker or lighter than the surrounding image. A large number of motion picture defects belong to this category. In order to build a mathematical model for local random defects some basic definitions are needed.

Flat zone.   The flat zone C of image $\mathbf{I}$ at point $P \in \mathbb{Z}^n$ is the largest connected component of $\mathbb{Z}^n$ which contains P and where $\mathbf{I}$ is constant. Note that a pixel whose gray level value is different from the gray level values of all its neighbors will be a flat zone on its own.

Connected operators [44].   An operator ρ is connected if and only if each flat zone of any gray level image $\mathbf{I}$ is included in a flat zone of image $\rho(\mathbf{I})$. In other words, connected operators only produce fusions of flat zones. As a consequence connected operators do not create new contours on the image.

Connected operators have proved to be very powerful image analysis tools. They have been applied to 2-D image analysis successfully, especially for image segmentation [40]. Afterward they were applied also to image sequences [36, 29, 39]. When we talk about gray level images, we tend to use a vocabulary that does not have a strict mathematical definition. The main examples are the "objects" that we can see in a scene. They have a meaning for us because we apply our semantic models, but they are very difficult to define as geometrical entities. This is the reason why we are going to introduce the definition of light and dark objects.

Light and dark objects. A subset Ω of $\mathbb{Z}^n$ is a light object of gray level image $\mathbf{I}$ if and only if there is an integer i such that Ω is a connected component of the threshold of $\mathbf{I}$ at level i, denoted by $\mathcal{T}^{(i)}(\mathbf{I})$. Let Ω be a light object of $\mathbf{I}$. There might be several i such that Ω is a connected component of $\mathcal{T}^{(i)}(\mathbf{I})$; the smallest of those values will be called the gray level of Ω. The dark objects of $\mathbf{I}$ are the light objects of the inverse of $\mathbf{I}$.

Light and dark objects are somehow close to our semantic models, and they are easy to manipulate with the morphological geodesic operators, as we will see in the following sections. In order to avoid confusion, from now on if the term "object" is used alone it will refer to semantic objects; otherwise, the terms "light" and "dark" objects will be used.

As was noticed before, local random defects are in practice either darker or lighter than the surrounding image. They will be called dark and light defects, respectively. Local random defects are noncorrelated in time. As long as their density is not too high, the probability that the intersection between defects of consecutive frames is not void is very small. The first model is based on the two previous remarks:

First model. A dark defect of frame $\mathbf{I}_k$ is a dark object Ω of frame $\mathbf{I}_k$ such that there is no dark object Ω' of frames $\mathbf{I}_{k-1}$ or $\mathbf{I}_{k+1}$ of same or lower gray level such that $\Omega \cap \Omega' \neq \emptyset$. Light defects are modeled in an analogous way, using this time light objects. But, if the defect density is too high, the model above becomes too strict. A second, more general model, is proposed:

Second model. A dark defect of length L (the length is temporal, measured along the time axis) of frame $\mathbf{I}_k$ is a dark object Ω_k of frame $\mathbf{I}_k$ such that there are no $L + 1$ dark objects $\Omega_{k_0}, \Omega_{k_0+1}, \ldots, \Omega_{k_0+L}$ belonging to frames $\mathbf{I}_{k_0}, \ldots, \mathbf{I}_{k_0+L}$ of same or lower gray level than Ω_k, such that k belongs to $\{k_0, \ldots, k_0 + L\}$ and such that $\bigcap_{k \in \{k_0 \ldots k_L\}} \Omega_k \neq \emptyset$. Light defects are modeled in an analogous way, using this time light objects.

Of course this model, as all models, does not perfectly describe the reality. For example, if a defect is in contact with an object of similar gray level in the same frame, then it will not be detected. Moreover, given that the model does not take motion into account, objects that get disconnected along time because of motion can produce false detections. This typically happens with small moving objects. However, this model gives good results.

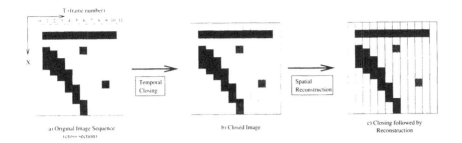

Fig. 4.9 Example of image sequence processing with morphological operators. Regions in light gray show the differences between the current image and the original one. (a) Cross section of the original image sequence along the X and T axis. (b) Closing of the original image sequence with B_k. (c) Spatial reconstruction of the closed image.

4.7.2 Detection of Local Random Defects

Now that a mathematical model for local random defects has been proposed, mathematical morphology will be used to detect them. The main tools will be the openings and closings by reconstruction.

Image sequences will be modeled as gray level images of $\mathbb{Z}^3$; the third dimension will correspond to time. Each pixel in the sequence will be identified by its coordinates (n_1, n_2, k). However, the temporal dimension will not be treated in the same way as the spatial dimensions, so we will use a spatial 4-connectivity: the neighbors of pixel (n_1, n_2, k) will be $\{(n_1, n_2, k), (n_1 + 1, n_2, k), (n_1, n_2 + 1, k), (n_1 - 1, n_2, k), (n_1, n_2 - 1, k)\}$. Therefore the following spatial structuring element will be used for the reconstructions:

$$B_s = \{(0,0,0), (+1,0,0), (0,+1,0), (-1,0,0), (0,-1,0)\}.$$

The following structuring element will be used for temporal operations:

$$B_k = \{(0,0,-L_1), \dots, (0,0,0), \dots, (0,0,+L_2)\},$$

where L_1 and L_2 are two positive integers, that will depend on the temporal length L of the defects that are to be detected ($L_1 + L_2 = L$).

The main property that will be used is the following:
A temporal closing (resp. opening) with structuring element B_k followed by a spatial reconstruction by erosion (resp. by dilation) with structuring element B_s will erase all dark (resp. light) defects.

Thanks to the reconstruction, the resulting sequence of operations, called Ψ, is a connected operator with respect to the spatial 4-connectivity. Moreover the resulting operator Ψ is increasing and idempotent, so it is a morphological filter [36]. This sequence of operations is illustrated in Fig. 4.9.

Since the aim is to obtain a binary image sequence, where the binary components correspond to the defects, the difference between the original image I_k and $\Psi(I_k)$ will be calculated and the result will be binarized. It is shown below how this is done.

Sliding window.   The number of frames in a motion picture is huge. Therefore a 3-D image cannot be simply built with all of them. A sliding window is used. One frame $\mathbf{I}_k$ at a time is treated, but in order to apply the temporal filters, p border frames are taken ($p = max\{L_1, L_2\}$) on each side of $\mathbf{I}_k$ and a 3-D image $\mathbf{F}_p$ is built by means of the frames $\{\mathbf{I}_{k-p}, \mathbf{I}_{k-p+1}, \ldots, \mathbf{I}_k, \ldots, \mathbf{I}_{k+p}\}$. As the processing is carried out, the sliding window is shifted by one frame. The central frame of the 3-D image $\Psi(\mathbf{F}_p)$ is denoted by $\Psi(\mathbf{I}_k)$.

Binarization.   Once that $\mathbf{J}_k = |\Psi(\mathbf{I}_k) - \mathbf{I}_k|$ is calculated, it is binarized using a double threshold, also called the hysteresis threshold. First the low and high threshold images $\mathcal{T}^{(low)}(\mathbf{J}_k)$ and $\mathcal{T}^{(high)}(\mathbf{J}_k)$ are calculated. Then a geodesic reconstruction by dilation is applied to $\mathcal{T}^{(high)}(\mathbf{J}_k)$ using as mask $\mathcal{T}^{(low)}(\mathbf{J}_k)$.

Hierarchical queues.   Reconstruction operators used in the openings/closings by reconstruction can be efficiently implemented using hierarchical queues [30].

The resulting binary mask corresponds to the (light or dark) local random defects of the image. Before moving on to the interpolation phase, note that if the shape of the defects is known a priori, then this information can be used to treat the binary mask in order to keep only the local random defects whose shape corresponds to the defect model.

4.7.3   Interpolation

As local random defects are detected, they have to be erased. Most of the time these defects have completely erased the color information that was present at their location. Therefore we are faced with a missing information problem.

There are two interpolation strategies. The first is a spatial strategy: the reconstruction is undertaken by interpolating the texture within the same image. The information in the neighboring frames is not considered. The second strategy is temporal: the texture information in the neighboring frames is used to interpolate the defects in the current frame.

Both strategies have been studied. Spatial interpolation methods might work correctly as long as no unique feature has been lost due to the defect (e.g., if a character had lost an eye because of a local random defect, it would be impossible to reconstruct it based only on local spatial texture information). In practice, spatial methods work when the defects are small. For example, rational interpolators (see Chapter 3) have given interesting results in this context [18, 19]. Otherwise, temporal methods give better results: if the same region of the scene has not been lost in the

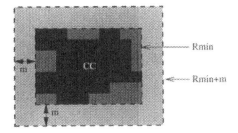

Fig. 4.10   Illustration of the definition of $R_{min}(CC)$ and $R_{min+m}(CC)$.

neighboring frames, then it should be possible to use that information to reconstruct the missing data in the current frame. In addition temporal approaches are a perfect complement to the detection method that has been used: if a small moving object has been considered a defect by the detection phase, the erroneous detection can be corrected by the interpolation phase. Consequently it has been decided to use only temporal interpolation methods to correct local random defects.

The temporal interpolation method is now described; it is called *copy and paste temporal interpolation*. The current frame being $\mathbf{I}_k$, we suppose that the preceding frame $\mathbf{I}_{k-1}$ has been restored, giving $\mathbf{I}^*_{k-1}$. Therefore image $\mathbf{I}^*_{k-1}$ is used in order to reconstruct missing pixels in $\mathbf{I}_k$. Observe that the interpolation method gives a first interpolator: the result of the opening/closing by reconstruction. However, in practice, the resulting texture is not satisfactory.

During the development of the local random defects detection algorithm, motion compensation was avoided as much as possible. It was accepted only to compensate for camera motion. But now that the defects, which represent a small subset of the original image, have been detected, we can afford to use motion compensation to interpolate them.

A block matching algorithm [16] has been adapted to perform the temporal interpolation. From now on only a pure spatial connectivity will be considered. Let CC be a connected subset that corresponds to a local random defect of frame $\mathbf{I}_k$. Let $R_{min}(CC)$ be the smallest horizontal rectangle containing CC, and $R_{min+m}(CC)$ be the smallest rectangle containing CC dilated m times in 8-connectivity (i.e., $R_{min+m}(CC)$ is a bounding rectangle of CC with a margin of m). Figure 4.10 illustrates these definitions. For simplicity, $R_{min+m}(CC)$ will be denoted by R.

Let $\mathbf{v}$ be a displacement vector with integer coordinates. The *block matching error without defect* of rectangle R with displacement $\mathbf{v}$ between frames $\mathbf{I}_k$ and $\mathbf{I}^*_{k-1}$ is defined as

$$E(R, \mathbf{v}, \mathbf{I}_k, \mathbf{I}^*_{k-1}) = \sum_{P \in R, P \ni CC} |\mathbf{I}_k(P) - \mathbf{I}^*_{k-1}(P + \mathbf{v})|. \qquad (4.8)$$

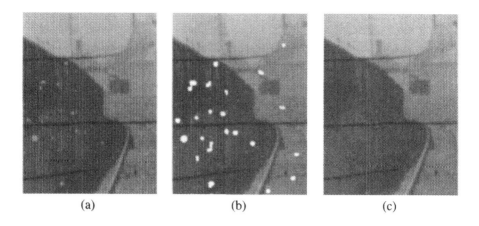

(a)   (b)   (c)

Fig. 4.11   Example of restoration of light local defects. (a) Original image (detail). (b) Detection. (c) Interpolation.

Let $D_{\mathbf{v}}$ be the domain of possible values for $\mathbf{v}$. Then the problem is to find the displacement vector $\mathbf{v}_0$ that minimizes $E(R, \mathbf{v}, \mathbf{I}_k, \mathbf{I}^*_{k-1})$. This minimum is computed by means of an exhaustive search: all possible $E(R, \mathbf{v}, \mathbf{I}_k, \mathbf{I}^*_{k-1})$ when $\mathbf{v}$ varies within $D_{\mathbf{v}}$ are computed. If the minimum is reached for different values of $\mathbf{v}$, then the smallest displacement vector is kept. The domain $D_{\mathbf{v}}$ is defined by a constant d:

$$\mathbf{v} = (\mathbf{v}_x, \mathbf{v}_y) \in D_{\mathbf{v}} \Leftrightarrow |\mathbf{v}_x| \le d, |\mathbf{v}_y| \le d. \tag{4.9}$$

Once that $\mathbf{v}_0$ has been calculated, it is used to interpolate the defect:

$$\forall P \in CC, \ \ \mathbf{I}^*_k(P) = \mathbf{I}^*_{k-1}(P + \mathbf{v}_0). \tag{4.10}$$

The name of the interpolation method is justified by this equation: a region is copied from the previous frame and pasted on the defect.

In Fig. 4.11 an interpolation example obtained with the copy and paste method applied to some images of the film *Arrivée d'un train en gare de Sidney* by the Lumière brothers is shown. A detail of size 251×335 of the images is pictured. The interpolation parameters were $d = 10$ and $m = 2$. Image (a) is the original one. Image (b) gives the result of the detection of light local random defects. Image (c) shows the result of the interpolation.

The computation time is small: one block matching is performed for each detected defect. It can be seen that the interpolation quality is good. However, in a few places a texture discontinuity can be observed around the interpolated regions. This phenomenon appears more clearly in image (b) of Fig. 4.12. A simple solution for this problem is proposed. A convolution is applied on the pixels along the border of the restored regions. In Fig. 4.12(c) it is shown that after this simple post-treatment the discontinuity disappears.

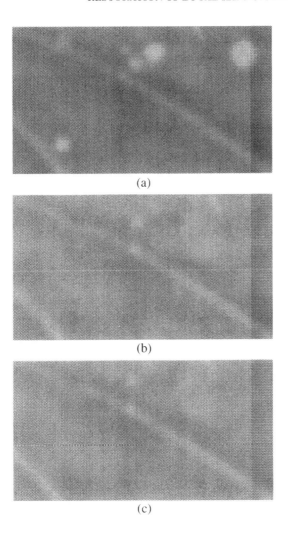

Fig. 4.12   Temporal interpolation of light local random defects. (a) Original image. (b) Interpolation: observe the texture discontinuity around the restored regions. (c) Interpolation followed by smoothing along the borders. The size of the images is 166×92.

4.7.4   Restoration Quality Measure

The minimal block matching error without defect $E(R, \mathbf{v}_0, \mathbf{I}_k, \mathbf{I}^*_{k-1})$ computed during the interpolation phase gives an interesting measure of the quality of the restoration of the treated defect. If this error is high, then there was a problem during the interpolation: no good matching was found for the missing region. However, the opposite is not true. If the error is small, we still can have a detection error caused by a fast moving object.

4.7.5   Conclusion

In this section a local random defects restoration algorithm was presented. It is composed of two main phases: a detection step based on the morphological opening and closing by reconstruction, and a temporal interpolation step using block matching.

The detection step can be preceded by a camera motion compensation algorithm. Such a global motion compensation algorithm can be relatively fast. At this stage more precise motion compensation algorithms are not used because they are slow and lack accuracy in a general framework.

The detection phase, thanks to the reconstruction step, allows to take into account most displacements. However, fast moving small objects can erroneously be considered as defects.

The interpolation phase plays a double role: first, it reconstructs defects, and second, it can correct a false detections produced by small fast moving objects. In addition this algorithm is fast, and the quality of the restored images is high.

Both algorithms (detection and interpolation) have been described for gray level images. Their application to color images is straightforward. The detection algorithm is applied to the luminance of the original images. During the interpolation step, block matching is computed on the luminance channel, and the final pixel reconstruction and the smoothing are applied to each color component.

There remains the problem of the choice of the restoration parameters. Even if the number of parameters is small, it would be interesting to choose most of them automatically. For example, if the maximal displacement between two frames could be computed with a simple method, then d, the parameter that defines the search domain for the displacement vector in the interpolation phase, could automatically be chosen.

4.8   CONCLUSION: THE RESTORATION SYSTEM

All these algorithms have been embedded in a restoration system called *SARSA*, which is described below.

4.8.1   Combining the Restoration Algorithms

It is impossible to treat all the images in the sequence at the same time. An iterative approach has been adopted. One frame at a time is treated, and it is supposed that the previous frames have already been restored. Moreover many algorithms (e.g., the local random defects restoration algorithm) use neighbouring frames in the processing of the current frame. All algorithms are applied to the current frame before saving it and moving to the next one.

Restoration algorithms are applied sequentially to each frame. In order to classify them, the following rule is used: apply first those algorithms that facilitate more the

Original Image Sequence

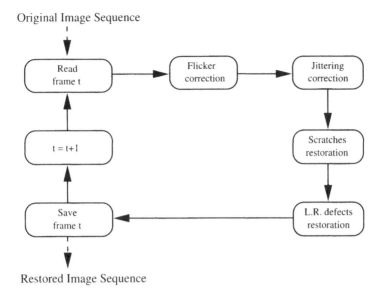

Fig. 4.13   Scheme of the restoration system.

work of the following algorithms. For example jitter is very annoying when detecting local random defects; as a consequence it should be corrected before treating local random defects. By this rule, the restoration algorithms are ordered:

1. flicker,

2. jitter,

3. scratches,

4. local random defects.

The resulting scheme is illustrated in Fig. 4.13.

The flicker correction algorithm works efficiently provided that the frames are not saturated and that flicker affects homogeneously each frame. The jitter correction algorithm gives good results, but it is too slow because of the exhaustive global motion estimation phase. This shortcoming can be easily overcome. The vertical scratch restoration algorithm is very fast and detects correctly perfectly vertical scratches. However, the texture interpolation phase can be improved. Finally, the local random defects restorations algorithm is very efficient: during the detection phase, thanks to the reconstruction step, only global camera motion compensation is needed. More accurate motion estimation algorithms are used during the interpolation phase, but they are only applied within damaged regions.

4.8.2 Performance

As explained in Section 4.3, algorithm speed was not one of the main concerns of this project. However, given the huge amount of data that has to be processed, algorithms cannot afford to be too slow. The current version of the prototype conforms with this requisite, except for the jitter correction algorithm. As faster global motion estimation techniques are developed, they should solve this problem.

An efficient way of improving the performance of a digital image sequence restoration system is parallelization. Probably the easiest way of achieving this is by treating each sequence of the film separately. A transition detection algorithm would identify scene changes, like cuts or dissolves. Each sequence would then be sent to a separate machine, with the corresponding restoration parameters. In fact commercial companies are building restoration systems based on this approach.

But the main improvement brought by this work is automation. SARSA claims to be an automatic restoration system. What does the term "automatic" mean in this context? SARSA takes as input an image sequence of any length and produces, without any external intervention, an output restored sequence. This is the reason why it is called "automatic". However, it needs a set of parameters in order to process the sequence, and they are actually chosen by an operator. Ideally this selection is done according to the contents of the image sequence. Could it be done automatically? We believe that in most cases this will be very difficult. Instead, we are trying to reduce the number of parameters. It should also be noted that these algorithms have proved to be quite robust: a small modification of the restoration parameters will not modify significantly the resulting restoration.

4.8.3 Verification Phase

To achieve high-quality restorations, there must be a verification phase following the application of the restoration algorithms. For example, some objects can be mistaken for defects and be erroneously erased. In order to make easier this verification phase, we have introduced the notion of *restoration quality measure*. If this measure is low, then the probability that the restoration is incorrect is high.

If a restoration error is observed, then the user should be able to easily cancel that correction. This is possible by storing the computed restoration information (jitter vectors, α_k and β_k parameters for flicker, and the binary masks for scratches and local random defects), and, of course, the original images. This verification phase is not fully implemented within SARSA.

4.8.4 Past, Present, and Future

SARSA has been used to restore many sequences from old motion pictures and videos presenting different kinds of defects. The results are very interesting, even for very damaged sequences. This restoration system is also flexible: new restoration algorithms can be included easily.

The restoration algorithms presented in this chapter have been used as basis for the development of an industrial prototype restoration system.

REFERENCES

1. G.R. Arce. Multistage order statistics filters for image sequence processing. *IEEE Trans. Acoustics, Speech, and Signal Processing*, 39(5): 1146-1163, 1991.

2. J.H. Chenot. New tools for digital restoration of television archives. In *Proc. of the IAB Seminar Television Archives Preservation and Creative Use*, Montreux, Switzerland, April 1998.

3. M.N. Chong, P. Liu, W.B. Goh, and D. Krishnan. A new spatiotemporal MRF model for the detection of missing data in image sequences. In *Proc. of 1997 IEEE Int. Conf. Acoustics, Speech, and Signal Processing*, Vol. 4, Munich, Germany, April 1997, pp. 2977-2980.

4. E. Decencière Ferrandière. Motion picture restoration using morphological tools. In P. Maragos, R.W. Schafer, and M.A. Butt, eds., *Mathematical Morphology and Its Applications to Signal Processing*. Kluwer Academic, Norwell, MA, 1996, pp. 361-368.

5. E. Decencière Ferrandière. Restoration of old motion pictures. *Microscopy, Microanalysis, Microstructures*, 7(5/6): 311-316, October-December 1996.

6. E. Decencière Ferrandière. *Restauration Automatique de Films Anciens*. Ph.D. thesis, Ecole Nationale Supérieure des Mines de Paris, December 1997.

7. E. Decencière Ferrandière, S. Marshall, and J. Serra. Application of the morphological geodesic reconstruction to image sequence analysis. *IEE Proc. Vision, Image and Signal Processing*, 144(6): 339-344, December 1997.

8. B. Despas and J.L. Fournier. *Guide de la Conservation des Films*. Commission Supérieure Technique de l'Image et du Son (CST), Paris, 1995.

9. E. Dubois and S. Sabri. Noise reduction in image sequences using motion compensated temporal filtering. *IEEE Trans. Communications*, 32(7): 826-831, 1984.

10. M. Friend. Film/digital/film. In *AMIA Conference*. AMIA, October 1994.

11. S. Geman, D.E. McClure, and D. Geman. A nonlinear filter for film restoration and other problems in image processing. *Computer Vision, Graphics, and Image Processing: Graphical Models and Image Processing*, 54(4): 281-289, July 1992.

12. S. Godsill and A.C. Kokaram. Restoration of image sequences using a causal spatio-temporal model. *The Art and Science of Bayesian Image Analysis*, Leeds, U.K., July 1997, pp. 189-194.

13. S.J. Godsill and A.C. Kokaram. Joint interpolation, motion and parameter estimation for image sequences with missing data. In *Signal Processing VIII: Theories and Applications*, Vol. I, Trieste, Italy, September 1996, pp. 1-4.

14. W.B. Goh, M.N. Chong, S. Kalra, and D. Krishnan. Bi-directional 3D autoregressive model approach to motion picture restoration. In *Proc. 1996 IEEE Int. Conf. Acoustics, Speech, and Signal Processing*, Vol. 4, May 1996, pp. 2275-2278.

15. T.S. Huang and Y.P. Hsu. Image sequence enhancement. In T.S. Huang, ed. *Image Sequence Analysis*. Springer, Berlin, 1981, ch.4, pp. 289-309.

16. J.R. Jain and A.K. Jain. Displacement measurement and its application in interframe image coding. *IEEE Trans. Communications*, 29(12): 1799-1808, 1981.

17. S. Kalra, M.N. Chong, and D. Krishan. A new auto-regressive (AR) model-based algorithm for motion picture restoration. In *Proc. 1997 IEEE Int. Conf. Acoustics, Speech, and Signal Processing*, Vol. 4, Munich, Germany, April 1997, pp. 2557-2560.

18. L. Khriji, M. Gabbouj, G. Ramponi, and E. Decencière Ferrandière. Old movie restoration using rational spatial interpolators. In *Proc. 6th IEEE Int. Conf. Electronics, Circuits and Systems*, Vol. 2, Paphos, Cyprus, September 1999, pp. 1151-1154.

19. L. Khriji, M. Gabbouj, S. Marsi, G. Ramponi, and E. Decencière Ferrandière. Nonlinear interpolators for old movie restoration. In *Proc. 1999 IEEE Int. Conf. Image Processing*, Vol. 3, Kobe, Japan, October 1999, pp. 169-173.

20. R.P. Kleihorst, R.L. Lagendijk, and J. Biemond. Noise reduction of image sequences using motion compensation and signal decomposition. *IEEE Trans. Image Processing*, 4(3): 274-284, 1995.

21. A. Kokaram. Detection and removal of line scratches in degraded motion picture sequences. In *Signal Processing VIII: Theories and Applications*, Trieste, Italy, September 1996.

22. A. Kokaram. *Motion Picture Restoration*. Springer, Berlin, 1998.

23. A.C. Kokaram. *Motion Ricture Restoration*. Ph.D. thesis, Cambridge University, May 1993.

24. A.C. Kokaram and J.G. Godsill. A system for reconstruction of missing data in image sequences using sampled 3D AR models and MRF motion priors. In

F. Buxton and R. Cipolla, eds., *Computer Vision–ECCV'96*. Cambridge, UK, April 1996, pp. 613-624.

25. A.C. Kokaram, R.D. Morris, W.J. Fitzgerald, and P.J.W. Rayner. Detection of missing data in image sequences. *IEEE Trans. Image Processing*, 4(11): 1496-1508, 1995.

26. A.C. Kokaram, R.D. Morris, W.J. Fitzgerald, and P.J.W. Rayner. Interpolation of missing data in image sequences. *IEEE Trans. Image Processing*, 4(11): 1508-1519, 1995.

27. A.C. Kokaram and P. Rayner. Removal of replacement noise in motion picture sequences using 3D auto-regressive modelling. In *Signal Processing VII: Theories and Applications*, Vol. 3, Edinburgh, U.K., September 1994, pp. 1780-1783.

28. D. Lyon. Real-time archive restoration. In *Proc. IAB Seminar Television Archives Preservation and Creative Use*, Montreux, Switzerland, April 1998.

29. B. Marcotegui. *Segmentation de Séquences d' Images en Vue du Codage*. Ph.D. thesis, Ecole Nationale Supérieure des Mines de Paris, 1996.

30. F. Meyer. Un algorithme optimal pour la ligne de partage des eaux. In *8ème Congrès de Reconnaissance des Formes et Intelligence Artificielle*, Vol. 2, Lyon, France, November 1991, pp. 847-857.

31. R.D. Morris and W.J. Fitzgerald. Replacement noise in image sequences– Detection and interpolation by motion field segmentation. In *Proc. 1994 IEEE Int. Conf. Acoustics, Speech, and Signal Processing*, Vol. V, Adelaide, Australia, April 1994, pp. 245-248.

32. R.D. Morris and W.J. Fitzgerald. Stochastic and deterministic methods in motion picture restoration. In *Proc. Int. Workshop in Image Processing*, Budapest, Hungary, June 1994.

33. R.D. Morris and W.J. Fitzgerald. Stochastic and deterministic methods in motion picture restoration. *J. Communications*, 45:17-21, July-August 1994.

34. R.D. Morris. *Image Sequence Restoration via Gibbs Distributions*. Ph.D. thesis, Trinity College, 1994.

35. M.K. Ozkan, M.I. Sezan, and A.M. Tekalp. Adaptive motion-compensated filtering of noisy image sequences. *IEEE Trans. Circuits and Systems for Video Technology*, 3(4): 277-290, 1993.

36. M. Pardàs, J. Serra, and L. Torrés. Connectivity filters for image sequences. In *Proc. SPIE Image Algebra and Morphological Image Processing III*, Vol. 1769, San Diego, July 19-24, 1992, pp. 318-329.

37. V. Pinel. La restauration des films. In J. Aumont, A. Gaudreault, and M. Marie, eds. *Histoire du Cinéma: Nouvelles Approches (Colloque de Cerisy)*. Publications de la Sorbonne, August 1989.

38. P. Richardson and D. Suter. Restoration of historic film for digital compression: A case study. In *Proc. 1995 IEEE Int. Conf. Image Processing*, Vol. 2, May 1995, pp. 49-52.

39. P. Salembier. Practical extensions of connected operators. In P. Maragos, R.W. Schafer, and M.A. Butt, eds., *Mathematical Morphology and Its Applications to Signal Processing*. Kluwer Academic, Norwell, MA, 1996, pp. 97-100.

40. P. Salembier and J. Serra. Multiscale image segmentation. In *Proc. SPIE: Visual Communications and Image Processing*, Vol. 1818, Boston, November 1992, pp. 620-631.

41. M. Schmitt and J. Mattioli. *Morphologie Mathématique*. Masson, Paris, 1993.

42. J. Serra. *Image Analysis and Mathematical Morphology–Vol. I*. Academic Press, New York, 1982.

43. J. Serra. *Image Analysis and Mathematical Morphology–Vol. II: Theoretical Advances*. Academic Press, London, 1988.

44. J. Serra and P. Salembier. Connected operators and pyramids. In *Proc. SPIE: Nonlinear Algebra and Morphological Image Processing*, Vol. 2030, San Diego, CA, July 1993, pp. 65-76.

45. A.G. Setos. The fox movietone news preservation project: An introduction. *SMPTE J.*, 105(9): 532-536, 1996.

46. P. Soille. *Morphological Image Analysis*. Springer-Verlag, Berlin, 1999.

47. R. Storey. Electronic detection and concealment of film dirt. *SMPTE J.*, 642-647, June 1985.

48. P. van Roosmalen, S.J.P. Westen, R.L. Lagendijk, and J. Biemond. Noise reduction for image sequences using an oriented pyramid thresholding technique. In *Proc. 1996 IEEE Int. Conf. Image Processing*, Vol. 1, Lausanne, Switzerland, September 1996, pp. 375-378.

49. P.M.B. van Roosmalen. Improved blotch detection by postprocessing. In *Proc. of IEEE Benelux Signal Processing Chapter*. Leuven, Belgium, March 1998.

50. P.M.B. van Roosmalen. *Restoration of Archived Film and Video*. Ph.D. thesis, Delft University of Technology, Delft, The Netherlands, 1999.

51. T. Vlachos and G. Thomas. Motion estimation for the correction of twins-lens telecine flicker. In *1996 IEEE Int. Conf. Image Processing*, Vol. 1, Lausanne, Switzerland, September 1996, pp. 109-112.

5 Adaptive Order Statistic Filtering of Still Images and Video Sequences

C. KOTROPOULOS and I. PITAS

Aristotle University of Thessaloniki
Department of Informatics
Artificial Intelligence and Information Analysis Laboratory[1]
Box 451, Thessaloniki 540 06, Greece

5.1 INTRODUCTION

Adaptive signal processing has shown considerable development in the past two decades. Adaptive filters have been applied in a wide variety of problems including system identification, channel equalization, echo cancellation in telephone channels, suppression of narrowband interference in wideband signals and adaptive arrays [87, 33, 9, 38]. The most widely known adaptive filters are linear ones having the form of either a finite impulse response (FIR) or lattice structure. All the just mentioned problems involve one-dimensional (1-D) signals. However, progress has been much slower in the development of adaptive algorithms for problems involving two-dimensional (2-D) signals (e.g., image filtering). The first extension of the popular 1-D least mean squares (LMS) algorithm to two dimensions was reported by [30] in 1988. An optimality criterion governing the choice of the adaptation step-size for the 2-D LMS (TDLMS) algorithm is developed by [56]. A local-mean estimation procedure is incorporated into the TDLMS filter so that the filter preserves the image edges [52]. However, linear filters are not suitable for applications where the noise is impulsive or where the signal is strongly nonstationary (e.g., in image processing).

Nonlinear filters have become a very important tool in signal processing, and especially in image analysis and computer vision. For a review of the nonlinear

---

[1] http://poseidon.csd.auth.gr

filter classes the reader may consult [70]. One of the best known nonlinear families is based on order statistics [72]. Order statistics uses the concept of data ordering. There are now a plethora of nonlinear filters based on data ordering. Three major classes of order statistic filters, namely the L-filter [12], the $L\ell$-filter [64, 27], and the ranked-order filters, are the fundamental filter structures (i.e., the nonlinear models) under discussion. The first two fundamental filter structures have the form of a linear combination of the observations, and they exploit either the rank or the combined rank and location information inherent in the observations. Although these filter structures have been proved to be highly effective in suppressing noise when processing signals with nonstationary mean levels and abrupt changes, they lead unavoidably to signal blurring. Such cases are customary in image processing where important visual cues provided by the edges and fine details should be preserved. The solution is to use ranked-order filters. Adaptive designs for the three fundamental filter structures are described subsequently in this chapter.

Two broad classes of adaptive nonlinear filters are known in the literature. The first class consists of adaptive filters whose coefficients are determined by *iterative algorithms* (e.g., least mean squares, recursive least squares) for the minimization of the mean squared error (MSE) between the filter output and the desired response. We will confine ourselves to the use of the least mean squares (LMS) algorithm in the design of the nonlinear filters reviewed in this chapter. The motivation behind using LMS is its ability to cope with nonstationary and/or time varying signals. Several authors have used the LMS algorithm to design nonlinear filters. For example, it has been used in the design of L-filters [71, 43, 88, 47], $L\ell$-filters [73], permutation filters [40], and order statistic filter banks [1]. A survey of adaptive order statistic filters appearing in the literature up to 1993 can be found in [45]. Here, our discussion will be devoted to the most recent advances. The so-called signal-adaptive filters form the second adaptive nonlinear filter class. They are filters that change their smoothing properties at each image pixel according to the local image content. They have been used in image processing applications where impulsive or signal-dependent noise is present [50, 53].

Adaptive filter designs are described for the L-filter [47] (Section 5.3) and the $L\ell$-filter [73] (Section 5.5). The performance of the above-mentioned filter structures in noise smoothing is studied for both still images and video sequences. Important issues in the filter design are treated: (1) the extension of adaptive L-filters to the multichannel case for filtering color images/video sequences using the marginal sub-ordering principle [48] (Section 5.4); (2) the incorporation of structural constraints in the filter design and their impact on filtering both gray scale and color images (Sections 5.3.1 and 5.4.3); (3) the selection of the adaptation step-size (Section 5.3.2); (4) the use of either LMS or LMS-Newton algorithm in the filter design (Sections 5.4.2.3 and 5.4.3.2). Besides the fundamental nonlinear filter structures we also review their powerful extensions and generalizations. For example, extensions of the $L\ell$-filter are the order statistic filter banks [1] (Section 5.6), the permutation filter lattices [40] (Section 5.7) and the $\ell + L$-filters [92] (Section 5.9).

The LMS algorithm is also used for the design of adaptive ranked-order filters that are reviewed in the chapter, i.e., the weighted-order statistic lattices [2] (Section 5.7.6), the morphological-rank-linear filter [66] (Section 5.8), the weighted-order statistic affine FIR filter [23] (Section 5.10) and the weighted myriad filters [36] (Section 5.11).

Most of the signal-adaptive filters offer maximal noise smoothing in homogeneous regions, whereas they change their coefficients close to image edges or impulses, so their performance becomes similar to that of the median filter. The fundamental filter in this class is the *signal-adaptive median* filter [10]. In Section 5.12 we describe a signal-dependent adaptive L-filter structure that employs two adaptive L-filters, one applied to homogeneous regions and a second filter applied close to image edges. Furthermore the adaptation of the window in the signal-adaptive median filter by employing morphological operations is described. The inclusion of signal-adaptive filters in the chapter serves as a bridge between the filters that stem from the L- and $L\ell$-filters and those stemming from the ranked-order filters.

Before proceeding to the description of the filter classes, we elaborate on the fundamentals of order statistic filtering.

5.2   ORDER STATISTIC FILTERING

Let us consider that the observed image $x(\mathbf{n})$ can be expressed as a sum of an arbitrary image $s(\mathbf{n})$ plus zero-mean 2-D additive white noise:

$$x(\mathbf{n}) = s(\mathbf{n}) + v(\mathbf{n}), \tag{5.1}$$

where $\mathbf{n} = (n_1, n_2)$ denotes the pixel coordinates. In image processing, a neighborhood is defined around each pixel $\mathbf{n}$. Our purpose is to design a filter in this neighborhood (to be called the *filter window* hereafter) that aims at estimating the noise-free central image pixel value $s(\mathbf{n})$ by minimizing a certain criterion. Without loss of generality, among the several filter masks (e.g., cross, $\times$-shape, square, circle) that are used in digital image processing [70], we will rely upon the rectangular window of dimensions $\mathcal{I}_1 \times \mathcal{I}_2$ where $\mathcal{I}_1 = 2\iota_1 + 1$ and $\mathcal{I}_2 = 2\iota_2 + 1$:

$$\mathbf{X}(\mathbf{n}) = \begin{bmatrix} x(n_1 - \iota_1, n_2 - \iota_2) & x(n_1 - \iota_1, n_2 - \iota_2 + 1) & \cdots & x(n_1 - \iota_1, n_2 + \iota_2) \\ \vdots & & & \vdots \\ x(n_1 + \iota_1, n_2 - \iota_2) & x(n_1 + \iota_1, n_2 - \iota_2 + 1) & \cdots & x(n_1 + \iota_1, n_2 + \iota_2) \end{bmatrix}. \tag{5.2}$$

Let us rearrange the $\mathcal{I}_1 \times \mathcal{I}_2$ filter window in a lexicographic order (i.e., row-wise) to a $N \times 1$ vector

$$\begin{aligned} \mathbf{x}(\mathbf{n}) &= (x(n_1 - \iota_1, n_2 - \iota_2), \ldots, \\ &\quad x(n_1 - \iota_1, n_2 + \iota_2), \ldots, x(n_1 + \iota_1, n_2 + \iota_2))^T \end{aligned} \tag{5.3}$$

where $N = \mathcal{I}_1\mathcal{I}_2$. We assume that the filter window is sliding over the image in a raster or prime-row scan fashion. If N_1 and N_2 denote the image rows and columns, respectively, in a raster scan, a (scalar) running index $n = (n_1 - 1)N_2 + n_2$ can be used instead of the pixel coordinates $\mathbf{n}$, where $1 \le n_1 \le N_1$ and $1 \le n_2 \le N_2$. For a prime-row scan the corresponding relationship is

$$n = \begin{cases} (n_1 - 1)N_2 + N_1 + 1 - n_2 & \text{if } n_1 \text{ is even} \\ (n_1 - 1)N_2 + n_2 & \text{if } n_1 \text{ is odd.} \end{cases} \tag{5.4}$$

Henceforth the following 1-D notation is adopted for simplicity:

$$\mathbf{x}_\ell(n) = (x_1(n), x_2(n), \dots, x_N(n))^T. \tag{5.5}$$

Let $\mathbf{x}_L(n)$ be the ordered input vector at n given by

$$\mathbf{x}_L(n) = \left(x_{(1)}(n), x_{(2)}(n), \dots, x_{(N)}(n)\right)^T, \tag{5.6}$$

where $x_{(1)}(n) \le x_{(2)}(n) \le \dots \le x_{(N)}(n)$ are the ordered pixel values (i.e., observations). The vector $\mathbf{x}_L(n)$ is commonly referred as the vector of the *order statistics* of $\mathbf{x}_\ell(n)$. Note that any vector $(x_{j_1}, x_{j_2}, \dots, x_{j_N})^T$, where $(j_1, j_2, \dots, j_N)$ is a permutation of $(1, 2, \dots, N)$, has the same vector of order statistics $\mathbf{x}_L(n)$. Hence $\mathbf{x}_L(n)$ looses the *spatial/temporal order* of observations. Let us define the *rank indicator vector*

$$\boldsymbol{\rho}_j = (\rho_{j1}, \rho_{j2}, \dots, \rho_{jN})^T, \tag{5.7}$$

with

$$\rho_{ji} = \begin{cases} 1 & \text{if } x_j \leftrightarrow x_{(i)} \\ 0 & \text{otherwise,} \end{cases} \tag{5.8}$$

where $x_j \leftrightarrow x_{(i)}$ denotes that the jth spatial/temporal sample occupies the ith order statistic. Let ϱ_j denote the *rank* of x_j. According to Eq. (5.8), $\rho_{j\varrho_j} = 1$. If a tie occurs in the observations, such that $x_{j_1} = x_{j_2}$ for some $j_1 < j_2$, then the corresponding ranks will be assigned values with the restriction $\varrho_{j_1} < \varrho_{j_2}$. This principle is known as *stable ordering* and guarantees only one rank vector for each $\mathbf{x}_\ell$ [64, 27].

In this section, we omit n for notation simplicity. Let us consider a generic filter that linearly weighs either the observation vector, $\mathbf{x}_\ell$, or the vector of the order statistics $\mathbf{x}_L$. Let y be the corresponding filter output. If the filter is an L-filter, then the output y is given by [12]

$$y = \mathbf{a}^T \mathbf{x}_L = \sum_{i=1}^{N} a_i x_{(i)}, \tag{5.9}$$

where $\mathbf{a} = (a_1, a_2, \dots, a_N)^T$ is the L-filter coefficient vector. The output of Eq. (5.9) is based solely on the order statistics of the input observations within the filter window.

It can be rewritten as

$$y = \left(\boldsymbol{\rho}_1^T \mathbf{a} \mid \boldsymbol{\rho}_2^T \mathbf{a} \mid \cdots \mid \boldsymbol{\rho}_N^T \mathbf{a}\right) \mathbf{x}_\ell = \sum_{j=1}^{N} a_{\varrho_j}\, x_j, \qquad (5.10)$$

by employing the observations x_j and data-dependent filter coefficients a_{ϱ_j}. That is, the L-filter incorporates explicitly x_j and data-dependent coefficients that are used in such a way that the spatial/temporal information becomes irrelevant [27].

Similarly let us define the *spatial/temporal location vector* [40]

$$\boldsymbol{\xi}_{(i)} = (\xi_{i1}, \xi_{i2}, \dots, \xi_{iN})^T \qquad (5.11)$$

where

$$\xi_{ij} = \begin{cases} 1 & \text{if } x_{(i)} \leftrightarrow x_j \\ 0 & \text{otherwise.} \end{cases} \qquad (5.12)$$

Let $\ell_{(i)}$ denote the location of $x_{(i)}$ in $\mathbf{x}_\ell$. A linear FIR filter produces an output

$$y = \mathbf{b}^T \mathbf{x}_\ell = \left(\boldsymbol{\xi}_{(1)}^T \mathbf{b} \mid \boldsymbol{\xi}_{(2)}^T \mathbf{b} \mid \cdots \mid \boldsymbol{\xi}_{(N)}^T \mathbf{b}\right) \mathbf{x}_L = \sum_{i=1}^{N} b_{\ell_{(i)}}\, x_{(i)}. \qquad (5.13)$$

From Eq. (5.13) it can be seen that the output of a linear FIR filter can be expressed in terms of $\mathbf{x}_L$ and data-dependent coefficients $b_{\ell_{(i)}}$. Equations (5.10) and (5.13) can be rewritten in a compact form as follows [27]:

$$y = \sum_{i=1}^{N} c(\ell_{(i)}, i)\, x_{(i)} = \sum_{j=1}^{N} c(j, \varrho_j)\, x_j = \mathbf{1}_N^T\, (\mathcal{C} \circ \mathcal{P})\, \mathbf{x}_\ell, \qquad (5.14)$$

where $\mathbf{1}_N$ is the $N \times 1$ unitary vector, meaning that $\mathbf{1}_N = (1, 1, \dots, 1)^T$, $\mathcal{C}$ is an $N \times N$ coefficient matrix, $\mathcal{P}$ is an $N \times N$ data-dependent permutation matrix and $\circ$ denotes the elementwise (Shur) product. Equation (5.14) defines the output of a *combination filter* (C-filter) [27]. The permutation matrix $\mathcal{P}$ in Eq. (5.14) is given by

$$\mathcal{P} = \begin{bmatrix} \boldsymbol{\xi}_{(1)}^T \\ \boldsymbol{\xi}_{(2)}^T \\ \vdots \\ \boldsymbol{\xi}_{(N)}^T \end{bmatrix} = [\boldsymbol{\rho}_1 \mid \boldsymbol{\rho}_2 \mid \cdots \mid \boldsymbol{\rho}_N]. \qquad (5.15)$$

From Eq. (5.15) it is seen that the ith row of $\mathcal{P}$ characterizes the observation with rank i, whereas the jth column vector characterizes the observation with spatial/temporal order j.

The coefficient vector $(\mathcal{C} \circ \mathcal{P})^T \mathbf{1}$ in Eq. (5.14) is a function of $\mathbf{x}_\ell$, since, for different realizations of ϱ_j, $j = 1, 2, \dots, N$, or $\ell_{(i)}$, $i = 1, 2, \dots, N$, different $\mathcal{C}$ coefficients are used in determining the output.   Let Ω_C be the collection of all

C-filters that can arise from a given C matrix. Clearly, the cardinality of the set Ω_C is $|\Omega_C| = N!$. Therefore, the output of a C-filter in each filter window can be treated as being produced by one of these $N!$ FIR filters. However, although the coefficient vector $(C \circ P)^T \mathbf{1}$ may vary for different P, its components are chosen from the restricted set of the N^2 elements of matrix C.

Trimming can be introduced in a C-filter by replacing the vector $\mathbf{1}$ with $\boldsymbol{\tau}$ [27]:

$$y = \boldsymbol{\tau}^T (C \circ P) \mathbf{x}_\ell, \tag{5.16}$$

where $\boldsymbol{\tau} = (\tau_1, \tau_2, \ldots, \tau_N)^T$ contains the trimming information and will be referred to as the *trimming vector*. Its ith component, τ_i, corresponds to the ith order statistic $x_{(i)}$. The presence of either an 1 or 0 at this position determines whether $x_{(i)}$ will be trimmed or not. More general weighting than the zero-one weighting can also be considered.

Another approach in generalizing Eqs. (5.10) and (5.13) has been proposed in [64]. Let $\mathbf{x}_{L\ell}$ denote the following $N^2 \times 1$ vector:

$$\mathbf{x}_{L\ell} = \left(x_{(1)} \boldsymbol{\xi}_{(1)}^T \mid x_{(2)} \boldsymbol{\xi}_{(2)}^T \mid \cdots \mid x_{(N)} \boldsymbol{\xi}_{(N)}^T \right)^T, \tag{5.17}$$

and let $\mathbf{c}$ be the N^2-dimensional vector

$$\mathbf{c} = \left(c_{(1)1}, \ldots, c_{(1)N} \mid c_{(2)1}, \ldots, c_{(2)N} \mid \cdots \mid c_{(N)1}, \ldots, c_{(N)N} \right)^T. \tag{5.18}$$

It can easily be verified that $x_{(i)} \boldsymbol{\xi}_{(i)}$ is an N-dimensional vector having $x_{(i)}$ as the sole nonzero element in its $\ell_{(i)}$th place. An estimate $\widehat{s}$ of the original (noise-free) image can be obtained by

$$y = \widehat{s} = \mathbf{c}^T \mathbf{x}_{L\ell}. \tag{5.19}$$

Equation (5.19) defines the so-called N^2-$L\ell$-*estimator* [64]. It is also possible to define an N^2-dimensional vector

$$\mathbf{x}_{\ell L} = \left(x_1 \boldsymbol{\rho}_1^T \mid x_2 \boldsymbol{\rho}_2^T \mid \cdots \mid x_N \boldsymbol{\rho}_N^T \right)^T \tag{5.20}$$

which is similar to the vector defined by Eq. (5.17). We can still estimate the original (noise-free) image using Eqs. (5.19) and (5.20). The vector defined in Eq. (5.20) was used in the definition of $L\ell$-filters and their generalizations in [40, 1]. However, it has been proved that there is a one-to-one correspondence between the elements of the vector $\mathbf{x}_{L\ell}$ defined in Eq. (5.17) and the elements of the vector $\mathbf{x}_{\ell L}$ defined in Eq. (5.20) [40]. This fact can be expanded to a one-to-one mapping between the coefficient vectors that would be employed.

It is worth noting that the decomposition $\mathbf{x}_\ell \in \mathbb{R}^N \leftrightarrow \mathbf{x}_{L\ell} \in \mathbb{R}^{N^2}$ is a one-to-one mapping. The vector $\mathbf{x}_\ell$ can be reconstructed from $\mathbf{x}_{L\ell}$ as

$$\mathbf{x}_\ell = \left(\mathbf{1}_N^T \otimes \mathbf{I}_N \right) \mathbf{x}_{L\ell}, \tag{5.21}$$

where $\mathbf{I}_N$ is the $N \times N$ identity matrix and $\otimes$ denotes the Kronecker product. We can also obtain $\mathbf{x}_L$ from $\mathbf{x}_{L\ell}$ as

$$\mathbf{x}_L = \left(\mathbf{I}_N \otimes \mathbf{1}_N^T\right) \mathbf{x}_{L\ell}. \qquad (5.22)$$

By Eqs. (5.21) and (5.22), it can be proved that a linear FIR filter with coefficient vector $\mathbf{b}$ is equivalent to an N^2-$L\ell$-filter with coefficient vector

$$\mathbf{c} = \mathbf{1}_N \otimes \mathbf{b}, \qquad (5.23)$$

and that an L-filter with coefficient vector $\mathbf{a}$ is equivalent to an N^2-$L\ell$-filter with coefficient vector

$$\mathbf{c} = \mathbf{a} \otimes \mathbf{1}_N. \qquad (5.24)$$

Further generalizations are also possible [64]. For example, let $\Omega = \{\omega_i : i = 1, 2, \ldots, Q, Q \leq N!\}$ be a subset of all possible permutations ω_i induced by the ranking operation on the elements of $\mathbf{x}_\ell$. The Q-$L\ell$-estimator results if

$$\mathbf{x}_{Q-L\ell} = \underbrace{\left(\mathbf{0}_N^T \mid \mathbf{0}_N^T \mid \cdots \mid \mathbf{0}_N^T \mid \underbrace{\mathbf{x}_\ell^T}_{i\text{th block}} \mid \mathbf{0}_N^T \mid \cdots \mid \mathbf{0}_N^T \right)^T}_{Q\text{-blocks}} \qquad (5.25)$$

is linearly weighted [64]. If we partition the observation vector $\mathbf{x}_\ell$ into not necessarily disjoint subsets and apply the above-described definition to the subsets, we can obtain the partitioned N^2-$L\ell$-estimator [64].

It is possible to derive a simplified filter design for N^2-$L\ell$-estimators and partitioned ones, where the N^2-dimensional coefficient vector $\mathbf{c}$ results as a Kronecker product of two N-dimensional coefficient vectors $\mathbf{a}$ and $\mathbf{b}$:

$$\mathbf{c} = \mathbf{a} \otimes \mathbf{b}. \qquad (5.26)$$

By comparing Eq. (5.26) with Eqs. (5.23) and (5.24), we deduce that $\mathbf{a}$ accounts for the ranks of the observations in $\mathbf{x}_L$ and, accordingly, is referred to as the *nonlinear part* of the estimator and $\mathbf{b}$ accounts for the locations of the observations in $\mathbf{x}_\ell$ and is referred to as the *linear part* . The estimate of the original (noise-free) image, i.e., filter output, produced by

$$y = \widehat{s} = (\mathbf{a} \otimes \mathbf{b})^T \mathbf{x}_{L\ell} \qquad (5.27)$$

defines the *Kronecker product form of the $L\ell$-estimator*, and for brevity it will be referred to as *Kronecker $L\ell$-filter*.

From the preceding analysis, it becomes evident that $L\ell$- and C-filters combine information contained in the *location-rank (LR) orderings*, where location refers to space/time. This LR information results from a crisp relation between the location and rank orderings and contains no information on sample values or their spread [7]. Indeed, let $\mathcal{X}$ be the set that has as elements the components of $\mathbf{x}_\ell$, that is, the observations. To relate the location and rank orderings of the samples in $\mathcal{X}$, we can define a crisp or binary relation [7]

$$\Re = \left\{ (x_j, x_{(i)}) \mid x_j \leftrightarrow x_{(i)}; x_j, x_{(i)} \in \mathcal{X} \right\}. \tag{5.28}$$

The relationship defined by Eq. (5.28) can be represented by a crisp LR matrix

$$\mathcal{P}^T = \begin{bmatrix} \rho_1^T \\ \rho_2^T \\ \vdots \\ \rho_N^T \end{bmatrix}, \tag{5.29}$$

where $\rho_j, j = 1, 2, \ldots, N$, is the rank indicator vector defined by Eq. (5.7) and $\mathcal{P}$ is the permutation matrix defined by Eq. (5.15). We can obtain the time and rank order indexes as follows:

$$
\begin{aligned}
\underline{\ell} &= \left(\ell_{(1)}, \ell_{(2)}, \ldots, \ell_{(N)} \right)^T = \mathcal{P} \left(1, 2, \ldots, N \right)^T, \\
\varrho &= \left(\varrho_1, \varrho_2, \ldots, \varrho_N \right)^T = \mathcal{P}^T \left(1, 2, \ldots, N \right)^T,
\end{aligned}
\tag{5.30}
$$

where $\ell_{(i)}$ is the location of $x_{(i)}$ in $\mathbf{x}_\ell$ and ϱ_j is the rank of x_j in $\mathbf{x}_L$. It can also be shown that

$$\mathbf{x}_\ell = \mathcal{P}^T \mathbf{x}_L \quad \text{and} \quad \mathbf{x}_L = \mathcal{P} \mathbf{x}_\ell. \tag{5.31}$$

The relationship between the location and rank ordering can be generalized to the real domain through fuzzy set theory [7]:

$$\widetilde{\Re} = \left\{ (x_j, x_{(i)}), m_{\widetilde{\Re}}(x_j, x_{(i)}) \mid x_j, x_{(i)} \in \mathcal{X} \right\}, \tag{5.32}$$

where $m_{\widetilde{\Re}}(x_j, x_{(i)})$ gives the degree of membership between x_j and $x_{(i)}$. The relation defined by Eq. (5.32) can be represented by a fuzzy LR matrix

$$\widetilde{\mathcal{P}} = \begin{bmatrix} m_{\widetilde{\Re}}(x_1, x_{(1)}) & \cdots & m_{\widetilde{\Re}}(x_1, x_{(N)}) \\ \vdots & & \vdots \\ m_{\widetilde{\Re}}(x_N, x_{(1)}) & \cdots & m_{\widetilde{\Re}}(x_N, x_{(N)}) \end{bmatrix}. \tag{5.33}$$

In order to incorporate information on the spread of samples into the fuzzy LR matrix, the membership function should satisfy the following properties [7]:

1. $m_{\widetilde{\Re}}(\alpha, \beta) = m_{\widetilde{\Re}}(\alpha - \beta)$.

2. $m_{\widetilde{\Re}}(\alpha, \beta) = m_{\widetilde{\Re}}(\beta, \alpha)$.

3. $\lim_{\beta \to \alpha} m_{\widetilde{\Re}}(\alpha, \beta) = 1$.

4. $\lim_{|\alpha - \beta| \to \infty} m_{\widetilde{\Re}}(\alpha, \beta) = 0$.

Suitable membership functions that satisfy these four properties are the Gaussian function

$$m_G(\alpha, \beta) = \exp\left(-\frac{(\alpha - \beta)^2}{2\sigma^2}\right) \tag{5.34}$$

and the triangular one

$$m_T(\alpha, \beta) = \begin{cases} 1 - \dfrac{|\alpha - \beta|}{\sigma} & \text{if } |\alpha - \beta| \leq \sigma \\ 0 & \text{otherwise.} \end{cases} \tag{5.35}$$

In replacing $\mathcal{P}$ with $\widetilde{\mathcal{P}}$ in Eq. (5.30), we obtain the fuzzy location and rank order indexes. Similarly we obtain fuzzy spatial/time and rank-ordered samples by replacing $\mathcal{P}$ with $\widetilde{\mathcal{P}}$ in Eq. (5.31). To restrict the fuzzy indexes and samples to the same range as their crisp counterparts, we should appropriately normalize the rows or columns of $\widetilde{\mathcal{P}}$ so that their elements sum up to 1.

Having defined the fundamentals of order statistic filtering, we proceed to the description of the adaptive order statistic filters reviewed in this chapter.

5.3   SINGLE-CHANNEL *L*-FILTERS

The output of the L-filter is defined as a linear combination of the order statistics of the input signal samples [12]. The L-filters have found extensive application in digital signal and image processing because they have a well-defined design methodology when they are used as estimators that minimize the mean-squared error (MSE) between the filter output and the noise-free signal [62, 57, 44, 46]. A design of L-filters that relies on a noniterative minimization of the MSE between the filter output and the desired response yields very tedious expressions for computing the marginal and joint cumulative functions of the order statistics (cf. [57]). Consequently the design of adaptive L-filters has proved to have appeal in order to avoid the computational burden of the noniterative methods.

The adaptation of the coefficients employed in order statistic filters by using linear adaptive signal processing techniques has received much attention in the literature [45, 63, 69, 71, 76, 43, 19, 15, 88]. In this section we will confine ourselves to the design of adaptive L-filters. Many authors have tried to design adaptive L-filters either by using the least mean squares (LMS) or the recursive least squares (RLS) algorithm [63, 69, 71] or by using constrained LMS adaptive algorithms [76, 43, 19, 15, 88].

We examine the case where images are corrupted by additive white noise. Several adaptive L-filters are designed and their performance in noise suppression is compared. The properties of the developed adaptive L-filters are studied as well. Another primary goal is to establish links between the adaptive L-filters under study and other algorithms developed elsewhere. All the adaptive L-filters stem from the same algorithm, the LMS that is used to minimize a cost function [e.g., MSE, mean absolute error (MAE), overall output power].

First, the location-invariant LMS L-filter for a nonconstant signal corrupted by zero-mean additive white noise is described. Our interest in deriving the location-invariant LMS L-filter relies on the observation that this filter structure can be modified so that it performs even where a reference signal is not available. We demonstrate that the location-invariant LMS L-filter can be described in terms of the general structure proposed by Griffiths and Jim [29]. Next, the normalized LMS (NLMS) L-filter is studied. Our motivation in studying the NLMS L-filter is justified in three ways: (1) This LMS variant provides a way to automate the choice for the step-size parameter and speed up the convergence of the algorithm [38]. (2) In practice, its design is based on quite limited knowledge of the input-signal statistics [79]. (3) It is able to track the varying signal statistics [9]. The derivation of the majority of adaptive filter algorithms relies on the minimization of MSE criterion. Another optimization criterion that is encountered in image processing is the MAE criterion. The so-called signed error LMS L-filter that minimizes the MAE between the filter output and the desired response is derived. The signed error LMS L-filter shares the benefits of the signed error LMS algorithm, that is, the reduced numerical requirements due to the elimination of any multiplication in the coefficient updating equation. A modified LMS L-filter with nonhomogeneous step-sizes is introduced in order to accelerate the rate of convergence of the adaptive L-filter by allowing the convergence of each L-filter coefficient to be controlled by a separate step-size parameter. An ad-hoc and a theoretically justified step-size selection is described. Finally, an adaptive L-filter structure that is reminiscent of the cellular network topology is reviewed.

5.3.1   Location-Invariant LMS L-filter

We describe the location-invariant LMS L-filter design in the case of a nonconstant signal corrupted by zero-mean additive white noise. Let us assume that $x_L(n)$, defined in Eq. (5.6), and $s(n)$ are jointly stationary stochastic signals. This assumption, which is implicitly made throughout this chapter, implies that the original (noise-free) image $s(n)$ is edge-free [27], a condition that is hardly met in reality. However, since we aim at designing adaptive filters, we rely on the tracking capabilities of the filters for nonstationary signals as images are. We are seeking the L-filter whose output at pixel n given by

$$y(n) = \mathbf{a}^T(n)\, \mathbf{x}_L(n), \tag{5.36}$$

minimizes the MSE

$$
\begin{aligned}
\varepsilon(n) &= \mathrm{E}\left\{(y(n) - s(n))^2\right\} \\
&= \mathrm{E}\left\{s^2(n)\right\} - 2\mathbf{a}^T(n)\,\mathbf{p}_L + \mathbf{a}^T(n)\,\mathbf{R}_L\,\mathbf{a}(n)
\end{aligned}
\tag{5.37}
$$

subject to the constraint

$$\mathbf{1}_N^T\,\mathbf{a}(n) = 1. \tag{5.38}$$

The constraint of Eq. (5.38) ensures that the filter preserves the zero-frequency or dc signals [12, 57]. In Eq. (5.37), $\mathbf{R}_L = \mathrm{E}\left\{\mathbf{x}_L(n)\,\mathbf{x}_L^T(n)\right\}$ is the correlation matrix of the observed ordered image pixel values and $\mathbf{p}_L = \mathrm{E}\{s(n)\,\mathbf{x}_L(n)\}$ denotes the crosscorrelation vector between the ordered input vector $\mathbf{x}_L(n)$ and the desired image pixel value $s(n)$. Let $\nu = (N+1)/2$. In employing Eq. (5.38), we can partition the *L*-filter coefficient vector as follows:

$$\mathbf{a}(n) = \left(\mathbf{a}_1^T(n)|a_\nu(n)|\mathbf{a}_2^T(n)\right)^T, \tag{5.39}$$

where $\mathbf{a}_1(n)$ and $\mathbf{a}_2(n)$ are $(N-1)/2 \times 1$ vectors given by

$$\mathbf{a}_1(n) = (a_1(n),\ldots,a_{\nu-1}(n))^T \quad \mathbf{a}_2(n) = (a_{\nu+1}(n),\ldots,a_N(n))^T. \tag{5.40}$$

The coefficient for the median input sample is evaluated then according to

$$a_\nu(n) = 1 - \mathbf{1}_{\nu-1}^T \mathbf{a}_1(n) - \mathbf{1}_{\nu-1}^T \mathbf{a}_2(n). \tag{5.41}$$

Similarly the ordered input vector can be rewritten as

$$\mathbf{x}_L(n) = \left(\mathbf{x}_{L1}^T(n)|x_{(\nu)}(n)|\mathbf{x}_{L2}^T(n)\right)^T. \tag{5.42}$$

Let $\mathbf{a}'(n)$ be the reduced *L*-filter coefficient vector

$$\mathbf{a}'(n) = \left(\mathbf{a}_1^T(n)|\mathbf{a}_2^T(n)\right)^T \tag{5.43}$$

and $\mathbf{x}'_L(n)$ be the $(N-1) \times 1$ vector

$$\mathbf{x}'_L(n) = \begin{bmatrix} \mathbf{x}_{L1}(n) - x_{(\nu)}(n)\mathbf{1}_{\nu-1} \\ \mathbf{x}_{L2}(n) - x_{(\nu)}(n)\mathbf{1}_{\nu-1} \end{bmatrix}. \tag{5.44}$$

Following the analysis in [43] it can be proven that the LMS recursive relation for updating the reduced *L*-filter coefficient vector is given by

$$\widehat{\mathbf{a}}'(n+1) = \widehat{\mathbf{a}}'(n) + \mu\,e(n)\,\mathbf{x}'_L(n), \tag{5.45}$$

where $e(n)$ is the estimation error at pixel n, that is, $e(n) = s(n) - y(n)$. Equations (5.45) and (5.41) constitute the *location-invariant LMS L-filter*. The convergence properties of the location-invariant LMS *L*-filter depend on the eigenvalue distribution of the correlation matrix of the vector $\mathbf{x}'_L(n)$.

Several researchers have proposed adaptive algorithms for the design of *L*-filters that fulfill structural constraints, such as the location-invariance, that is discussed in this paper, as well as the unbiasedness [76, 43, 19, 15, 88]. The design of constrained adaptive LMS algorithms has been a hot research topic for two decades in adaptive array beamforming[24, 84]. Two are the main classes of constrained LMS algorithms: the *gradient projection algorithm* and *Frost's algorithm*. A generalized structure has been proposed by Griffiths and Jim [29] that includes Frost's algorithm as well as other linearly constrained adaptive algorithms. The algorithms reported in [76, 19, 15, 88] utilize Frost's algorithm. The possibility of using the gradient

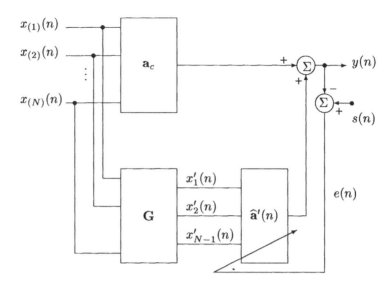

Fig. 5.1   Generalized linearly constrained adaptive processing structure (reprinted by copyright permission from IEEE).

projection algorithm has also been proposed in [15]. Between those two, Frost's algorithm is preferable because it does not accumulate errors as opposed to gradient projection algorithm [24, 29, 77, 16, 84]. We will demonstrate that Eq. (5.45) falls into *Griffiths's and Jim's general structure* shown in Fig. 5.1. This general structure consists of two paths. In the upper path, the input observations are filtered by a fixed filter whose coefficients have sum 1. In the lower path, the input observations are multiplied by matrix preprocessor G of dimensions $(N - 1) \times N$ at most, and the resulted data are fed to an adaptive filter whose coefficients are updated according to the unconstrained LMS algorithm. In addition the rows of matrix G should sum up to zero. In our case both the fixed filter and the adaptive one are L-filters that are driven by the observed ordered image pixel values. If we choose a_c to be a median filter and a preprocessor matrix of the form

$$G = \begin{bmatrix} 1 & 0 & \ldots & 0 & -1 & 0 & \ldots & 0 \\ 0 & 1 & \ldots & 0 & -1 & 0 & \ldots & 0 \\ \vdots & & & & & & & \vdots \\ 0 & 0 & \ldots & 1 & -1 & 0 & \ldots & 0 \\ 0 & 0 & \ldots & 0 & -1 & 1 & \ldots & 0 \\ \vdots & & & & & & & \vdots \\ 0 & 0 & \ldots & 0 & -1 & 0 & \ldots & 1 \end{bmatrix}, \tag{5.46}$$

the location-invariant LMS *L*-filter results.

If a desired image is not available, Eq. (5.45) is modified as

$$\widehat{\mathbf{a}}'(n+1) = \widehat{\mathbf{a}}'(n) - \mu\, y(n)\, \mathbf{x}'_L(n). \tag{5.47}$$

The resulting filter minimizes the overall output power

$$\varepsilon(n) = \mathbf{a}^T(n)\mathbf{R}_L\mathbf{a}(n) \tag{5.48}$$

subject to the constraint defined in Eq. (5.38).

5.3.2   Variants of the Unconstrained LMS *L*-filter

In this section we deal with the unconstrained LMS adaptive *L*-filter [69, 71] whose coefficients are updated by using the following recursive formula:

$$\widehat{\mathbf{a}}(n+1) = \widehat{\mathbf{a}}(n) + \mu\, e(n)\, \mathbf{x}_L(n). \tag{5.49}$$

Three variants of the unconstrained LMS adaptive *L*-filter are discussed: the NLMS *L*-filter, the signed error LMS *L*-filter, and the modified LMS *L*-filter with nonhomogeneous step-sizes.

5.3.2.1   Normalized LMS L-filter   A difficult problem frequently met in the design of adaptive filters based on the LMS algorithm, such as the algorithm of Eq. (5.49), is the selection of the step-size parameter μ. Although, in theory, sufficient conditions on μ exist that guarantee the convergence of the LMS algorithm, these conditions depend on the knowledge of the eigenvalues of the correlation matrix $\mathbf{R}$ [33, 34, 22]. In the case of the LMS adaptive *L*-filter defined by Eq. (5.49), $\mathbf{R}$ is the correlation matrix of the ordered input vectors. For example, the necessary and sufficient condition for the average *L*-filter coefficient vector to be convergent is

$$0 < \mu < \frac{2}{\lambda_{\max}}, \tag{5.50}$$

where $\lambda_{\max}$ denotes the maximal eigenvalue of matrix $\mathbf{R}_L$ [33]. Furthermore μ should satisfy the following stricter condition:

$$0 < \mu < \frac{2}{3\,\mathrm{tr}[\mathbf{R}]} = \frac{2}{3 \times \text{ total input power}} \tag{5.51}$$

in order to achieve convergence of the average mean-squared error to a steady-state value [22]. In Eq. (5.51), $\mathrm{tr}[\mathbf{R}]$ is the trace of the matrix $\mathbf{R}$. In addition, when the adaptive filter is operated in a nonstationary environment (as is the case in image processing), the inequalities defined in Eqs. (5.50) and (5.51) become useless, since the correlation matrix $\mathbf{R}$ is time/space varying. It is then reasonable to employ a time/space varying step-size parameter $\mu(n)$. Let us evaluate the a posteriori estimation error at pixel n, defined as follows:

$$e'(n) = s(n) - \widehat{\mathbf{a}}^T(n+1)\mathbf{x}_L(n). \tag{5.52}$$

If $\mu(n)$ is chosen to be

$$\mu(n) = \frac{1}{\mathbf{x}_L^T(n)\mathbf{x}_L(n)} = \frac{1}{||\mathbf{x}_L(n)||^2}, \tag{5.53}$$

then $e'(n)$ becomes zero. A step-size sequence of the form Eq. (5.53) motivated us to modify Eq. (5.49) as follows:

$$\widehat{\mathbf{a}}(n+1) = \widehat{\mathbf{a}}(n) + \frac{\mu_0}{||\mathbf{x}_L(n)||^2} e(n)\,\mathbf{x}_L(k). \tag{5.54}$$

Equation (5.54) describes the adaptation of the coefficients of the *normalized LMS L-filter*. It may be interpreted as an ordinary LMS updating formula that operates on the normalized ordered noisy-input observations $\mathbf{x}_L(n)/||\mathbf{x}_L(n)||$. Therefore, according to Eq. (5.51), μ_0 should be chosen to satisfy the inequality

$$0 < \mu_0 \leq \frac{2}{3}. \tag{5.55}$$

In practice, it has been found that Eq. (5.55) is still conservative, since the best results are obtained for $\mu_0 = 0.8$. However, if μ_0 is chosen to be equal to 2/3, we are still close to its optimal value.

5.3.2.2 *Signed error LMS L-filter*

Another criterion that is frequently encountered in nonlinear image processing is the MAE criterion (also called least mean absolute value criterion [9]). The motivation to use MAE instead of the MSE stems from the fact that the median is the optimal MAE estimator of location for the double-exponential distribution (i.e., the Laplacian distribution) [70]. The MAE criterion to be minimized is defined as

$$\varepsilon'(n) = \mathrm{E}\left\{|s(n) - y(n)|\right\}. \tag{5.56}$$

In this case the coefficient vector $\mathbf{a}(n)$ must be updated at n so that Eq. (5.56) is minimized. The method of steepest-descent yields the following recursive relation for updating the filter coefficients:

$$\mathbf{a}(n+1) = \mathbf{a}(n) + \mu\left[-\nabla\varepsilon'(n)\right], \tag{5.57}$$

where

$$\nabla\varepsilon'(n) = \frac{\partial\varepsilon'(n)}{\partial\mathbf{a}(n)} = \frac{\partial}{\partial\mathbf{a}(n)}\,\mathrm{E}\left\{|s(n) - \mathbf{a}^T(n)\mathbf{x}_L(n)|\right\}. \tag{5.58}$$

An unbiased estimate for the gradient $\nabla \varepsilon'(n)$ can be obtained, if we drop out the expectation operator:

$$\widehat{\nabla} J'(n) = \frac{\partial}{\partial \mathbf{a}(n)} |s(n) - \mathbf{a}^T(n)\mathbf{x}_L(n)| = -\mathrm{sgn} \left[s(n) - \mathbf{a}^T(n)\mathbf{x}_L(n)\right] \mathbf{x}_L(n),$$

(5.59)

where $\mathrm{sgn}[\cdot]$ denotes the sign function

$$\mathrm{sgn}[x] = \begin{cases} 1 & \text{if } x > 0 \\ -1 & \text{if } x < 0. \end{cases}$$

(5.60)

Therefore the updating formula for the coefficients of the *signed error LMS L-filter* is

$$\widehat{\mathbf{a}}(n+1) = \widehat{\mathbf{a}}(n) + \mu \, \mathrm{sgn}[e(n)] \, \mathbf{x}_L(n).$$

(5.61)

Equation (5.61) is equivalent to the linear least mean absolute value LMS algorithm [9]. The only difference is that Eq. (5.61) employs the vector of the ordered observed image pixel values instead of the input vector itself. The step-size parameter can be chosen as follows: The a posteriori estimation error at n is simply given by

$$e'(n) = s(n) - \widehat{\mathbf{a}}^T(n+1)\mathbf{x}_L(n) = e(n) - \mu \, \mathrm{sgn}[e(n)] \, \mathbf{x}_L^T(n)\mathbf{x}_L(n).$$

(5.62)

If the step-size sequence $\mu(n)$ is chosen as

$$\mu(n) = \frac{\mu_0 |e(n)|}{\mathbf{x}_L^T(n)\mathbf{x}_L(n)}, \qquad 0 < \mu_0 < 1,$$

(5.63)

then $|e'(n)|$ will be $(1 - \mu_0)|e(n)|$, that is, the a posteriori absolute estimation error is smaller that the (a priori) absolute estimation error.

5.3.2.3 Modified LMS L-filter with nonhomogeneous step-sizes

It is well known that the rate of convergence of the LMS algorithm is one order of magnitude slower than that of other adaptive algorithms (e.g., the RLS or the RLS Lattice algorithm) [33]. This slow rate of convergence can be attributed to the fact that only one parameter, the step-size μ, controls the convergence of all the filter coefficients. On the contrary, in the case of the RLS algorithm, the convergence of each filter coefficient is controlled by a separate element of the Kalman gain vector. In addition, at each time instant, the Kalman gain vector is updated utilizing all the information contained in the input data, extending back to the time instant when the algorithm is initiated. This observation motivated us to employ different step-size parameters for the various LMS L-filter coefficients in the recursive relation that describes their adaptation:

$$\widehat{\mathbf{a}}(n+1) = \widehat{\mathbf{a}}(n) + e(n) \, \mathbf{M}(n)\mathbf{x}_L(n)$$

(5.64)

where $\mathbf{M}(n)$ denotes the following diagonal matrix:

$$\mathbf{M}(n) = \mathrm{diag} [\mu_1(n), \mu_2(n), \dots, \mu_N(n)].$$

(5.65)

Ad hoc nonhomogeneous step-size selection.   We have found, by experiment, that the following step-size sequence

$$\mu_i(n) = \mu_0 \frac{\sum_{j=0}^{n} x_{(i)}(n-j)}{\sum_{j=0}^{n} x_{(1)}(n-j)} \tag{5.66}$$

gives results comparable to those obtained by using the normalized LMS L-filter algorithm. The selection of the step-size sequence according to Eq. (5.66) guarantees a smaller a posteriori estimation error for every n. Indeed, since for nonnegative input observations the following inequality holds:

$$x_{(i)}(l) \geq x_{(1)}(l) \; \forall l, \tag{5.67}$$

we have

$$e'(n) = e(n) \left\{ 1 - \sum_{i=1}^{N} \mu_i(n) x_{(i)}^2(n) \right\} < e(n) \left\{ 1 - \mu_0 \sum_{i=1}^{N} x_{(i)}^2(n) \right\}. \tag{5.68}$$

Therefore, if μ_0 is chosen so that

$$0 < \mu_0 < \frac{1}{\max_n \{\sum_{i=1}^{N} x_{(i)}^2(n)\}} = \frac{1}{\text{peak input power}}, \tag{5.69}$$

then $e'(n) < e(n)$. Another property that the modified LMS L-filter, defined by Eq. (5.64), possesses is drawn from Eq. (5.68). The right-hand side of the inequality in Eq. (5.68) is identified to be the a posteriori error at n of an LMS L-filter that uses a constant step-size parameter $\mu = \mu_0$. Thus the modified LMS L-filter given by Eq. (5.64) produces always a smaller a posteriori estimation error than the ordinary LMS L-filter that employs a constant step-size $\mu = \mu_0$.

Theoretically motivated step-size selection.   The nonhomogeneous step-size selection described previously was clearly an ad hoc selection. In the following we describe a variable step-size selection mechanism that is reminiscent of the one proposed in [17]. It has been found that such a variable step-size selection can accelerate the convergence of the filter coefficients toward the optimal ones, especially at the beginning of the filtering session, without deteriorating the noise reduction achieved by the filter at convergence (i.e., the mean squared error at convergence) [49]. Moreover, the proposed variable step-size selection algorithm yields always a stable adaptive filtering algorithm, which is not the case with NLMS algorithm or the method proposed in [17].

We have seen in Eq. (5.51) that the adaptation step should satisfy a certain inequality so that the average mean squared error converges to a steady-state value. The upper bound in Eq. (5.51) depends on the trace of the correlation matrix of the ordered input vectors. Note that the trace of the correlation matrix of the ordered input vectors is equal to the trace of the correlation matrix of the input observations (without ordering).

Let $\mathbf{R}_L = \mathbf{U}\mathbf{\Lambda}\mathbf{U}^T$, where $\mathbf{U} = \{U_{ij}\}$ is the modal matrix of $\mathbf{R}_L$, whose jth column is the eigenvector associated with the jth eigenvalue of $\mathbf{R}_L$, and $\mathbf{\Lambda}$ is a diagonal matrix composed of the eigenvalues of $\mathbf{R}_L$. Let $\phi(n) = \mathbf{a}(n) - \mathbf{a}^o$ denote the coefficient-error vector at n, where $\mathbf{a}^o$ is the optimal L-filter coefficient vector

$$\mathbf{a}^o = \mathbf{R}_L^{-1}\mathbf{p}_L \tag{5.70}$$

and $\mathbf{p}_L = \mathrm{E}\{s(n)\mathbf{x}_L(n)\}$. It is more convenient to work with the transformed coefficient-error vector, $\tilde{\phi}(n) = \mathbf{U}^T\phi(n)$. Following similar lines to [17], we can show that

$$\tilde{\phi}(n+1) = (\mathbf{I}_N - \widehat{\mathbf{M}}\mathbf{\Lambda})\tilde{\phi}(n), \tag{5.71}$$

where $\widehat{\mathbf{M}} = \mathbf{U}^T\mathbf{M}\mathbf{U}$ and $\mathbf{I}_N$ is the $N \times N$ identity matrix. It can be easily seen that $\widehat{\mathbf{M}}$ is no more diagonal. Its ij-th element is

$$\widehat{\mu}_{ij} = \sum_{k=1}^{N} \mu_k U_{ki} U_{kj}. \tag{5.72}$$

The evolution of the coefficient-error covariance matrix results in [33, 17]

$$\begin{aligned} \mathbf{K}(n+1) = \quad & \mathbf{K}(n) - \widehat{\mathbf{M}}\mathbf{\Lambda}\mathbf{K}(n) - \mathbf{K}(n)\mathbf{\Lambda}\widehat{\mathbf{M}} \\ & + \widehat{\mathbf{M}}\mathbf{\Lambda}\mathrm{tr}\left[\mathbf{\Lambda}\mathbf{K}(n)\right]\widehat{\mathbf{M}} + \varepsilon_{\min}\widehat{\mathbf{M}}\mathbf{\Lambda}\widehat{\mathbf{M}}, \end{aligned} \tag{5.73}$$

where $\varepsilon_{\min}$ denotes the minimum MSE. For a moment we will assume that $\mathrm{E}\{\phi_i\phi_j\} = \sigma_\phi^2\delta_{ij}$ with δ_{ij} denoting a Kronecker delta:

$$\delta_{ij} = \begin{cases} 1 & \text{if } i = j \\ 0 & \text{otherwise.} \end{cases} \tag{5.74}$$

Such an assumption implies that $\mathbf{K}(n) = \sigma_\phi^2\mathbf{I}_N$. Furthermore, let us consider $\widehat{\mathbf{M}} = \mu\mathbf{I}_N$. Equation (5.73) can be simplified to a more tractable form:

$$\begin{aligned} \mathbf{K}(n+1) = \quad & \mathbf{K}(n)\left(\mathbf{I}_N - 2\mathbf{\Lambda}\widehat{\mathbf{M}}\right) \\ & + \mathrm{tr}\left[\mathbf{\Lambda}\mathbf{K}(n)\right]\widehat{\mathbf{M}}\mathbf{\Lambda}\widehat{\mathbf{M}} + \varepsilon_{\min}\widehat{\mathbf{M}}\mathbf{\Lambda}\widehat{\mathbf{M}}. \end{aligned} \tag{5.75}$$

The step-sizes can be chosen so that the excess mean squared error is minimized. Such a minimization problem has been solved by Bersad [11], when Eq. (5.75) holds and all the eigenvalues are equal. Clearly, such an assumption does not hold for the correlation matrix of the order statistics $\mathbf{R}_L = \mathrm{E}\{\mathbf{x}_L(n)\mathbf{x}_L^T(n)\}$, and it can only be considered as a design assumption. To minimize the excess mean squared error, the diagonal elements of $\mathbf{K}(n)$ should be minimized. Following similar reasoning to

[17] it can be shown that $\widehat{\mu}_{ii}$ should be chosen as

$$\widehat{\mu}_{ii} = \frac{1}{\text{tr}[\mathbf{R}]}. \tag{5.76}$$

In the remaining analysis all the assumptions that yield Eq. (5.75) will be dropped. By combining Eqs. (5.76) and (5.72), we obtain

$$\sum_{k=1}^{N} \mu_k U_{ki}^2 = \frac{1}{\text{tr}[\mathbf{R}]}. \tag{5.77}$$

One may write N equations as Eq. (5.77) for $i = 1, 2, \ldots, N$. Then, the set of N equations can be solved for μ_k. However, the solution of the set of equations does not guarantee that each μ_k is less than $1/\text{tr}[\mathbf{R}]$. On the contrary, it is trivial to show that the set of equations Eq. (5.77) implies that

$$\sum_{k=1}^{N} \mu_k = \frac{N}{\text{tr}[\mathbf{R}]}. \tag{5.78}$$

Accordingly we follow a different approach. Let us assume that the step-size μ_k that controls the adaptation of coefficient $a_k(n)$ (i.e., the weight of the kth order statistic $x_{(k)}(n)$) is given by

$$\mu_k = f(\mu_{k1}, \mu_{k2}, \ldots, \mu_{ki}, \ldots, \mu_{kN}), \tag{5.79}$$

where $f(\cdot)$ stands for an appropriate function and each μ_{ki} is chosen by taking into consideration only the ith eigenvector of $\mathbf{R}_L$. Let us denote by $\mathbf{R}_{L;[k]}$ the matrix

$$\mathbf{R}_{L;[k]} = \{R_{L_{ij}}\}, \quad i, j = 1, 2, \ldots, k; \quad k = 1, 2, \ldots, N. \tag{5.80}$$

Each μ_{ki} can be defined as follows:

$$\mu_{ki} = G_k \, \mu_{\min_i}, \quad \text{where } G_k = \frac{\text{tr}[\mathbf{R}]}{\text{tr}[\mathbf{R}_{L;[k]}]}. \tag{5.81}$$

By substituting Eq. (5.81) into Eq. (5.77), we obtain

$$\mu_{\min_i} = \left(\text{tr}^2[\mathbf{R}] \sum_{k=1}^{N} \frac{U_{ki}^2}{\text{tr}[\mathbf{R}_{L;[k]}]} \right)^{-1}. \tag{5.82}$$

Accordingly

$$\mu_{ki} = \begin{cases} \left(\text{tr}[\mathbf{R}] \text{tr}[\mathbf{R}_{L;[k]}] \sum_{j=1}^{N} \dfrac{U_{ji}^2}{\text{tr}[\mathbf{R}_{L;[j]}]} \right)^{-1} = \theta & \text{if } \theta \leq \dfrac{1}{\text{tr}[\mathbf{R}]} \\ \dfrac{1}{\text{tr}[\mathbf{R}]} & \text{otherwise.} \end{cases} \tag{5.83}$$

Having computed μ_{ki}, the step-size μ_k can be obtained, for example, by averaging μ_{ki}:

$$\mu_k = \frac{1}{N} \sum_{i=1}^{N} \mu_{ki}. \tag{5.84}$$

The major difficulties of the variable step-size selection analyzed above are the following: (1) It requires the computation of the eigenvalues and eigenvectors of the correlation matrix at every image pixel. Therefore it increases dramatically the computational complexity of the algorithm. (2) There is no guarantee that the $\mu_{\min_i}$ determined in Eq. (5.82) yields always a stable filter operation. We have to check if $\mu_{\min_i} < 1/\text{tr}[\mathbf{R}]$ before accepting the value computed in Eq. (5.82). It has been found that the most crucial term in the proposed variable step-size selection algorithm is the ratio G_k. Accordingly we propose to compute a sequence of variable step-sizes by employing a run-time estimate of G_k and a constant step-size $\mu_0 < 1/\text{tr}[\mathbf{R}]$ (e.g., $\mu_0 = 10^{-8}$) as follows:

$$\mu_k(n) = \begin{cases} \widehat{G}_k(n)\,\mu_0 & \text{if} \qquad \mu_k(n) < \mu_{\max} \\ \dfrac{\mu_0'}{\mathbf{x}^T(n)\mathbf{x}(n)} & \text{otherwise} \end{cases} \tag{5.85}$$

where $\mu_{\max}$ is an upper bound on the variation of μ, for example, 10^{-6}, μ_0' is a positive real number less than 1, and $\widehat{G}_k(n)$ is an estimate for the ratio G_k for every image pixel that is computed by

$$\begin{aligned} \widehat{G}_k(n) &= \frac{\sum_{i=1}^{N} Q_i(n)}{\sum_{i=1}^{k} Q_i(n)}, \\ Q_i(n) &= \frac{1}{n-N} \sum_{j=N}^{n} x_{(i)}^2(j), \quad i = 1, 2, \ldots, N; \quad n = N+1, \ldots. \end{aligned} \tag{5.86}$$

Obviously $Q_i(n)$ can be computed recursively.

5.3.3   Cellular LMS *L*-filters

In this section we review an adaptive *L*-filter structure that was proposed in [26]. It is reminiscent of the cellular neural network topology [18, 85]. It can cope more efficiently with the nonstationary nature of still images and image sequences due to the fact that all the interactions between the processing cells (which are adaptive *L*-filters themselves) are mainly confined to a finite neighborhood around each processing cell. A single adaptive *L*-filter is expected to track the nonstationarity only in environments of slowly varying statistics. However, in still images or image sequences, we find image region statistics that vary both spatially and in time. Furthermore, due to image scanning methods (e.g., row-wise scanning described in Section 5.2) abrupt transitions between regions having different statistics are regularly encountered in

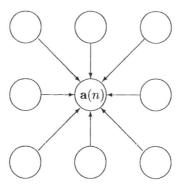

Fig. 5.2   Interactions within a neighborhood around the cell n.

the adaptive algorithm, thus preventing the filter coefficients from converging to the optimal ones.

An ordinary LMS adaptive L-filter updates its coefficients by using Eq. (5.49). It is seen that the adaptation of the L-filter coefficients at pixel n depends explicitly on the past values of the desired signal $s(l)$ and the ordered input vectors $\mathbf{x}_L(l)$ for $l < n$. Such a direct dependence on the past history in a nonstationary environment, in addition to the transitions introduced by the scanning method, may prevent the filter to converge locally to a steady-state one. This observation motivated us to adopt an approach where the direct interactions between the image pixels are limited within a finite local neighborhood. It is well known that cellular neural systems and cellular automata share such a property: all the interactions are local within a finite radius [18, 85]. Accordingly we borrow the cellular network topology in order to design an adaptive L-filter structure that is called *cellular adaptive L-filter* hereafter.

The cellular adaptive L-filter structure is a $N_1 \times N_2$ cellular network having $N_1 \times N_2$ processing cells, one assigned to each image pixel, arranged in N_1 rows and N_2 columns [26]. An example of a 3×3 cellular adaptive L-filter structure is depicted in Fig. 5.2. The cycles represent the processing cells assigned to each image pixel. The links between the cells indicate that there are interactions between the linked cells. Each processing cell is an adaptive LMS L-filter. A neighborhood is defined around each processing cell. Let $\mathbf{a}(n; k)$ denote the L-filter coefficients of the nth processing cell at the kth iteration. At $k = 0$, all processing cells are randomly initialized. At each iteration k, all processing cells evaluate synchronously an estimation error of the form

$$e(j; k) = s(j; k) - \mathbf{a}^T(j; k)\,\mathbf{x}_L(j; k), \quad j = 1, 2, \ldots, N_1 \cdot N_2. \tag{5.87}$$

The steepest descent method yields the following recursion for the filter coefficients:

$$\mathbf{a}(n; k+1) = \mathbf{a}(n; k) - \tfrac{1}{2}\mu\nabla\tilde{\varepsilon}(n; k), \tag{5.88}$$

where μ is the step-size and $\tilde{\varepsilon}$ denotes the mean-squared average error in the neighborhood of the nth cellular L-filter given by

$$\tilde{\varepsilon}(n;k) = \mathrm{E}\left\{\frac{1}{M}\left(\sum_{j \in \mathcal{M}_\wp(n)} e(j;k)\right)^2\right\}, \tag{5.89}$$

where M is the cardinality of the processing cells in the $\wp$-neighborhood of any processing cell, $\mathcal{M}_\wp(n)$. By differentiating $\tilde{\varepsilon}$ with respect to coefficients $\mathbf{a}(n;k)$, we get

$$\nabla\tilde{\varepsilon}(n;k) \overset{\triangle}{=} \frac{\partial\tilde{\varepsilon}}{\partial\mathbf{a}(n;k)} = -\frac{2}{M}\mathrm{E}\left\{\left(\sum_{j\in\mathcal{M}_\wp(n)} e(j;k)\right)\mathbf{x}_L(n;k)\right\}. \tag{5.90}$$

Let us estimate the expected value in Eq. (5.90) by

$$\mathrm{E}\left\{\left(\sum_{j\in\mathcal{M}_\wp(n)} e(j;k)\right)\mathbf{x}_L(n;k)\right\} \simeq \sum_{j\in\mathcal{M}_\wp(n)} e(j;k)\,\mathbf{x}_L(n;k) \tag{5.91}$$

as in the LMS algorithm. Accordingly, the following updating equation of the cellular L-filter coefficients is obtained:

$$\widehat{\mathbf{a}}(n;k+1) = \widehat{\mathbf{a}}(n;k) + \mu\,\bar{e}(n;k)\,\mathbf{x}_L(n;k), \tag{5.92}$$

where

$$\bar{e}(n;k) = \frac{1}{M}\sum_{j\in\mathcal{M}_\wp(n)} e(j;k). \tag{5.93}$$

Equations (5.92) and (5.93) imply that all processing cells adjacent to the pixel n, meaning all the processing cells inside the neighborhood of $\mathcal{M}_\wp(n)$, exchange their estimation errors, and the nth processing cell evaluates an averaged estimation error. Equation (5.92) is the *cell dynamic equation*. It can easily be understood that in the case of still images $s(n;k) = s(n)$ and $\mathbf{x}_L(n;k) = \mathbf{x}_L(n)$ for all iterations $k = 1, 2, \ldots$. In the case of image sequences, the index k could correspond either to the frame that is being filtered or to an iteration index. In the latter case the previous original/filtered frame can be used as the reference image. Unlike the ordinary LMS adaptive L-filter, which is a recursive algorithm that cannot be implemented in parallel, the cellular adaptive L-filter can easily be parallelized. Indeed, each processing cell exchanges its estimation error with the adjacent cells and all the remaining operations are local, put differently, they can be performed at each processing cell in parallel.

The cellular adaptive L-filter should converge to a constant steady-state filter after a transient period following initialization. We will briefly discuss the convergence properties of the cellular adaptive LMS L-filter. First, let us recast the cell dynamic equation (Eq. (5.92)) in the following form:

$$\widehat{a}(n; k+1) = \Upsilon(n)\widehat{a}(n; k) - h(n; k) + g(n), \tag{5.94}$$

where $\mathcal{M}^-(n)$ is the neighborhood of pixel n excluding this pixel, and

$$\Upsilon(n) = I_N - \frac{\mu}{M}x_L(n)x_L^T(n), \tag{5.95}$$

$$h(n; k) = \frac{\mu}{M}\left(\sum_{j\in\mathcal{M}^-(n)} x_L^T(j)\widehat{a}(j; k)\right) x_L(n), \tag{5.96}$$

$$g(n) = \frac{\mu}{M}\left(\sum_{j\in\mathcal{M}(n)} s(j)\right) x_L(n). \tag{5.97}$$

By mathematical induction it has been proved that [26]

$$\widehat{a}(n; k+1) = \Upsilon^k(n)\widehat{a}(i; 0) - \sum_{l=0}^{k-1}\Upsilon^{k-1-l}(n)\left(g(n) - h(n; l)\right) + \left(g(n) - h(n; k)\right). \tag{5.98}$$

The system defined by Eq. (5.98) is a discrete autonomous system. The system converges to a steady-state if the eigenvalues of matrix $\Upsilon(n)$ lie inside the unit circle. It can easily be proved that the eigenvalues of $\Upsilon(n)$ are

$$\lambda_1 = \lambda_2 = \ldots \lambda_{N-1} = 1, \quad \lambda_N = 1 - \frac{\mu}{M}x_L^T(n)x_L(n). \tag{5.99}$$

Accordingly the step-size μ must be chosen in such a way that the largest eigenvalue lies inside the unit circle:

$$0 < \mu < \frac{2M}{x_L^T(n)x_L(n)}. \tag{5.100}$$

Moreover it has been proved that the sum of cellular L-filter coefficients is upper bounded as follows [26]:

$$1_N^T\widehat{a}(n) \le \mu \max_i\{s(i)\} \left[1_N^Tx(n)\right]. \tag{5.101}$$

The above-mentioned bound is finite and independent of the iteration index k.

5.3.4 Experimental Results

We present several sets of experiments in order to assess the performance of the adaptive L-filters we have discussed thus far. First we describe the experimental setup and the quantitative criteria used in the experiments. The L-filter coefficients have been randomly initialized in the interval $(0, 1)$, and they have been normalized by their sum so as their sum equals unity. The entire noisy image has been used in

Table 5.1   Noise reduction and mean absolute error reduction (in dB) achieved by the various *L*-filters in the restoration of "Lenna" corrupted by mixed impulsive and additive Gaussian noise

Method	NR	MAER
Median 3×3	-8.756	-8.147
Location-invariant LMS *L*-filter 3×3 ($\mu = 5 \times 10^{-7}$)	-9.747	-9.192
Modified LMS *L*-filter 3×3 with nonhomogeneous step-sizes ($\mu_0 = 5 \times 10^{-7}$)	-11.216	-10.867
Normalized LMS *L*-filter 3×3 ($\mu_0 = 0.8$)	-11.281	-11.071

the training of the adaptive filter. A single run on the training noisy image has been performed. The coefficients derived during the operation of the algorithm on the last image row have been averaged and have been applied to filter the entire image. Two criteria have been employed, namely the *noise reduction index* (NR) defined as the ratio of the mean output noise power to the mean input noise power, or

$$\text{NR} = 10 \log \frac{\frac{1}{N_1 N_2} \sum_{n_1=1}^{N_1} \sum_{n_2=1}^{N_2} (y(n_1, n_2) - s(n_1, n_2))^2}{\frac{1}{N_1 N_2} \sum_{n_1=1}^{N_1} \sum_{n_2=1}^{N_2} (x(n_1, n_2) - s(n_1, n_2))^2} \quad (\text{in dB}), \quad (5.102)$$

and the *mean absolute error reduction* (MAER) defined as the ratio of the mean absolute error in the output to the mean absolute error in the input, or

$$\text{MAER} = 20 \log \frac{\frac{1}{N_1 N_2} \sum_{n_1=1}^{N_1} \sum_{n_2=1}^{N_2} |y(n_1, n_2) - s(n_1, n_2)|}{\frac{1}{N_1 N_2} \sum_{n_1=1}^{N_1} \sum_{n_2=1}^{N_2} |x(n_1, n_2) - s(n_1, n_2)|} \quad (\text{in dB}). \quad (5.103)$$

In Eqs. (5.102) and (5.103), $s(n_1, n_2)$ is the intensity of the original image at pixel (n_1, n_2), $x(n_1, n_2)$ denotes the intensity of the same pixel when the original image is corrupted by noise and $y(n_1, n_2)$ is the filter output at the same. N_1 and N_2 denote the number of image rows and columns, respectively.

5.3.4.1   Performance of location-invariant, normalized and modified adaptive
L-filters   Figure 5.3(a) shows "Lenna," an original image frequently encountered in image processing literature. The original image is severely corrupted by adding zero-mean additive white Gaussian noise of standard deviation $\sigma_n = 50$ plus impulsive noise with probability $p = 10\%$. The NR as well as the MAER achieved by the location-invariant LMS *L*-filter, the normalized LMS *L*-filter and the modified LMS *L*-filter with nonhomogeneous step-sizes, all of dimensions 3×3, are listed in Table 5.1. In the same table we have included the corresponding figures of merit for the 3×3 median filter. As can be seen, the condition for location-invariant estimation is strict enough and the resulting adaptive *L*-filter is only 1 dB better than the median filter with respect to both quantitative measures. The modified LMS *L*-filter with

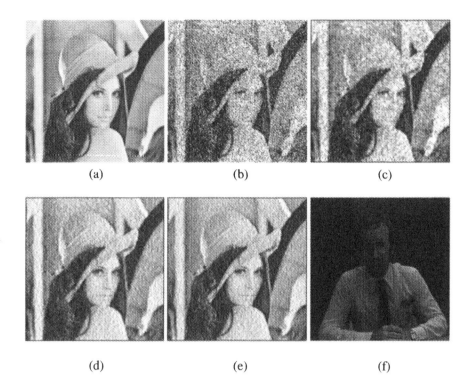

(a)   (b)   (c)

(d)   (e)   (f)

Fig. 5.3 Comparison between the adaptive LMS L-filters under study in suppressing mixed impulsive and additive Gaussian noise. (a) Original image "Lenna." (b) Original image corrupted by mixed impulsive and additive Gaussian noise. (c) Output of the 3×3 location-invariant LMS L-filter. (d) Output of the 3×3 modified LMS L-filter that employs nonhomogeneous step-sizes. (e) Output of the 3×3 normalized LMS L-filter. (f) Original image "Trevor White" (reprinted by copyright permission from IEEE).

nonhomogeneous step-sizes is the second best adaptive L-filter yielding an almost 2.5 dB better NR and MAER compared to the median filter. The normalized LMS L-filter achieves the best performance both in the MSE sense as well as in MAE sense.

The optimal value of parameter μ_0 has been found experimentally. Figure 5.4 shows the NR and the MAER achieved for several values of the parameter μ_0. Note that the optimal value of μ_0 is different for the two figures of merit. In our experiments we have used the value of μ_0 for which NR attains a minimum, $\mu_0 = 0.8$.

The original image corrupted by mixed impulsive and additive Gaussian noise is shown in Fig. 5.3(b). The output of the location-invariant LMS L-filter of dimensions 3×3 may be found in Fig. 5.3(c). The filtered image by using the modified LMS L-filter with nonhomogeneous step-sizes and the output of the normalized LMS L-filter are shown in Fig. 5.3(d) and Fig. 5.3(e), respectively.

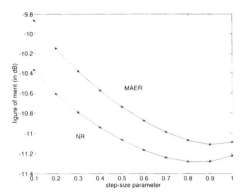

Fig. 5.4 Plot of the noise reduction (NR) and mean absolute error reduction (MAER) indexes achieved by the normalized LMS *L*-filter versus the step-size parameter μ_0 (reprinted by copyright permission from IEEE).

When a reference image is not available, the straightforward choice is to apply the location-invariant LMS *L*-filter, because only that design can be modified to work without a reference signal, as has already been discussed in Section 5.3.1. The first entry in Table 5.2 shows the noise reduction index achieved at the output of the modified location-invariant adaptive *L*-filter given by Eq. (5.47), when the original image "Lenna" is corrupted by mixed impulsive and additive Gaussian noise. In parentheses, we have included the same figure of merit for the median filter. The step-size parameter has been chosen according to

$$\mu(n) = \frac{\mu_0}{||\mathbf{x}_L(n)||^2}; \quad \mu_0 = 1 \times 10^{-10}. \tag{5.104}$$

Note that the modified location-invariant LMS *L*-filter defined by Eq. (5.47) is able to adapt to the noise distribution and to attain an almost identical performance to that of the maximum likelihood estimator of location for the noise model. Furthermore we have tested the robustness of the *L*-filter coefficients that are determined at the end of a training session and are applied to filter a noisy image that has been produced by corrupting a different reference image than the one used in the training session. More specifically, we have run the location-invariant LMS *L*-filtering algorithm given by Eq. (5.45) on "Trevor White," shown in Fig. 5.3(f), corrupted by impulsive ($p = 10\%$) and additive Gaussian noise ($\sigma_n = 50$) using as a reference image the original "Trevor White." Subsequently we have averaged the *L*-filter coefficients that were derived during the operation of the algorithm on the last image row. The resulting *L*-filter coefficients have been applied to filter the original image "Lenna" corrupted by the same mixed impulsive and additive Gaussian noise, as before. It has been found that only the location-invariant *L*-filter coefficients, defined in Eq. (5.45), are robust, in the sense that they do not depend on the reference image used in the training session.

Table 5.2 Noise reduction (in dB) achieved by the location-invariant adaptive L-filter when a reference image is not available

Filtered Image	NR	
"Lenna" corrupted by mixed impulsive and additive Gaussian noise	-8.40	(-8.756)
"Lenna" using the L-filter coefficients of "Trevor White"	-9.477	
"Trevor White" using the L-filter coefficients of "Lenna"	-9.293	

Table 5.3 Figures of merit in still image filtering achieved by an adaptive L-filter that employ different step-size selections

Step-size Selection	Running Coefficients NR (in dB)	MSE	Averaged Coefficients NR (in dB)	MSE
Constant	8.140	586.885	8.098	592.563
Normalized LMS	7.974	609.755	8.233	574.446
[17]	8.136	587.360	8.231	574.646
Eqs. (5.85) and (5.86)	8.131	588.147	8.281	568.107
Median filter	7.446	688.545		

The last two entries in Table 5.2 show the noise reduction achieved by filtering the original image "Lenna" by using the L-filter coefficients determined at the end of a training session on "Trevor White," and vice versa. It is seen that the attained noise reduction is close to the one achieved when the reference image is available.

5.3.4.2 *Performance of adaptive L-filters with variable step-sizes* We comment on the performance of the step-size selection algorithm described by Eqs. (5.85) and (5.86). First, we compare the performance of adaptive L-filters that employ the following step-size selections: (1) constant step-size ($\mu_0 = 10^{-8}$), (2) the normalized LMS algorithm [47], (3) the algorithm proposed in [17], and (4) the algorithm of Eqs. (5.85) and (5.86). A noisy version of the image "Airfield," depicted in Fig. 5.21(a), produced by adding mixed zero-mean Gaussian noise having standard deviation equal to 50, and impulsive noise having probability of impulse occurrence 10% has been used. Positive and negative impulses are equiprobable. The initial filter coefficients are set equal to zero. We also assume that the noise-free image is available. Performance results have been obtained either during the adaptation by using the coefficients determined at each image pixel or by using throughout the entire image the set of coefficients produced by averaging the filter coefficients computed in the last image row. For comparison purposes the same figures of merit for median filter are also included in Table 5.3. The inspection of Table 5.3 reveals that all step-size selection algorithms provide almost the same results with respect to the quantitative criteria computed. However, when the proposed algorithm is applied, the averaged filter coefficients yield the best results. Note that the proposed method does not rely on eigenvalue-eigenvector computations.

5.3.4.3 *Performance of cellular adaptive L-filters*

We demonstrate the performance of cellular adaptive LMS *L*-filters by employing the noise reduction index NR defined in Eq. (5.102). Let us use the second frame of the image sequence "Trevor White" as reference image. The reference image has been corrupted by mixed additive Gaussian and impulsive noise. The additive Gaussian noise has zero mean and standard deviation equal to 10. Negative and positive impulses occur with equal probability $p = 5\%$, resulting in an impulsive noise with total probability of impulse occurrence $p = 10\%$. A cellular network of 256×256 processing cells was employed. Each processing cell was directly connected to its two nearest horizontal neighboring cells; that is, the following neighborhood has been used:

$$\mathcal{M}(\mathbf{n}) = \{\mathbf{j} \mid j_1 = n_1, \; |j_2 - n_2| \leq 1\}. \tag{5.105}$$

The *L*-filter coefficients have been initialized as follows:

$$\widehat{\mathbf{a}}(n;0) = \widehat{\mathbf{R}}^{-1} \, \widehat{\mathbf{p}}, \tag{5.106}$$

with

$$\widehat{\mathbf{R}} = \frac{1}{N_1 \cdot N_2 - i_0 + 1} \sum_{i=i_0}^{N_1 \cdot N_2 - i_1} \mathbf{x}_L(i) \, \mathbf{x}_L^T(i),$$

$$\widehat{\mathbf{p}} = \frac{1}{N_1 \cdot N_2 - i_0 + 1} \sum_{i=i_0}^{N_1 \cdot N_2 - i_1} s(i) \, \mathbf{x}_L(i), \tag{5.107}$$

where i_0 and i_1 depend on the dimensions of the filter window. In the subsequent iterations, the cellular adaptive *L*-filter is initialized with the coefficients obtained at the end of the previous iteration. The value of step-size $\mu_0 = 10^{-5}$ was found experimentally to be optimal. The scanning of the image was performed in the *prime-row* manner.

Table 5.4 shows the NR obtained at the end of each iteration. No further performance improvement is obtained after the sixth iteration. From the inspection of Table 5.4, it can be seen that the NR reaches an upper limit within a few iterations. The number of iterations depends heavily on the initial coefficients. If the ordinary adaptive *L*-filter defined by Eq. (5.49) were used with the same initial conditions, an NR of 16.17 dB would be obtained. The cellular adaptive *L*-filter yields a 1.5 dB higher NR already after the first iteration. A 3 dB higher NR is attained at the end of the sixth iteration. The superior performance of cellular adaptive *L*-filters is based on the fact that the adaptation of the cellular *L*-filter coefficients at pixel n matches the local statistics. On the contrary, in the case of ordinary adaptive *L*-filters the

Table 5.4 Noise reduction in dB achieved by the cellular adaptive L-filter in the restoration of the second frame of "Trevor White" corrupted by mixed impulsive and additive Gaussian noise

Iteration	NR
1	17.535
2	18.159
3	18.722
4	19.009
5	19.146
6	19.17

Table 5.5 Noise reduction in dB achieved by the cellular adaptive L-filter and the ordinary adaptive L-filter in the restoration of several frames of "Trevor White" corrupted by mixed impulsive and additive Gaussian noise

Frame	Cellular Adaptive L-Filter	Ordinary Adaptive L-Filter
Second	19.146	16.17
Third	17.508	16.052
Fourth	17.19	16.088

adaptation of the coefficients at pixel n depends explicitly on all the past values of the desired signal $s(j)$ and the ordered input vector $\mathbf{x}_L(j)$ for $j < n$.

We repeated next the above-described experiment with the first four frames of the sequence. Table 5.5 shows the NR achieved by the cellular adaptive L-filters after having processed each frame for five iterations. In Table 5.5, the NR achieved by the adaptive L-filter given by Eq. (5.49) is also listed. Note that cellular adaptive L-filter outperforms the ordinary adaptive L-filter.

5.4 MULTICHANNEL L-FILTERS

The nonlinear processing of vector-valued signals is closely related to color image processing. It is well known that there is no unambiguous or universally accepted method to order multivariate data. Therefore several *subordering principles* have emerged: the *marginal ordering* (M-ordering), *reduced ordering* (R-ordering), *partial ordering*, and *conditional ordering* [8]. The M-ordering and the R-ordering principles are the most popular. The marginal median [67], the marginal α-trimmed mean [68], and the multichannel L-filters [46] are a few examples of filters based on the M-ordering principle. The following multichannel filters based on R-ordering have been proposed: the vector median [3], the ranked-order type estimator $\mathcal{R}_E$

[31], *L*-filters based on radial medians [39], the multichannel modified trimmed mean (MTM) and its double window extension (DWMTM) [68], the weighted vector median, and the α-trimmed vector median [86]. Other multivariate extensions of the MTM filter and *M*-filters stemming from *M*-estimators are investigated in [42], where the influence function of the mentioned filters is derived as well. A multichannel *L*-filter based on R-ordering was proposed in [59]. New filters that utilize Bayesian techniques and nonparametric methodologies to adapt to local data in image processing are proposed in [74]. A new approach to vector median filtering can be found in [75], where, first, absolute sorting of the vectorial space based on Peano space filling curves is used and, second, a scalar median filtering operation is applied.

5.4.1 Marginal *L*-filters

Let $\mathbf{x}_1, \mathbf{x}_2, \ldots, \mathbf{x}_N$ be N random d-dimensional observations. Each vector-valued observation $\mathbf{x}_j = (x_{j1}, x_{j2}, \ldots, x_{jd})^T$ belongs to a d-dimensional space denoted by $I\!R^d$. The M-ordering scheme orders the vector components independently, thus yielding

$$x_{i(1)} \leq x_{i(2)} \leq \cdots \leq x_{i(N)} \qquad i = 1, 2, \ldots, d. \qquad (5.108)$$

The output of a d-channel *L*-filter of length N operating on a sequence of d-dimensional vectors $\{\mathbf{x}(n)\}$ for N odd is given by [46]

$$\mathbf{y}(n) \overset{\triangle}{=} \mathbf{T}[\mathbf{x}(n)] = \sum_{i=1}^{d} \mathbf{A}_i \mathbf{x}_{Li}(n), \qquad (5.109)$$

where $\mathbf{A}_i$ is a $(d \times N)$ coefficient matrix and $\mathbf{x}_{Li}(n) = \left(x_{i(1)}(n), \ldots, x_{i(N)}(n)\right)^T$ is the $(N \times 1)$ vector of the order statistics along the ith channel, $i = 1, 2, \ldots, d$. Let $\mathbf{a}_{il}^T, l = 1, 2, \ldots, d$, denote the lth row of matrix $\mathbf{A}_i$.

We assume that the observed d-dimensional signal $\{\mathbf{x}(n)\}$ can be expressed as a sum of a d-dimensional noise-free signal $\{\mathbf{s}(n)\}$ and a vector-valued noise sequence $\{\mathbf{v}(n)\}$ having a zero mean vector, i.e., $\mathbf{x}(n) = \mathbf{s}(n) + \mathbf{v}(n)$. The noise vector $\mathbf{v}(n) = (v_1(n), \ldots, v_d(n))^T$ is a d-dimensional vector of random variables characterized by the joint probability density function (pdf) of its components. The noise vector components are assumed to be uncorrelated in the general case. In addition we assume that the noise vectors at different values of the spatial index n are independent identically distributed (i.i.d.) random vectors and that at each value of the spatial index n the signal vector $\mathbf{s}(n)$ and the noise vector $\mathbf{v}(n)$ are uncorrelated. We want to find the multichannel *L*-filter coefficient matrices $\mathbf{A}_i$, $i = 1, 2, \ldots, d$, that minimize the MSE between the filter output $\mathbf{y}(n)$ and the noise-free signal $\mathbf{s}(n)$. Strictly speaking, we define MSE as the trace of the MSE matrix. Let

$$\underline{\mathbf{a}}_i = \left(\mathbf{a}_{1i}^T \mid \mathbf{a}_{2i}^T \mid \cdots \mid \mathbf{a}_{di}^T\right)^T \qquad (5.110)$$

be the composite coefficient vector created by concatenating the ith row from all matrices $\mathbf{A}_j$, $j = 1, 2, \ldots, d$, and let

$$\underline{x}_L(n) = \left(\mathbf{x}_{L1}^T(n) \mid \mathbf{x}_{L2}^T(n) \mid \ldots \mid \mathbf{x}_{Ld}^T(n)\right)^T \tag{5.111}$$

be the composite vector of the order statistics created by concatenating the vectors of the order statistics given by Eq. (5.108) from all the d channels. Following reasoning similar to that in [46], but without invoking the assumption of a constant signal $\mathbf{s}$, it can be shown that the MSE is expressed as

$$\begin{aligned} \varepsilon &= \mathrm{E}\left\{[\mathbf{y}(n) - \mathbf{s}(n)]^T [\mathbf{y}(n) - \mathbf{s}(n)]\right\} \\ &= \sum_{i=1}^{d} \left[\underline{a}_i^T \mathbf{R}_d \underline{a}_i - 2\underline{a}_i^T \underline{p}_i\right] + \mathrm{E}\left\{\mathbf{s}^T(n)\mathbf{s}(n)\right\}, \end{aligned} \tag{5.112}$$

where

$$\mathbf{R}_d = \mathrm{E}\left\{\underline{x}_L(n)\underline{x}_L^T(n)\right\}, \tag{5.113}$$

$$\underline{p}_i = \mathrm{E}\left\{s_i(n)\underline{x}_L(n)\right\}, \quad i = 1, 2, \ldots, d. \tag{5.114}$$

In Eqs. (5.113) and (5.114) we take for granted that $\underline{x}_L(n)$ and $s_i(n)$ are jointly stationary stochastic processes. It can easily be seen that $\mathbf{R}_d$ is a composite matrix that consists of the correlation matrices of the ordered input samples from the same channel -for example, $\mathbf{R}_{ii} = \mathrm{E}\{\mathbf{x}_{Li}(n)\mathbf{x}_{Li}^T(n)\}$- as well as from different channels -for example, $\mathbf{R}_{ij} = \mathrm{E}\{\mathbf{x}_{Li}(n)\mathbf{x}_{Lj}^T(n)\}$, $i \neq j$. Put differently,

$$\mathbf{R}_d = \begin{bmatrix} \mathbf{R}_{11} & \mathbf{R}_{12} & \cdots & \mathbf{R}_{1d} \\ \mathbf{R}_{12}^T & \mathbf{R}_{22} & \cdots & \mathbf{R}_{2d} \\ \vdots & & \ddots & \vdots \\ \mathbf{R}_{1d}^T & \mathbf{R}_{2d}^T & \cdots & \mathbf{R}_{dd} \end{bmatrix}. \tag{5.115}$$

Minimizing the error ε, defined in Eq. (5.112), over $\underline{a}_i$ is a quadratic minimization problem that has a unique solution under the condition that $\mathbf{R}_d$ is positive definite. It is easily deduced that the minimum MSE coefficient vector is

$$\underline{a}_i^o = \mathbf{R}_d^{-1}\underline{p}_i. \tag{5.116}$$

Equation (5.116) explicitly yields the filter coefficients provided that we are able to calculate the moments of the order statistics from univariate populations that appear in $\mathbf{R}_{ii}$, as well as the product moments of the order statistics from bivariate populations that appear in $\mathbf{R}_{ij}$, $i \neq j$ and $i = 1, 2, \ldots, d$. This is fairly easy for i.i.d. input variates, that is, in the case of a constant signal $\mathbf{s}(n) = \mathbf{s}$ as demonstrated in [46]. Even for independent, nonidentically distributed input variates, the mathematical framework tends to become very complicated (cf. [46]). The difficulties are increased in color image processing, where the observations $\underline{x}(n)$ and the desired signal $\mathbf{s}(n)$

are strongly nonstationary. To overcome these obstacles, we resort to the use of iterative algorithms for the minimization of $\varepsilon(n)$ in Eq. (5.112).

5.4.2   Unconstrained Minimization of the MSE

In this section three adaptive multichannel *L*-filters are reviewed that iteratively minimize the MSE given by Eq. (5.112) without imposing any constraints on the filter coefficients. These algorithms are: (1) the LMS, (2) the NLMS, and (3) the LMS Newton (LMSN) filters.

5.4.2.1   LMS adaptive multichannel L-filter   The filter coefficient vectors $\underline{a}_i$, $i = 1, 2, \ldots, d$, that minimize the MSE given by Eq. (5.112), can be computed recursively using the steepest descent algorithm as follows:

$$\underline{a}_i(n + 1) = \underline{a}_i(n) + \tfrac{1}{2}\mu[-\nabla\varepsilon_i(n)], \tag{5.117}$$

where $\nabla\varepsilon_i(n) = \partial\varepsilon/\partial\underline{a}_i(n)$ and μ denotes the adaptation step-size parameter. Using $\underline{x}_L(n)\underline{x}_L^T(n)$ and $s_i(n)\underline{x}_L(n)$ as instantaneous estimates of $\mathbf{R}_d$ and $\underline{p}_i$, respectively, the LMS adaptive multichannel *L*-filter is obtained:

$$\begin{aligned}
\widehat{\underline{a}}_i(n + 1) &= \widehat{\underline{a}}_i(n) + \mu\left[s_i(n) - \underline{x}_L^T(n)\widehat{\underline{a}}_i(n)\right]\underline{x}_L(n) \\
&= \widehat{\underline{a}}_i(n) + \mu\, e_i(n)\,\underline{x}_L(n).
\end{aligned} \tag{5.118}$$

The bracketed term in Eq. (5.118) is the a priori estimation error $e_i(n)$ between the *i*th component of the desired signal and the filter output, since $y_i(n) = \underline{x}_L^T(n)\widehat{\underline{a}}_i$. We see that the LMS algorithm yields a very simple recursive relation for updating the *L*-filter coefficients. This is the rationale underlying its choice for minimizing the MSE. The convergence properties of the adaptive LMS multichannel *L*-filter depend on the eigenvalue distribution of the composite correlation matrix $\mathbf{R}_d$. However, an important question arises: Which is the appropriate range of the adaptation step μ that guarantees the convergence of the adaptive filter in both the mean and the mean-square sense? Up-to-date results are available only for stationary environments and for simple cases, such as time-variant system identification in nonstationary environments [87, 33]. To obtain an optimal sequence of adaptation step parameters, we follow the approach proposed in [56]. Let $\mathbf{M}_i$ be a diagonal matrix of dimensions $(dN \times dN)$ associated with the updating equation for the coefficient vector $\underline{a}_i(n)$, namely

$$\widehat{\underline{a}}_i(n + 1) = \widehat{\underline{a}}_i(n) + \tfrac{1}{2}\mathbf{M}_i[-\nabla\varepsilon_i(n)]. \tag{5.119}$$

The MSE $\varepsilon(n)$ can be approximated by its instantaneous value, $\varepsilon(n) = \sum_{i=1}^{p} e_i^2(n)$. The step-size sequence, $\mu_{i,jj}$, that minimizes the squared a priori error at $n + 1$, $e_i^2(n + 1)$, is given by [48]

$$\mu_{i,jj}^o(n) = \frac{|\chi_j(n)|}{\sum_{l=1}^{dN} |\chi_l(n)|^3} \tag{5.120}$$

where $\chi_j(n)$ is the jth element of the composite vector of the ordered observations $\underline{X}_L(n)$. We see that Eq. (5.120) does not depend on index i that refers to channels. Therefore the same step-size sequence can be applied to all equations that update the filter coefficients.

5.4.2.2 NLMS adaptive multichannel L-filter Let $\mu_{i,jj}(n) = \mu(n)$ be a single adaptation step-size parameter for all the elements of coefficient vectors $\underline{a}_i$. The derivation of the step-size sequence that minimizes the a priori estimation error at $n + 1$ yields [48]

$$\mu(n) = \frac{1}{\underline{x}_L^T(n)\underline{x}_L(n)}. \tag{5.121}$$

The substitution of Eq. (5.121) into Eq. (5.118) yields the updating equations for the coefficients of the normalized LMS adaptive multichannel L-filter, namely

$$\hat{\underline{a}}_i(n + 1) = \hat{\underline{a}}_i(n) + \frac{\mu_0}{\underline{x}_L^T(n)\underline{x}_L(n)} e_i(n)\underline{x}_L(n), \qquad i = 1, 2, \ldots, d, \tag{5.122}$$

where $\mu_0 \in (0, 1]$ is a parameter that is introduced for additional control.

5.4.2.3 LMSN adaptive multichannel L-filter The eigenvalue spread of the composite correlation matrix $\mathbf{R}_d$ is large in principle. In such a case the LMSN algorithm is a powerful alternative to LMS [21]. The LMSN algorithm is an approximate implementation of Newton's method for minimizing a cost function of several variables. It employs computationally efficient estimates for the autocorrelation matrix of the input signal (in our case, of the composite vector of the ordered observations) and for the gradient of the objective function (i.e., the MSE). The updating formula for the LMSN multichannel L-filter is given by

$$\begin{aligned} \hat{\underline{a}}_i(n + 1) &= \hat{\underline{a}}_i(n) + \tfrac{1}{2}\mu\,\hat{\mathbf{R}}_d^{-1}(n)\left[-\hat{\nabla}\varepsilon_{(i)}(n)\right] \\ &= \hat{\underline{a}}_i(n) + \mu\hat{\mathbf{R}}_d^{-1}(n)e_i(n)\underline{x}_L(n), \quad i = 1, 2, \ldots, d. \end{aligned} \tag{5.123}$$

Equation (5.123) is similar to the coefficient vector recursion described in [87], but it encompasses an estimate for the composite correlation matrix and the composite vector of the ordered observations instead of the correlation matrix of the input signal and the input multichannel signal itself. An estimate of the composite matrix $\mathbf{R}_d$ can be calculated by using the Robbins-Monro procedure that solves the equation $\mathrm{E}\left\{\underline{x}_L(n)\underline{x}_L^T(n) - \mathbf{R}_d\right\} = 0$ [21]. An iterative solution of this equation is given by $\hat{\mathbf{R}}_d(n) = \hat{\mathbf{R}}_d(n - 1) + \zeta\left[\underline{x}_L(n)\underline{x}_L^T(n) - \hat{\mathbf{R}}_d(n - 1)\right]$. Using the matrix inversion lemma, we obtain

$$\hat{\mathbf{R}}_d^{-1}(n) = \frac{1}{1 - \zeta}\left[\hat{\mathbf{R}}_d^{-1}(n - 1) - \frac{\hat{\mathbf{R}}_d^{-1}(n - 1)\underline{x}_L(n)\underline{x}_L^T(n)\hat{\mathbf{R}}_d^{-1}(n - 1)}{\left(\frac{1-\zeta}{\zeta}\right) + \underline{x}_L^T(n)\hat{\mathbf{R}}_d^{-1}(n - 1)\underline{x}_L(n)}\right]. \tag{5.124}$$

Equation (5.124) is initialized with $\widehat{\mathbf{R}}_d^{-1}(0) = \eta^{-1}\mathbf{I}_{dN}$, where η is a suitably chosen constant. A detailed analysis of the computational complexity of the unconstrained adaptive multichannel *L*-filters reviewed can be found at [48].

In the following section, we study the design of constrained adaptive multichannel *L*-filters that minimize MSE given by Eq. (5.112), subject to structural constraints imposed on the filter coefficients.

5.4.3   Constrained Minimization of the MSE

Frequently structural constraints are imposed on the filter coefficients. For example, in the single-channel case, the sum of the filter coefficients must be equal to one. This is true for both adaptive linear filters, such as the TDLMS [30, 52] and nonlinear adaptive filters such as the location-invariant LMS *L*-filter [47]. Moreover, as we see in Eqs. (5.118), (5.122), and (5.123), all the unconstrained adaptive filters derived so far depend on the knowledge of a reference signal $\mathbf{s}(n)$. In certain cases a reference signal can easily be found. For example, in an image sequence one can always choose a previous noise-free frame as a reference image. In cases where this is not possible, the need to develop an adaptive filter structure that does not rely on a reference signal emerges. It will be shown that the adaptive location-invariant multichannel *L*-filter can be modified so that it does not depend on a reference signal similarly to the single channel case. Moreover the experiments demonstrate that the performance of the adaptive location-invariant multichannel *L*-filters is practically independent of the reference signal that is used. Therefore they possess robustness properties in this sense.

In the multichannel case the optimal nonadaptive location-invariant *L*-filter was derived in [46]. Let us recall the definition of the location-invariant multichannel *L*-filter first. A multichannel marginal *L*-filter is said to be *location-invariant* if its output is able to track small perturbations of its input. The definition of a location-invariant multichannel *L*-filter yields the following set of constraints imposed on the filter coefficients:

$$\mathbf{G}^T\underline{\mathbf{a}}_i = \boldsymbol{\kappa}_i, \qquad i = 1, 2, \ldots, d. \tag{5.125}$$

In Eq. (5.125) $\boldsymbol{\kappa}_i$ is the *i*th basis vector in $\mathbb{R}^d$, i.e., a vector whose elements are zero except the *i*th element which equals 1 and $\mathbf{G}^T$ is a $(d \times dN)$ matrix that has the structure

$$\mathbf{G}^T = \begin{bmatrix} \mathbf{1}_N^T & \mathbf{0}_N^T & \cdots & \mathbf{0}_N^T \\ \mathbf{0}_N^T & \mathbf{1}_N^T & \cdots & \mathbf{0}_N^T \\ \vdots & & \ddots & \vdots \\ \mathbf{0}_N^T & \mathbf{0}_N^T & \cdots & \mathbf{1}_N^T \end{bmatrix} \tag{5.126}$$

where $\mathbf{1}_N$ denotes the $(N \times 1)$ unitary vector and $\mathbf{0}_N$ is a $(N \times 1)$ vector of zeros. In the following, we derive two constrained adaptive multichannel *L*-filters based on LMS and LMSN algorithms.

5.4.3.1   LMS location-invariant multichannel L-filter   We are seeking the L-filter whose output minimizes the MSE given by Eq. (5.112) subject to Eq. (5.125). A well-established methodology for minimizing a cost function subject to constraints was proposed by Frost [24]. This approach is adopted in our analysis. The problem under study is formulated as the minimization of the following Lagrangian function:

$$H = \tfrac{1}{2}\varepsilon + \underline{\psi}^T \begin{bmatrix} \mathbf{G}^T\underline{a}_1 - \kappa_1 \\ \vdots \\ \mathbf{G}^T\underline{a}_d - \kappa_d \end{bmatrix}, \tag{5.127}$$

where ε is given by Eq. (5.112) and $\underline{\psi} = \left(\underline{\psi}_1^T \mid \cdots \mid \underline{\psi}_d^T\right)^T$ is the $(d^2 \times 1)$ vector of Lagrange multipliers. By differentiating H with respect to $\underline{a}_i$, we obtain

$$\frac{\partial H}{\partial \underline{a}_i} = \mathbf{R}_d\,\underline{a}_i - \underline{p}_i + \mathbf{G}\underline{\psi}_i. \tag{5.128}$$

Accordingly the steepest descent solution is given by

$$\underline{a}_i(n+1) = \underline{a}_i(n) - \mu\left[\mathbf{R}_d\,\underline{a}_i(n) - \underline{p}_i + \mathbf{G}\underline{\psi}_i\right]. \tag{5.129}$$

We demand $\underline{a}_i(n+1)$ to satisfy Eq. (5.125). By substituting Eq. (5.129) into Eq. (5.125) and solving for $\underline{\psi}_i$, we get

$$\underline{\psi}_i = \frac{1}{\mu}(\mathbf{G}^T\mathbf{G})^{-1}\left\{\mathbf{G}^T\underline{a}_i(n) - \mu\mathbf{G}^T\left[\mathbf{R}_d\underline{a}_i(n) - \underline{p}_i\right] - \kappa_i\right\}. \tag{5.130}$$

By combining Eqs. (5.129) and (5.130), the steepest descent solution is rewritten as follows:

$$\underline{a}_i(n+1) = \mathbf{\Pi}\left[\underline{a}_i(n) + \mu\left(\underline{p}_i - \mathbf{R}_d\,\underline{a}_i(n)\right)\right] + \mathbf{f}_i, \quad i = 1, 2, \ldots, d, \tag{5.131}$$

where $\mathbf{\Pi}$ is the *projection matrix* of dimensions $(dN \times dN)$ defined by

$$\mathbf{\Pi} = \left[\mathbf{I}_{dN} - \mathbf{G}(\mathbf{G}^T\mathbf{G})^{-1}\mathbf{G}^T\right] = \left(\mathbf{I}_{dN} - \frac{1}{N}\mathbf{G}\mathbf{G}^T\right), \tag{5.132}$$

and $\mathbf{f}_i$ is the $(dN \times 1)$ vector given by

$$\mathbf{f}_i = \mathbf{G}(\mathbf{G}^T\mathbf{G})^{-1}\kappa_i = \frac{1}{N}\mathbf{G}\kappa_i, \quad i = 1, 2, \ldots, d. \tag{5.133}$$

The algorithm is initialized by

$$\underline{a}_i(0) = \mathbf{f}_i, \quad i = 1, 2, \ldots, d. \tag{5.134}$$

By using instantaneous estimates for $\mathbf{R}_d$ and $\underline{p}_i$, the LMS location-invariant multichannel *L*-filter is obtained

$$\hat{\underline{a}}_i(n+1) = \mathbf{\Pi} \left[\hat{\underline{a}}_i(n) + \mu e_i(n) \underline{x}_L(n) \right] + \mathbf{f}_i, \quad i = 1, 2, \ldots, d. \tag{5.135}$$

Next we consider the case where a reference signal is not available. In such a case the criterion to be minimized is the total output power subject to constraints of Eq. (5.125) that prevent the filter coefficients from becoming identically zero. It can easily be shown that the LMS location-invariant *L*-filter coefficients are now updated as follows:

$$\hat{\underline{a}}_i(n+1) = \mathbf{\Pi} \left[\hat{\underline{a}}_i(n) - \mu y_i(n) \underline{x}_L(n) \right] + \mathbf{f}_i, \quad i = 1, 2, \ldots, d. \tag{5.136}$$

It is evident that by replacing $e_i(n)$ by $-y_i(n)$ the same filter structure can be used for the minimization of the total output power subject to the constraints of Eq. (5.125).

5.4.3.2   LMSN location-invariant multichannel L-filter   The vast majority of constrained adaptive algorithms rely on the LMS algorithm. To the authors' knowledge, no attempt has been made to design constrained adaptive filters based on other adaptive algorithms, such as the recursive least squares (RLS) or the LMSN algorithm. The case is much simpler for the LMSN than the RLS algorithm, because LMSN shares the same framework with LMS in the sense that LMSN employs the gradient of the error function given by Eq. (5.128). By premultiplying both sides of Eq. (5.128) by $\mathbf{R}_d^{-1}$, we obtain

$$\mathbf{R}_p^{-1} \frac{\partial H}{\partial \underline{a}_i(n)} = \underline{a}_i(n) - \mathbf{R}_d^{-1} \underline{p}_i + \mathbf{R}_d^{-1} \mathbf{G} \psi_i, \quad i = 1, 2, \ldots, d. \tag{5.137}$$

We identify that

$$\underline{a}_i^o = \mathbf{R}_d^{-1} \left[\underline{p}_i - \mathbf{G} \psi_i \right], \quad i = 1, 2, \ldots, d, \tag{5.138}$$

is the solution to the minimization of the cost function given by Eq. (5.127). Eq. (5.138) can be rewritten as

$$\underline{a}_i^o(n) = \mathbf{a}_{(i)}(n) - \mathbf{R}_d^{-1} \frac{\partial H}{\partial \underline{a}_i(n)}. \tag{5.139}$$

The steepest descent solution is obtained from Eq. (5.139) by adding another step-size parameter μ:

$$\underline{a}_i(n+1) = \underline{a}_i(n) + \mu \left[\underline{a}_i^o(n) - \underline{a}_i(n) \right] = \underline{a}_i(n) - \mu \mathbf{R}_d^{-1} \frac{\partial H}{\partial \underline{a}_i(n)}. \tag{5.140}$$

By substituting $\mathbf{R}_d^{-1}(n)$ with the estimate given by Eq. (5.124) and by using instantaneous estimates for the expected values involved in the gradient of H with respect to $\underline{a}_i(n)$, the following recursions result:

$$\widehat{\underline{a}}_i(n+1) = \widehat{\underline{a}}_i(n) + \mu \widehat{\mathbf{R}}_d^{-1}(n) \left[e_i(n)\underline{x}_L(n) - \mathbf{G}\psi_i \right], \quad i = 1, 2, \ldots, d. \quad (5.141)$$

The coefficients, given by Eq. (5.141), should satisfy the constraints defined by Eq. (5.125), accordingly

$$\psi_i = \left[\mathbf{G}^T \widehat{\mathbf{R}}_d^{-1}(n)\mathbf{G} \right]^{-1} \left\{ \frac{1}{\mu} \left[\mathbf{G}^T \widehat{\underline{a}}_i(n) - \kappa_i \right] + \mathbf{G}^T \widehat{\mathbf{R}}_d^{-1}(n)e_i(n)\underline{x}_L(n) \right\}. \quad (5.142)$$

The substitution of Eq. (5.142) into Eq. (5.141) reveals that the structure of the LMSN location-invariant multichannel L-filter is the same with that of the LMS location-invariant one, but with a different matrix $\mathbf{\Pi}$ and a different vector $\mathbf{f}_i$. The new matrix $\mathbf{\Pi}'$ and vector $\mathbf{f}_i'$ are now given by

$$\mathbf{\Pi}' = \mathbf{I}_{dN} - \widehat{\mathbf{R}}_d^{-1}(n)\mathbf{G} \left[\mathbf{G}^T \widehat{\mathbf{R}}_d^{-1}(n)\mathbf{G} \right]^{-1} \mathbf{G}^T, \quad (5.143)$$

$$\mathbf{f}_i' = \widehat{\mathbf{R}}_d^{-1}(n)\mathbf{G} \left[\mathbf{G}^T \widehat{\mathbf{R}}_p^{-1}(n)\mathbf{G} \right]^{-1} \kappa_i, \quad i = 1, 2, \ldots, d. \quad (5.144)$$

The recursion is initialized by

$$\widehat{\underline{a}}_i(0) = \mathbf{f}_i', \quad i = 1, 2, \ldots, d. \quad (5.145)$$

5.4.4 Experimental Results

An observed color image is denoted by $\mathbf{x}(\mathbf{n})$, where $\mathbf{n} = (n_1, n_2)^T$ is the 2-D vector of the pixel coordinates. Accordingly $\mathbf{x}(\mathbf{n})$ is a three-channel 2-D signal. It can refer to any color space, such as RGB, XYZ, and $U^*V^*W^*$. Our goal is to test the ability of the proposed adaptive multichannel L-filters in suppressing the noise in color images. Both unconstrained and constrained adaptive multichannel L-filters are considered. We compare the noise reduction capability of the adaptive multichannel L-filters under study to the other multichannel nonlinear filters as well as to their single-channel counterparts. The following nonlinear filters were considered: the vector median [3], the marginal median [67, 68], the marginal α-trimmed mean [67, 68], the multichannel MTM [68], the multichannel DWMTM [68] and the ranked-order estimator $\mathcal{R}_E$ [31]. Whenever the filter window is not specified explicitly, a window of dimensions 3×3 is assumed.

The multichannel DWMTM filter [68] uses two window sizes 3×3 and 5×5, as in the single-channel case [51]. The multichannel MTM filter is a generalization of the single-channel one proposed in [51]. Its output is evaluated as for the DWMTM filter with the exception that only one window of size 3×3 is used [68]. The trimming parameter for α-trimmed mean filters has been 0.2 in all experiments. For the $\mathcal{R}_E$ filter [31], only the best result is tabulated. We have also included the arithmetic mean in the comparative study, because it is a straightforward choice for noise filtering in many practical applications. The performance of three adaptive single-channel *L*-filters that are used to filter the noise in each primary color component (i.e., channel) was independently taken into consideration as well. A multichannel extension of the 2-D LMS (TDLMS) algorithm (i.e., adaptive multichannel linear filter) and the TDLMS (i.e., adaptive single-channel linear filter) [30] that is used in each primary color component separately were included in the comparative study. We employed the NR index defined by Eq. (5.102) as an objective figure of merit in the performance comparisons. Moreover the visual quality of the filtered images was used as a subjective figure of merit.

In all experiments the adaptive linear/nonlinear filter coefficients were initialized by zero. The LMS and LMSN location-invariant multichannel *L*-filters are initialized by Eq. (5.134) and Eq. (5.145), respectively. Moreover the filter coefficients determined recursively by the adaptive algorithm at each color image pixel were used to filter the noisy input image. Note that all adaptive filters considered in this section depend explicitly on the knowledge of a reference image [e.g., the original image $s(n)$], as we can see in the updating equations for the filter coefficients. However, a reference image is seldom available in practice. In the context of an image sequence (e.g., video), we can assume that a previous noise-free frame can act as a reference image for a number of subsequent image frames. Furthermore in such a context, it is also possible to exploit motion compensation (MC) in determining the desired response at each pixel assuming that the displacement vectors between the frame acting as a reference image and the actual noise-free image are known beforehand. Let $s_L(\mathbf{n};\, k)$ and $s_L(\mathbf{n};\, k-l)$ be the luminance components (i.e., the Y component in XYZ color space) of two color image frames that are l frames apart. Motivated by the success of motion-compensated filtering [13], we propose the following choice of a reference image $\widehat{\mathbf{s}}(\mathbf{n};\, k) = \mathbf{s}(\mathbf{n} - \mathbf{n}_D^*;\, k - l)$, where $\mathbf{n}_D^*$ is the displacement vector that minimizes a prediction error of the form $PE(\mathbf{n}_D) = \sum_{\mathcal{S}} |s_L(\mathbf{n};\, k) - s_L(\mathbf{n} - \mathbf{n}_D;\, k-l)|$ in a neighborhood $\mathcal{S}$ around each pixel $\mathbf{n}$. A block of 8×8 pixels was used in the estimation of the displacement vectors between the 50th frame and frames 45th through 49th of "Trevor White." The displacement vector field between the 50th and 48th frames produced by the block-matching algorithm, is shown in Fig. 5.5(a). The displacement vector field between the 50th and the 45th frames is depicted in Fig. 5.5(b).

Another point that requires some further clarification is the choice of the color space where the performance comparisons are to be made. It is well known that color distances are not Euclidean in the RGB primary system [35]. Color distances are approximated by Euclidean distances in the so called *uniform color spaces*, such as, the C.I.E. uniform chromaticity scale (UCS) system, the $L^*a^*b^*$, the $L^*u^*v^*$, and the $U^*V^*W^*$ (also known as modified UCS system) [35]. To guarantee that the measured NR indexes correspond to perceived color differences, we felt the need to

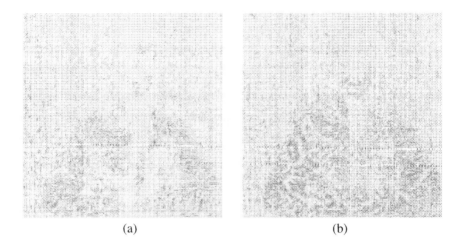

(a)   (b)

Fig. 5.5 Displacement vector fields (a) between the 50th and the 48th noise-free frames of image sequence "Trevor White"; (b) between the 50th and the 45th noise-free frames of the same image sequence.

test the performance of the several filters in a uniform color space. We chose the $U^*V^*W^*$ space for this purpose.

5.4.4.1 Filtering in the RGB color space   Let us consider the 50th frame of the color image sequence "Trevor White." This frame is corrupted by additive white trivariate contaminated Gaussian noise having the probability distribution $(1 - \epsilon)\mathcal{N}(\mathbf{0} \; ; \mathbf{C}_1) + \epsilon\mathcal{N}(\mathbf{0} \; ; \mathbf{C}_2)$ for $\epsilon = 0.1$ plus impulsive noise such that 6 % of the image pixels in each primary color component are replaced by impulses of value 0 or 255 equiprobably. $\mathbf{C}_i$, $i = 1, 2$, denotes the covariance matrix of each trivariate joint Gaussian distribution. The following covariance matrices were used:

$$\mathbf{C}_1 = \begin{bmatrix} 100 & 100 & 210 \\ 100 & 400 & 180 \\ 210 & 180 & 900 \end{bmatrix}, \quad \mathbf{C}_2 = \begin{bmatrix} 900 & -300 & -210 \\ -300 & 400 & 60 \\ -210 & 60 & 100 \end{bmatrix}. \quad (5.146)$$

The NR achieved in RGB color space by the filters under study is given in Table 5.6. The adaptation step-size parameter μ_0 that yields the best result in terms of the visual quality of the filtered image has been used in NLMS algorithms. It is in cluded in the corresponding entry of Table 5.6. For the LMSN algorithm, μ, ζ, and η were set to 0.0008, 0.001, and 0.01, respectively. Note that the best $\mathcal{R}_E$ estimator corresponds to index $J = 2$ [31]. From the examination of Table 5.6 we conclude:

1. Nonlinear filters are ranked as the four best filters, namely the multichannel DWMTM filter, the MC LMSN adaptive multichannel L-filter, the MC NLMS multichannel L-filter, and the MC LMSN location-invariant multichannel

Table 5.6   Noise reduction (in dB) achieved in (NTSC) RGB color space by several filters in the restoration of the 50th color frame of "Trevor White" corrupted by mixed additive white trivariate contaminated Gaussian plus impulsive noise (filter window 3×3)

Filter	NR	Rank
Marginal median	-11.748	[7]
Vector median L_1	-10.043	[13]
Vector median L_2	-8.508	[18]
$\mathcal{R}_E$-filter	-8.696	[16]
α-Trimmed mean ($\alpha = 0.2$)	-11.250	[11]
Arithmetic mean	-8.514	[17]
Multichannel MTM filter	-11.425	[10]
Multichannel DWMTM filter	-13.362	[1]
MC NLMS adaptive multichannel L-filter ($\mu_0 = 0.5$)	-13.228	[3]
MC LMSN adaptive multichannel L-filter	-13.320	[2]
MC LMS location-invariant multichannel L-filter ($\mu_0 = 0.2$)	-12.076	[6]
MC LMSN location-invariant multichannel L-filter	-12.403	[4]
MC normalized TDLMS multichannel filter ($\mu_0 = 0.1$)	-11.600	[8]
MC TDLMS-Newton multichannel filter	-11.169	[12]
MC NLMS adaptive single-channel L-filters ($\mu_0 = 0.05$)	-11.560	[9]
MC LMSN adaptive single-channel L-filters	-12.234	[5]
MC normalized TDLMS filters ($\mu_0 = 0.05$)	-9.264	[15]
MC TDLMS-Newton filters	-9.823	[14]

L-filter. The MC LMSN/NLMS adaptive multichannel *L*-filter attains an almost identical performance to that of the multichannel DWMTM filter. Recall, however, that the multichannel DWMTM filter employs a window of dimensions 5 × 5.

2. The adaptive multichannel *L*-filters outperform the adaptive multichannel linear filters by approximately 1.6 dB in the case of MC NLMS algorithm and by 2.1 dB in the case of MC LMSN algorithm.

3. The multichannel techniques are found to be superior to single-channel ones by 1.668 dB in MC NLMS adaptive *L*-filters and by 1.086 dB in MC LMSN adaptive *L*-filters.

4. The multichannel techniques yield a much better NR index than single-channel ones in linear filtering as well as in nonlinear filtering.

5. MC LMSN adaptive algorithms give identical results to MC NLMS ones with respect to NR index. However, it has been found that the LMSN adaptive multichannel *L*-filter without MC yields an almost 2 dB higher NR than the NLMS adaptive multichannel *L*-filter without MC.

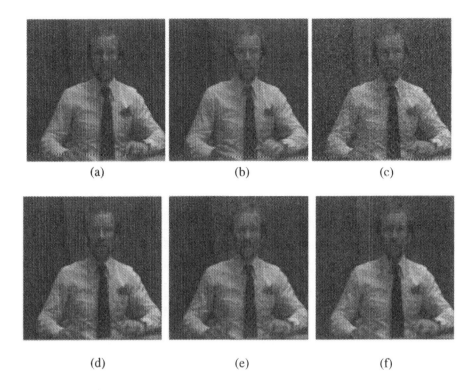

(a)   (b)   (c)

(d)   (e)   (f)

Fig. 5.6 Output of various filters when smoothing the mixed impulsive and additive white trivariate contaminated Gaussian noise that corrupts the 50th frame of color image sequence "Trevor White." Image filtering has been performed in RGB color space. (a) Original color image; (b) reference image (48th frame); (c) noisy input image; (d) output of the 3×3 MC NLMS adaptive multichannel L-filter; (e) output of the 3×3 MC LMSN adaptive multichannel L-filter; (f) output of the 3×3 marginal median filter. (The color images of this figure are available in Postscript/Adobe Portable Document Format via anonymous FTP from `ftp://ftp.wiley.com/public/sci_tech_med/image_video/`.)

Figure 5.6(a) shows the noise-free 50th frame of color image sequence "Trevor White" in RGB color space that is used as a test image. The 48th frame of "Trevor White" that is used as a reference image is shown in Fig. 5.6(b). The test image corrupted by mixed impulsive and additive trivariate contaminated Gaussian noise is depicted in Fig. 5.6(c). The output of the MC NLMS and the MC LMSN adaptive multichannel L-filter is shown in Fig. 5.6(d) and Fig. 5.6(e), respectively. We argue that MC NLMS gives a slightly superior filtered image than the MC LMSN in terms of the visual quality. For comparison purposes, the output of the marginal median filter is depicted in Fig. 5.6(f). The marginal median filter preserves the edges but fails to remove the noise in the homogeneous regions. On the contrary, in homogeneous

Table 5.7 Noise reduction (in dB) achieved in $U^*V^*W^*$ color space by several filters in the restoration of the 50th color frame of "Trevor White" corrupted by mixed additive white trivariate contaminated Gaussian plus impulsive noise (filter window 3×3)

Filter	NR	Rank
Marginal median	-11.270	[12]
Vector median L_1	-9.775	[15]
Vector median L_2 ($\mathcal{R}_E$-filter)	-8.553	[17]
α-Trimmed mean ($\alpha = 0.2$)	-10.904	[13]
Arithmetic mean	-9.212	[16]
Multichannel MTM filter	-11.291	[11]
Multichannel DWMTM filter	-13.920	[4]
MC NLMS adaptive multichannel L-filter ($\mu_0 = 0.5$)	-14.412	[1]
MC LMSN adaptive multichannel L-filter (fixed $\mu = 0.001$)	-14.159	[2]
MC LMS location-invariant multichannel L-filter ($\mu_0 = 0.1$)	-12.020	[9]
MC LMSN location-invariant multichannel L-filter	-12.183	[8]
MC normalized TDLMS multichannel filter ($\mu_0 = 0.5$)	-13.922	[3]
MC TDLMS-Newton multichannel filter ($\mu_0 = 0.1$)	-12.889	[7]
MC NLMS adaptive single-channel L-filters ($\mu_0 = 0.5$)	-13.513	[5]
MC LMSN adaptive single-channel L-filters (fixed $\mu = 0.001$)	-13.477	[6]
MC normalized TDLMS filters ($\mu_0 = 0.5$)	-10.668	[14]
MC TDLMS-Newton filters (fixed $\mu = 0.001$)	-11.501	[8]

regions, the MC NLMS/MC LMSN multichannel L filter performs better than the marginal median filter.

5.4.4.2 Filtering in the $U^*V^*W^*$ color space The performance of the several filters included in the comparisons has been measured in $U^*V^*W^*$ color space as well. As before, the adaptation step-size parameter that yields the best result in terms of the visual quality of the filtered image has been used in NLMS algorithms. It can be found in the corresponding entry of Table 5.7. For LMSN algorithms, one may use the parameters $\eta = 0.01$ and $\zeta = 0.001$ and either the fixed step size $\mu = 5 \times 10^{-4}$ or a variable step-size $\mu(n)$ given by $\mu(n) = \mu_0/(\underline{x}_L^T(n)\widehat{\mathbf{R}}_d^{-1}(n)\underline{x}_L(n))$ as in algorithm II [21]. In the latter case the parameter μ_0 used is given in the corresponding entry of Table 5.7. The inspection of Table 5.7 reveals the following facts:

1. The adaptive multichannel L-filters are ranked as the best filters.

2. They outperform the multichannel linear techniques by 0.5 dB in the case of the MC NLMS algorithm and by 1.27 dB in the case of the MC LMSN algorithm.

3. The MC NLMS adaptive multichannel L-filter is superior to the MC NLMS single-channel L-filters by 0.9 dB. The MC LMSN adaptive multichannel

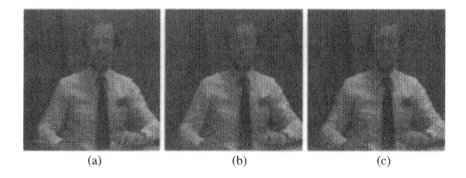

(a)   (b)   (c)

Fig. 5.7   Output of various filters when smoothing the mixed impulsive and additive white trivariate contaminated Gaussian noise that corrupts the 50th frame of color image sequence "Trevor White". Image filtering has been performed in $U^* V^* W^*$ color space. (a) Output of the 3×3 MC NLMS adaptive multichannel L-filter; (b) output of the 3×3 MC LMSN adaptive multichannel L-filter; (c) output of the 3×3 marginal median filter. (The color images of this figure are available in Postscript/Adobe Portable Document Format via anonymous FTP from ftp://ftp.wiley.com/public/sci_tech_med/image_video/.)

L-filter is slightly better than using three separate LMSN adaptive single-channel L-filters by 0.682 dB.

4. Multichannel techniques yield a much higher NR than the single-channel techniques in linear filtering.

5. Three out of the four best filtering techniques are found to be multichannel and nonlinear. It is surprising that the MC multichannel normalized TDLMS linear filtering has proved to be efficient in $U^* V^* W^*$ color space.

The output of several filters included in our comparative study was transformed from $U^* V^* W^*$ to RGB for display purposes. The output of the MC NLMS and MC LMSN adaptive multichannel L-filter is shown in Fig. 5.7(a) and Fig. 5.7(b), respectively. In comparing Fig. 5.7(a) and Fig 5.7(b) with Fig. 5.6(d), and Fig. 5.6(e), we conclude that by processing the color image in a uniform color space, color information is preserved better than by processing the image in RGB color space. This is self-evident in the uniform greenish background of the image. Moreover the inability of the marginal median to smooth the noise in homogeneous regions is clearly depicted in Fig. 5.7(c). Therefore, for a color image corrupted by mixed additive white trivariate contaminated Gaussian plus impulsive noise, very good performance in terms of NR was obtained by employing MC adaptive multichannel L-filters. Furthermore the aforementioned measured improvement in terms of the NR index corresponds to a perceived subjective improvement as manifested by the experiments performed in the perceptually uniform color space $U^* V^* W^*$.

Table 5.8 Noise reduction (in dB) achieved by the adaptive location-invariant multichannel L-filters in the restoration of the 50th color frame of "Trevor White" corrupted by mixed additive white trivariate contaminated Gaussian plus impulsive noise

	NR (dB) in RGB		NR (dB) in $U^*V^*W^*$	
Reference Image	48	45	48	45
Filter				
LMS location-invariant	-11.376	-10.220	-12.033	-11.028
MC LMS location-invariant	-12.076	-12.052	-12.020	-11.926
LMSN location-invariant	-11.935	-10.783	-12.265	-12.078
MC LMSN location-invariant	-12.403	-12.335	-12.183	-12.134

5.4.4.3 Performance of LMS/LMSN location-invariant multichannel L-filters
Finally we examine the performance of the LMS/LMSN adaptive location-invariant multichannel L-filter in more detail. Table 5.8 summarizes the NR achieved by these two filter structures in both color spaces with/without MC when either the 48th frame or the 45th one is chosen as the reference image. Note that results with/without motion compensation (MC) in color spaces RGB and $U^*V^*W^*$ are included. Either the 48th frame or the 45th one is used as the reference image. From the inspection of Table 5.8, we see that the MC adaptive location-invariant multichannel L-filters are proved to be robust adaptive filter structures because there is almost no deterioration in NR when the reference image changes. This point is valid in both color spaces.

5.5 $L\ell$-FILTERS

In this section we discuss adaptive designs based on the LMS algorithm for the N^2-$L\ell$-filter and the Kronecker $L\ell$-filter.

5.5.1 N^2-$L\ell$-Filter

Let us assume that $\mathbf{x}_{L\ell}(n)$ defined in Eq. (5.17) and $s(n)$ are jointly stationary stochastic signals. Based on the stationarity assumption we are interested in the derivation of the N^2-$L\ell$-filter that minimizes the MSE,

$$\begin{aligned}\varepsilon(n) &= \mathrm{E}\left\{(s(n) - \widehat{s}(n))^2\right\} \\ &= \mathbf{c}^T(n)\mathbf{R}_{L\ell}\mathbf{c}(n) - 2\mathbf{c}^T(n)\mathbf{p}_{L\ell} + \mathrm{E}\left\{s^2(n)\right\},\end{aligned} \quad (5.147)$$

where $\mathbf{R}_{L\ell} = \mathrm{E}\left\{\mathbf{x}_{L\ell}(n)\mathbf{x}_{L\ell}^T(n)\right\}$ is the correlation matrix of the vectors $\mathbf{x}_{L\ell}(n)$, and $\mathbf{p}_{L\ell} = \mathrm{E}\left\{s(n)\mathbf{x}_{L\ell}(n)\right\}$ is the crosscorrelation vector between the vector $\mathbf{x}_{L\ell}(n)$ and the desired response $s(n)$. Clearly, the optimal N^2-$L\ell$-filter coefficient vector

c^o is given by

$$\mathbf{c}^o = \mathbf{R}_{L\ell}^{-1}\mathbf{p}_{L\ell}. \tag{5.148}$$

In many practical cases (e.g., for incorporating a fixed trimming [27]) we require the minimization of the MSE defined by Eq. (5.147) subject to a set of constraints that is expressed as

$$\mathbf{G}\,\mathbf{c} = \boldsymbol{\kappa} \tag{5.149}$$

where $\mathbf{G}$ is an $(M \times N^2)$ matrix and $\boldsymbol{\kappa}$ is an M-dimensional vector with $M < N^2$. In using Lagrange multipliers, the optimal vector is found to be [64]

$$\widetilde{\mathbf{c}}^o = \mathbf{R}_{L\ell}^{-1}\mathbf{p}_{L\ell} + \mathbf{R}_{L\ell}^{-1}\mathbf{G}^T \left(\mathbf{G}\mathbf{R}_{L\ell}^{-1}\mathbf{G}^T\right)^{-1} \left(\boldsymbol{\kappa} - \mathbf{G}\mathbf{R}_{L\ell}^{-1}\mathbf{p}_{L\ell}\right). \tag{5.150}$$

Equation (5.150) clearly requires the matrices $\mathbf{R}_{L\ell}$ and $\mathbf{G}\mathbf{R}_{L\ell}^{-1}\mathbf{G}^T$ to be nonsingular. A necessary condition for $\mathbf{R}_{L\ell}$ to be nonsingular is [60]

$$\Pr\{x_{(i)} \leftrightarrow x_j\} \neq 0 \quad \forall i, j = 1, 2, \ldots, N, \tag{5.151}$$

where $\Pr\{\mathcal{Y}\}$ denotes the probability of event $\mathcal{Y}$. Moreover it has been shown that a sufficient condition for $\mathbf{R}_{L\ell}$ to be nonsingular is either $\mathbf{R}_\ell = \mathrm{E}\left\{\mathbf{x}_\ell(n)\mathbf{x}_\ell^T(n)\right\}$ or $\mathbf{R}_L = \mathrm{E}\left\{\mathbf{x}_L(n)\mathbf{x}_L^T(n)\right\}$ to be nonsingular. Therefore in practice, the singularity of the matrix $\mathbf{R}_{L\ell}$, just as in the case of linear filters, is not a problem. It is well known that $\mathbf{R}_\ell$ is positive definite, if the observations are linearly independent [65]. The singularity of the matrix $\mathbf{G}\mathbf{R}_{L\ell}^{-1}\mathbf{G}^T$ has to be checked case by case. However, if $\mathbf{G}$ is of full rank, then a unique solution can be computed that is given by Eq. (5.150) [27].

The elements of $\mathbf{R}_{L\ell}$ are difficult to be evaluated analytically due to the nonlinear rank-ordering operation producing $\mathbf{x}_{L\ell}(n)$ from $\mathbf{x}_\ell(n)$. They can be determined either numerically or estimated from sample averages by simulation.

An approach to determining the $L\ell$-filter coefficients $\mathbf{c}(n)$ using the LMS algorithm has been proposed in [27]. This approach is an alternative way of obtaining the filter coefficients when the analytical design is either intractable or computationally burdensome, as is our case. However, a major requirement is that a noise-free signal is available for the design. It is trivial to show that the coefficient update equation provided by the LMS algorithm for the minimization of the MSE given by Eq. (5.147) is

$$\mathbf{c}(n + 1) = \mathbf{c}(n) + \mu\,e(n)\,\mathbf{x}_{L\ell}(n). \tag{5.152}$$

However, only N of the N^2 coefficients of the $L\ell$-filter are updated at each iteration in Eq. (5.152). Indeed,

$$c_{(i)j}(n+1) = \begin{cases} c_{(i)\ell_{(i)}}(n) + \mu\,e(n)\,x_{(i)}(n) & \text{if } j = \ell_{(i)} \\ 0 & \text{otherwise,} \end{cases} \quad i = 1, 2, \ldots, N, \text{ or}$$

$$c_{(i)j}(n+1) = \begin{cases} c_{(\varrho_j)j}(n) + \mu\, e(n)\, x_j(n) & \text{if } i = \varrho_j \\ 0 & \text{otherwise,} \end{cases} \quad j = 1, 2, \ldots, N.$$

$$(5.153)$$

5.5.2 Kronecker $L\ell$-Filters

In the following, we address the optimal and adaptive design of Kronecker $L\ell$-filters that are expressed as in Eq. (5.27). We are seeking the Kronecker $L\ell$-filter that minimizes the MSE,

$$\begin{aligned} \varepsilon(n) &= \left(\mathbf{a}^T(n) \otimes \mathbf{b}^T(n)\right) \mathbf{R}_{L\ell}\left(\mathbf{a}(n) \otimes \mathbf{b}(n)\right) \\ &\quad -2\left(\mathbf{a}^T(n) \otimes \mathbf{b}^T(n)\right)\mathbf{p}_{L\ell} + \mathrm{E}\left\{s^2(n)\right\}. \end{aligned} \tag{5.154}$$

Let us suppose that two vectors $\mathbf{a}^o$ and $\mathbf{b}^o$ for which $\varepsilon(n)$ attains its unique minimum exist. Next, we briefly describe the algorithm proposed in [64] to find the optimal vectors $\mathbf{a}^o$ and $\mathbf{b}^o$. For a fixed vector $\mathbf{b}$ or $\mathbf{a}$ the MSE can be respectively expressed as

$$\varepsilon(\mathbf{a}; \mathbf{b}) = \mathbf{a}^T \mathbf{R}_{L\ell;b}\mathbf{a} - 2\mathbf{a}^T \mathbf{p}_{L\ell;b} + \mathrm{E}\left\{s^2(n)\right\}, \tag{5.155}$$

$$\varepsilon(\mathbf{b}; \mathbf{a}) = \mathbf{b}^T \mathbf{R}_{L\ell;a}\mathbf{b} - 2\mathbf{b}^T \mathbf{p}_{L\ell;a} + \mathrm{E}\left\{s^2(n)\right\}, \tag{5.156}$$

where $\mathbf{R}_{L\ell;b}$ and $\mathbf{R}_{L\ell;a}$ are $N \times N$ matrices with elements given by

$$\begin{aligned} R_{L\ell;b}(i_1, i_2) &= \sum_{j_1=1}^{N}\sum_{j_2=1}^{N} b_{j_1} b_{j_2} \mathrm{E}\left\{x_{(i_1)j_1}(n)x_{(i_2)j_2}(n)\right\}, \\ &\quad 1 \leq i_1, i_2 \leq N, \end{aligned} \tag{5.157}$$

$$\begin{aligned} R_{L\ell;a}(j_1, j_2) &= \sum_{i_1=1}^{N}\sum_{i_2=1}^{N} a_{i_1} a_{i_2} \mathrm{E}\left\{x_{(i_1)j_1}(n)x_{(i_2)j_2}(n)\right\}, \\ &\quad 1 \leq j_1, j_2 \leq N, \end{aligned} \tag{5.158}$$

and $\mathbf{p}_{L\ell;b}$ and $\mathbf{p}_{L\ell;a}$ are $N \times 1$ vectors whose elements are

$$p_{L\ell;b}(i) = \sum_{j}^{N} b_j \mathrm{E}\left\{s(n)x_{(i)j}(n)\right\}, \quad i = 1, 2, \ldots, N, \tag{5.159}$$

$$p_{L\ell;a}(j) = \sum_{i=1}^{N} a_i \mathrm{E}\left\{s(n)x_{(i)j}(n)\right\}, \quad j = 1, 2, \ldots, N. \tag{5.160}$$

The algorithm is initialized by setting $\mathbf{a}$ equal to $\mathbf{a}_0$, where $\mathbf{a}_0$ is an appropriate choice and finding the vector $\mathbf{b}_0$ that minimizes the MSE for fixed $\mathbf{a}$. Next $\mathbf{b}$ is set to $\mathbf{b}_0$ and the vector $\mathbf{a}_1$ that minimizes the MSE for fixed $\mathbf{b}$ is determined and so on. The corresponding MSE will be the nonincreasing sequence

$$\varepsilon(\mathbf{b}_0; \mathbf{a}_0) \geq \varepsilon(\mathbf{a}_1; \mathbf{b}_0) \geq \varepsilon(\mathbf{b}_1; \mathbf{a}_1) \geq \varepsilon(\mathbf{a}_2; \mathbf{b}_1) \geq \ldots. \tag{5.161}$$

The procedure converges asymptotically if the matrix $\mathbf{R}_{L\ell}$ is nonsingular.

It can be easily shown that the Kronecker $L\ell$-filter output can be rewritten as

$$y(n) = \sum_{i=1}^{N} a_i(n) \, b_{\ell_{(i)}} \, x_{(i)}(n). \tag{5.162}$$

Let us define the scalar quantity

$$z_i(n) = b_{\ell_{(i)}} \, x_{(i)}(n) = \mathbf{b}_i^T(n)\mathbf{x}_l, \tag{5.163}$$

where $\mathbf{b}_i(n), i = 1, 2, \ldots, N$, is an N-dimensional vector having elements

$$b_{ij}(n) = b_j(n) \, \delta_{j \, \ell_{(i)}}, \quad j = 1, 2, \ldots, N, \tag{5.164}$$

with δ_{ij} denoting a Kronecker delta (Eq. (5.74)). The substitution of Eqs. (5.163) and (5.164) into Eq. (5.162) yields

$$y(n) = \mathbf{a}^T(n) \, \mathbf{z}(n). \tag{5.165}$$

The MSE between the filter output and the desired response $s(n)$ is then

$$\varepsilon(n) = \mathrm{E}\left\{e^2(n)\right\} = \mathrm{E}\left\{\left[s(n) - \mathbf{a}^T(n)\mathbf{z}(n)\right]^2\right\}. \tag{5.166}$$

The LMS updating equation for the coefficient vector $\mathbf{a}(n)$ can be easily derived as

$$a_i(n + 1) = a_i(n) + \mu_a \, e(n) \, b_{\ell_{(i)}} \, x_{(i)}, \quad i = 1, 2, \ldots, N. \tag{5.167}$$

Equation (5.167) is easily recognized as the updating equation of the nonlinear part of the Kronecker $L\ell$-filter. The derivation of the updating equations for the coefficients b_{ij} that correspond to the linear part of the $L\ell$-filter can be obtained by applying a *backpropagation algorithm* for deriving an estimate of the gradient of $\varepsilon(n)$ with respect to $\mathbf{b}_i(n)$ as in [73, 61]

$$\frac{\partial \hat{\varepsilon}(n)}{\partial b_{ij}(n)} = \frac{\partial \hat{\varepsilon}(n)}{\partial z_i(n)} \frac{\partial z_i(n)}{\partial b_{ij}(n)} = -2 \, e(n) \, a_i(n) \, x_j(n) \, \delta_{j\ell_{(i)}}. \tag{5.168}$$

Accordingly, the updating equations for $b_{ij}(n)$ are

$$b_{ij}(n + 1) = \begin{cases} b_{i\ell_{(i)}}(n) + \mu_b \, e(n) \, a_i(n) \, x_{\ell_{(i)}}(n) & \text{if } j = \ell_{(i)} \\ 0 & \text{otherwise.} \end{cases} \tag{5.169}$$

The updating equations derived in this section lead to the same equations as in [73], but in a more elegant way. The backpropagation algorithm was used in [73] to update the coefficients of the general nonlinear filter structure [70], but not for updating the coefficients of the Kronecker *Lℓ*-filters. For the latter filters the LMS algorithm was applied to find the vectors $\mathbf{a}(n)$ and $\mathbf{b}(n)$ that minimize Eqs. (5.155) and (5.156) used in [64] to determine the optimal coefficients. Another worth noting subtle distinction is that the Kronecker *Ll*-filters *cannot* be considered as a special case of the general nonlinear filter structure [70], because the sorting operation does not take place after linear weighting by b_{ij}.

We can introduce space/time-varying step-size parameters $\mu_a(n)$ and $\mu_b(n)$ that guarantee that at each step the a posteriori estimation error at n,

$$e'(n) = s(n) - \sum_{i=1}^{N} a_i(n+1)\, b_{\ell_{(i)}}(n+1) x_{(i)}(n), \qquad (5.170)$$

is less than the a priori error $e(n)$ [73]. A sufficient condition for $e'(n) \leq e(n)$ is

$$0 \leq |1 - \mu_a(n)\mathcal{A}(n) - \mu_b(n)\mathcal{B}(n)| \leq 1, \qquad (5.171)$$

where

$$\mathcal{A}(n) = \sum_{i=1}^{N} b_{\ell_{(i)}}^2(n) x_{(i)}^2(n), \qquad (5.172)$$

$$\mathcal{B}(n) = \sum_{i=1}^{N} a_i^2(n) x_{(i)}^2(n). \qquad (5.173)$$

If $\mu_a(n) = \mu_b(n) = \mu(n)$, we obtain

$$0 \leq \mu(n) \leq \frac{2}{\mathcal{A}(n) + \mathcal{B}(n)}. \qquad (5.174)$$

5.5.3 Experimental Results

We comment the performance of the *Lℓ*-filters in filtering the image sequence "Trevor White" that has been corrupted by mixed Gaussian (zero mean and standard deviation 10.0) and impulsive noise with probability of occurrence $p = 10\%$. The image sequence is filtered by an adaptive linear, *L*-, *Lℓ*- and Kronecker *Lℓ*-filter, respectively. The current frame is the 50th frame of the sequence, illustrated in Fig. 5.8(a), whereas its corrupted version can be seen in Fig. 5.8(b). Filters defined on a $3 \times 3 \times 3$ running window are considered. In the experiment reported, we consider the noisy frames #49, #50, and #51. The initial filter coefficients are set to zero and a constant step-size $\mu_0 = 10^{-6}$ is used. No motion compensation was employed. The NR and the MSE in the filtered output are tabulated in Table 5.9. It can be seen that the adaptive *L*-filter attains the best performance. The superiority of the adaptive *L*-filter

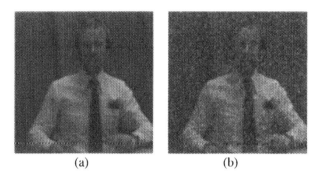

(a)　　　　　　　　　　　　(b)

Fig. 5.8 (a) Original 50th frame of image sequence "Trevor White". (b) Corrupted 50th frame by mixed impulsive and Gaussian noise.

Table 5.9 Simulation results obtained by filtering the 50th frame of image sequence "Trevor White" that is corrupted by mixed additive Gaussian and impulsive noise with 3-D adaptive linear, L-, Ll-, and Kronecker $L\ell$-filters

Filter	Noise Type	NR		MSE	
Linear	Mixed	9.911749		209.093853	
L	Gaussian and	15.608116	$\sqrt{}$	56.325423	$\sqrt{}$
$L\ell$	impulsive	14.455667		73.442836	
Kronecker $L\ell$		14.079106		80.095015	

over the $L\ell$-filter can be attributed to the significantly smaller number of L-filter coefficients to be adapted. On the contrary, the number of $L\ell$-filter coefficients is three times greater than that in still image filtering. Consequently the convergence of the filter coefficients to the optimal ones becomes much slower. This fact deteriorates filter performance. The latter remark is justified by observing the $L\ell$ filtered image, Fig. 5.9(c). The residual noise at the top rows is caused by the slower convergence of the filter coefficients. The filter performance gets better when the bottom rows are reached. As a remedy to the problem described, a Kronecker $L\ell$-filter has been used in order to lessen the number of coefficients, to reduce the $L\ell$-filter complexity and its execution time as well as its memory requirements. The filtered images by an adaptive linear, L-, $L\ell$-, and Kronecker $L\ell$- filter are shown in Fig. 5.9(a) to Fig. 5.9(d). The better subjective quality of the L-filtered frame is self-evident. The edges are better preserved (e.g., the glasses), and the noise is smoothed better. No residual noise exists in the top rows of the filtered frame. The performance of the adaptive linear filter is the worst of all.

(a)   (b)

(c)   (d)

Fig. 5.9   Filtered 50th frame of image sequence "Trevor White" that is corrupted by mixed Gaussian and impulsive noise by a 3-D adaptive: (a) Linear filter, (b) L-filter, (c) $L\ell$-filter, and (d) Kronecker $L\ell$-filter.

5.6   ORDER STATISTICS FILTER BANKS

A filter bank is a set of filters with a common input and a common output. It consists of two stages. The first stage is an *analysis bank* where the input signal $x(n)$ is split into N *subband signals* $y_k(n) = \mathcal{Z}^{-1}[H_k(z)X(z)]$, $k = 0, 1, \ldots, N - 1$, where $\mathcal{Z}^{-1}[\cdot]$ denotes the inverse z-transform and $H_k(z)$ is the system function of the kth filter in the analysis bank. The second stage of the system is the *synthesis bank* comprising of a set of filters $F_k(z)$ whose outputs are combined into a single response $\widehat{x}(n) = \mathcal{Z}^{-1}[\sum_{k=0}^{N-1} F_k(z)\widetilde{Y}_k(z)]$, where $\widetilde{Y}_k(z)$ denotes the z-transform of the filtered signal $y_k(n)$. Figure 5.10 depicts the general structure of a filter bank. For example, a uniform *discrete Fourier transform* (DFT) filter bank has analysis filters with system functions $H_k(z) = H_0(zW^{-k})$, where $W = e^{-j2\pi/N}$ and $H_0(z) = 1 + z^{-1} + \ldots + z^{-(N-1)}$. It can easily be verified that the magnitude of the subband $y_k(n)$ is the kth coefficient of the N-point DFT of the sequence $\{x(n), x(n - 1), \ldots, x(n - N + 1)\}$. Accordingly the uniform DFT filter bank is a spectrum analyzer that provides frequency information localized in time. Many generalizations

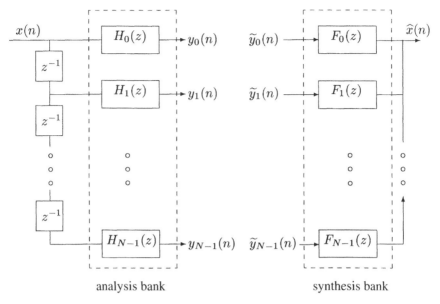

analysis bank   synthesis bank

Fig. 5.10   General structure of a filter bank.

of the uniform DFT filter bank have been proposed, such as *quadrature mirror filters.*

Filter banks have traditionally been developed within the framework of linear filters. Filter banks that provide frequency and rank information localized in time, called *order statistic filter banks*, are proposed in [1]. The order statistics filter banks can be used for multiresolution ranking in image and video processing applications. Two such applications are reported in [1]. Ranking at low resolutions may provide edge and signal trend information of large areas with low computational complexity. Order statistics filter banks can adapt the time-frequency space tilings according to the local edge characteristics. In the following, we briefly describe the design of the order statistics filter banks.

As was described in Section 5.2, the sorting operation can be formally considered as a permutation of the set of spatial/temporal indexes $(1, 2, \ldots, N)$ into the set of rank indexes $(\varrho_1, \varrho_2, \ldots, \varrho_N)$ [5, 40]. The permutation of the aforementioned set onto itself can be described in terms of a tree-structure showing all the $N!$ possible permutations. Let us denote each permutation of spatial/temporal indexes by $(\rho_1[k], \rho_2[k], \ldots, \rho_N[k]), k = 1, 2, \ldots, N!$, where k is an enumeration index that specifies a path in the tree. We define the N-dimensional vector $\mathbf{z}[k]$ as

$$\mathbf{z}[k] = \left(x_{1(\rho_1[k])}, x_{2(\rho_2[k])}, \ldots, x_{j(\rho_j[k])}, \ldots, x_{N(\rho_N[k])} \right)^T, \qquad (5.175)$$

where

$$x_{j(\rho_j[k])} = \begin{cases} x_j & \text{if } \rho_j[k] = \varrho_j \text{ in the } k\text{th permutation} \\ 0 & \text{otherwise.} \end{cases} \qquad (5.176)$$

The link between the vector $\mathbf{x}_{\ell L}$, defined in Eq. (5.20), and $\mathbf{z}[k]$ can be established as follows [1]:

$$\mathbf{z}[k] = \mathcal{Q}[k]\mathbf{x}_{\ell L}, \qquad (5.177)$$

where

$$\mathcal{Q}[k] = \begin{bmatrix} \mathbf{e}^T_{\rho_1[k]} & \mathbf{0}^T_N & \cdots & \mathbf{0}^T_N \\ \mathbf{0}^T_N & \mathbf{e}^T_{\rho_2[k]} & & \mathbf{0}^T_N \\ \vdots & & \ddots & \vdots \\ \mathbf{0}^T_N & \mathbf{0}^T_N & \cdots & \mathbf{e}^T_{\rho_N[k]} \end{bmatrix}, \qquad (5.178)$$

and $\mathbf{e}_{\rho_N[k]}$ is the $N \times 1$ indicator vector,

$$\mathbf{e}_{\rho_j[k]} = \left(\underbrace{0,\dots,0,}_{\rho_j[k]-1 \text{ elements}} \quad 1, \quad \underbrace{0,\dots,0}_{N-\rho_j[k] \text{ elements}} \right)^T. \qquad (5.179)$$

The inverse of Eq. (5.177) is given by [1]

$$\begin{aligned} \mathbf{x}_{\ell L} &= \frac{1}{(N-1)!} \left(\mathbf{e}^T_1 \sum_{k \in \Omega_{1(1)}} \mathbf{z}[k], \dots, \mathbf{e}^T_1 \sum_{k \in \Omega_{1(N)}} \mathbf{z}[k] \,\middle|\, \cdots \right. \\ &\qquad\qquad \left. \middle|\, \mathbf{e}^T_N \sum_{k \in \Omega_{N(1)}} \mathbf{z}[k], \dots, \mathbf{e}^T_N \sum_{k \in \Omega_{N(N)}} \mathbf{z}[k] \right)^T \\ &= \left((\mathbf{x}^1_{\ell L})^T \mid (\mathbf{x}^2_{\ell L})^T \mid \cdots (\mathbf{x}^N_{\ell L})^T \right)^T, \qquad (5.180) \end{aligned}$$

with $\Omega_{j(i)}$ denoting the set of permutation indexes k such that

$$\Omega_{j(i)} = \left\{ k : \quad x_{j(i)} \text{ is the } j\text{th element of } \mathbf{z}[k] \right\}. \qquad (5.181)$$

Clearly, $\mathbf{e}_i$ is the ith basis vector in $\mathbb{R}^N$. To derive the *analysis order statistics bank*, each $\mathbf{z}[k]$ is transformed via an $N \times N$ orthogonal matrix $\mathbf{U}$ to

$$\mathbf{y}[k] = \mathbf{U}\,\mathbf{z}[k], \quad k = 1, 2, \dots, N!, \qquad (5.182)$$

and the transformed vectors, whose enumeration index k is in the set $\Omega_{j(i)}$, are added to form the scalars

$$y_{j(i)} = \sum_{k \in \Omega_{j(i)}} \mathbf{u}^T_j \mathbf{z}[k], \qquad (5.183)$$

where $\mathbf{u}^T_j$ is the jth row of the orthogonal matrix $\mathbf{U}$. Let us create the vector $\mathbf{y}_{\ell L}$ having the structure of Eq. (5.20) using $y_{j(i)}$. It can be shown that [1]

$$\mathbf{y}_{\ell L}(n) = \mathbf{S}\,\mathbf{x}_{\ell L}(n), \qquad (5.184)$$

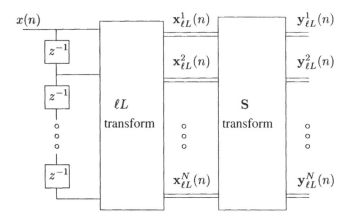

Fig. 5.11   Analysis order statistic bank.

where

$$\mathbf{S} = (N-1)\mathbf{U}_I \otimes \mathbf{I}_N + \mathbf{U}_J \otimes \mathbf{J}_N, \qquad (5.185)$$

$$\mathbf{U}_I = \operatorname{diag}(u_{11}, u_{22}, \ldots, u_{NN}), \qquad (5.186)$$

$$\mathbf{U}_J = \mathbf{U} - \mathbf{U}_I, \qquad (5.187)$$

$$\mathbf{J}_M = \mathbf{1}_N \mathbf{1}_N^T - \mathbf{I}_N. \qquad (5.188)$$

The analysis order statistic bank defined by Eq. (5.182) to Eq. (5.188) is an N-input to N^2-output nonlinear system. It is depicted in Fig. 5.11. It consists of an ℓL-transformation module and an appropriate module based on matrix $\mathbf{S}$ whose goal is to provide N^2-channel signals $y_{j(i)}$ that carry space/time-localized rank and frequency information of the sequence being processed. The N signals $y_{j(1)}, \ldots, y_{j(N)}$ that constitute the vector $\mathbf{y}_{\ell L}^j$ share the same spectral range or subband [1]. However, since the filters discussed are nonlinear and do not possess the additive superposition property, their frequency responses depend, in general, on the input signal. By applying the method proposed by Mallows[54], Arce and Tian prove that the subband characteristics of $y_{j(1)}$ and $y_{j(2)}$ and the jth subband of the corresponding linear *discrete cosine transform* (DCT) filter bank share the same spectral range for a zero-mean, discrete-space/time, Markov signal taking values in a symmetric set $\{-S, \ldots, S\}$. Furthermore the order statistic filter subbands offer better minimum stopband attenuation with respect to the zero frequency gain, but they exhibit a poorer transition bandwidth than the first DCT subband.

Another possible use of Eq. (5.184) is in accelerating the convergence of adaptive ℓL-filters by performing filtering on the subband domain, that is,

$$y_{\ell L}(n) = \mathbf{c}^T(n)\,\mathbf{S}\,\mathbf{x}_{\ell L}(n), \qquad (5.189)$$

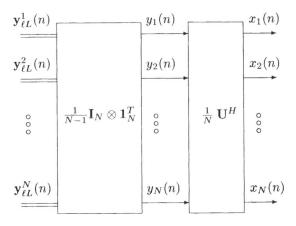

Fig. 5.12 Synthesis order statistic bank.

where $\mathbf{U}$ can be chosen as the DCT transformation matrix in Eq. (5.185).

The synthesis order statistic bank is given by

$$\mathbf{x}_\ell(n) = \frac{1}{N(N-1)} \mathbf{U}^H \left(\mathbf{I}_N \otimes \mathbf{1}_N^T\right) \mathbf{y}_{\ell L}(n) = \frac{1}{N(N-1)} \left(\mathbf{U}^H \otimes \mathbf{1}^T\right) \mathbf{y}_{\ell L}(n),$$
$$(5.190)$$

where $\mathbf{U}^H$ denotes the Hermitian transpose of the orthogonal matrix $\mathbf{U}$. It can be seen from Eq. (5.190) and Fig. 5.12 that the synthesis order statistic filter bank consists of the following:

1. A space/time-frequency-rank to space-frequency (i.e., subbands) transformation module defined by $\frac{1}{N-1}(\mathbf{I}_N \otimes \mathbf{1}^T)$.

2. A module that maps the subbands to the original spatial/temporal input sequence (i.e., the pre-multiplication by $\frac{1}{N}\mathbf{U}^H$).

Accordingly Eq. (5.190) defines a space/time-frequency-rank to space/time transformation.

5.7 PERMUTATION FILTER LATTICES

The class of order statistic filters was extended to permutation filters, which are flexible and modular like the polynomial filters [40]. Instead of polynomial expansions, we have permutation lattice expansions. At the simplest level in the lattice, permutation filters reduce to either a simple linear FIR or to an L-filter, but at the higher levels in the lattice the filters obtained can model complicated nonlinear systems more accurately than the aforementioned L-filter, while still preserving their robustness properties. The class of permutation filter lattices extends the $L\ell$-filter described in Section 5.5. We will slightly modify the notation used in [40] in order to maintain the same notation with Sections 5.2 and 5.5.

5.7.1 ℓL^k-Filters Based on Rank Permutations

In the context of this section let $\oplus$ denote the modulo N addition, that is,

$$j \oplus \alpha = (j \oplus \alpha) \bmod N, \tag{5.191}$$

where j and α are integers, and the operation $\oplus$ is defined in the set $\{1, 2, \ldots, N\}$ such that $N \bmod N = N$ and $(N + 1) \bmod N = 1$. A *reduced rank indicator* of $x_{j\oplus\alpha}$, ρ_j^α, is formed by removing the ϱ_j-th, $\varrho_{j\oplus1}$-th, $\ldots$, $\varrho_{j\oplus(\alpha-1)}$-th elements from the rank indicator vector $\boldsymbol{\rho}_{j\oplus\alpha}$ defined by Eq. (5.7). For example, if $\mathbf{x}_\ell = (6, 2, 10, 3, 1)^T$, then $\mathbf{x}_L = (1, 2, 3, 6, 10)^T$ and the rank parameters for all observations are $\varrho_1 = 4$, $\varrho_2 = 2$, $\varrho_3 = 5$, $\varrho_4 = 3$ and $\varrho_5 = 1$. The reduced indicator vector ρ_4^3 is obtained from $\boldsymbol{\rho}_2$ by deleting the third (ϱ_4), first (ϱ_5), and fourth (ϱ_1) elements

$$\boldsymbol{\rho}_4^3 = (\emptyset, 1, \emptyset, 0, \emptyset)^T = (1, 0)^T. \tag{5.192}$$

The reduced rank indicator ρ_j^k characterizes the rank information of the sample $x_{j\oplus k}$ when the rank information of the samples $(x_j, x_{j\oplus1}, x_{j\oplus(k-1)})$ is known.

Let us define the *rank permutation indicator* by

$$\varpi_j^k = \boldsymbol{\rho}_j \otimes \boldsymbol{\rho}_j^1 \otimes \cdots \otimes \boldsymbol{\rho}_j^{k-1}, \quad k = 1, 2, \ldots, N, \tag{5.193}$$

where $\otimes$ denotes the Kronecker product. It can be seen that ϖ_j^k is of dimensions $P_N^k \times 1$ where $P_N^k = N(N - 1) \ldots (N - k + 1)$. P_N^k represents the number of permutations choosing j samples from N distinct samples. Hence ϖ_j^0 does not unveil any rank information, ϖ_j^1 provides the rank information of x_j and ϖ_j^N accounts for the ranks of all input samples x_1 through x_N. In general, it can be easily seen that

$$\varpi_j^k = \varpi_j^{k-1} \otimes \boldsymbol{\rho}_j^{k-1}. \tag{5.194}$$

The rank permutation indicator forms the basis for the *rank permutation vector* $\mathbf{X}_k$ defined as the NP_N^k-dimensional vector

$$\mathbf{X}_k = \left(x_1 \left(\varpi_1^k\right)^T \mid x_2 \left(\varpi_2^k\right)^T \mid \cdots \mid x_N \left(\varpi_N^k\right)^T \right)^T \tag{5.195}$$

where each x_j is placed according to the rank of k-space/time ordered samples $(x_j, x_{j\oplus1}, \ldots, x_{j\oplus(k-1)})$. The ℓL^k *estimate* is given by

$$\hat{s}_{\ell L^k} = \mathbf{A}_k^T \mathbf{X}_k \quad \text{with} \quad \mathbf{A}_k = \left((\mathbf{a}_1^k)^T \mid (\mathbf{a}_2^k)^T \mid \cdots \mid (\mathbf{a}_N^k)^T\right)^T. \tag{5.196}$$

It is seen from Eq. (5.196) that the ℓL^k-filter employs a $NP_N^k \times 1$ coefficient vector. For $k = 0$, the permutation filter reduces to a linear FIR filter. For $k = 1$, the permutation filter is identical to the C-filter defined by Eq. (5.14), or ℓL-filter defined by Eq. (5.20), while for $k = N - 1$ we obtain the $N!$-ℓL-filter.

5.7.2 $L\ell^k$-Filters Based on Location Permutations

The time and ordering information can be considered as dual entities. Therefore, we can design a filter where the output is a weighted sum of the order statistic, but the weight applied to each ranked input sample, $x_{(i)}$, is determined by the temporal location of a set of consecutive ranked samples, as has been explained in Section 5.2. The definition of $L\ell^k$-*filters* is based on the temporal location indicator vector defined by Eq. (5.11). A *reduced location indicator* $\xi_{(i)}^\alpha$ can be defined by removing the $\ell_{(i)}$-th, $\ell_{(i\oplus 1)}$-th, ..., $\ell_{(i\oplus(\alpha-1))}$-th elements from $\xi_{(i\oplus\alpha)}$. Using the location indicator vectors, we can construct the *location permutation indicator vector* as

$$\varpi_{(i)}^k = \xi_{(i)} \otimes \xi_{(i)}^1 \otimes \xi_{(i)}^2 \otimes \cdots \otimes \xi_{(i)}^{k-1}, \qquad (5.197)$$

which provides the temporal location of order statistics $(x_{(i)}, x_{(i\oplus 1)}, \ldots, x_{(i\oplus(k-1))})$. The vectors $\varpi_{(i)}^k$ are then used to form the permuted observation vector

$$\mathbf{X}_{(k)} = \left(x_{(1)} (\varpi_{(1)}^k)^T \mid x_{(2)} (\varpi_{(2)}^k)^T \mid \cdots \mid x_{(N)} (\varpi_{(N)}^k)^T \right)^T \qquad (5.198)$$

that is exploited to derive the $L\ell^k$-*estimate*

$$\hat{s}_{l,\ell^k} = \mathbf{A}_{(k)}^T \mathbf{X}_{(k)} \quad \text{with} \quad \mathbf{A}_{(k)} = \left((\mathbf{a}_{(1)}^k)^T \mid (\mathbf{a}_{(2)}^k)^T \mid \cdots \mid (\mathbf{a}_{(N)}^k)^T \right)^T. \quad (5.199)$$

For $k = 0$, the $L\ell^0$-filter reduces to an L-filter. The performances of ℓL^k- and $L\ell^k$-filters are very similar when the same parameter k is used [40].

5.7.3 Building Permutation Order Statistic Lattices

5.7.3.1 $\ell^\alpha L^\beta$-filters A weight can be assigned to an input sample, x_j, according to the rank permutation of α consecutive spatial/temporal samples and also by the additional spatial/temporal location information of β consecutive rank ordered samples which were not previously utilized by the first set of temporal samples.

The rank permutation indicator ϖ_j^α not only provides the ranking of $(x_j, x_{j\oplus 1},$ ..., $x_{j\oplus(\alpha-1)})$ but also provides the location information of $(x_{(\varrho_j)}, x_{(\varrho_j\oplus 1)}, \ldots,$ $x_{(\varrho_{j\oplus(\alpha-1)})})$. Let $\mathbf{x}_j$ be the reduced observation spatial/temporal vector formed by removing the samples $(x_j, x_{j\oplus 1}, \ldots, x_{j\oplus(\alpha-1)})$ from $\mathbf{x}_\ell$. Let $\mathbf{x}_{\varrho j}$ be the following vector:

$$\mathbf{x}_{\varrho j} = \left(x_{(\varrho_j \oplus m_1)}, x_{(\varrho_j \oplus m_2)}, \ldots, x_{(\varrho_j \oplus m_{N-\alpha})} \right)^T, \qquad (5.200)$$

where m_l is the lth smallest positive integer such that $x_{(\varrho_j \oplus m_l)}$ is the lth sample away from $x_{(\varrho_j)}$ in $\mathbf{x}_L$, which is not coupled with ϖ_j^α. In Eq. (5.200) the samples are addressed from left to right and in a circular fashion if the end of the vector is reached. Let $\ell_{(j,\,l)}$ and $\boldsymbol{\xi}_{(j,\,l)}$ be the location parameter and the location indicator vector of $x_{(\varrho_j \oplus m_l)}$ in $\mathbf{x}_j$ for $l = 1, 2, \ldots, N - \alpha$, respectively. The reduced location indicator of $x_{(\varrho_j \oplus m_l)}$, denoted as $\boldsymbol{\xi}_{(j)}^l$, is obtained by removing the $\ell_{(j,\,1)}$-th, $\ell_{(j,\,2)}$-th, $\ldots$, $\ell_{(j,\,l-1)}$-th elements from $\boldsymbol{\xi}_{(j,\,l)}$. The location permutation of $(x_{(\varrho_j \oplus m_1)}, x_{(\varrho_j \oplus m_2)}, \ldots, x_{(\varrho_j \oplus m_{\beta-1})})$ is characterized by the *uncoupled location permutation indicator*, $\varpi_{(\varrho_j)}^{\alpha,\,\beta-1}$, defined by [40]

$$\varpi_{(\varrho_j)}^{\alpha,\,\beta-1} = \boldsymbol{\xi}_{(j)}^1 \otimes \boldsymbol{\xi}_{(j)}^2 \otimes \cdots \boldsymbol{\xi}_{(j)}^{\beta-1}. \tag{5.201}$$

We construct the following vector:

$$\mathbf{X}_{\alpha(\beta)} = \begin{bmatrix} x_1 \left(\varpi_1^\alpha \otimes \varpi_{(\varrho_1)}^{\alpha,\,\beta-1} \right) \\ x_2 \left(\varpi_2^\alpha \otimes \varpi_{(\varrho_2)}^{\alpha,\,\beta-1} \right) \\ \vdots \\ x_N \left(\varpi_N^\alpha \otimes \varpi_{(\varrho_N)}^{\alpha,\,\beta-1} \right) \end{bmatrix}. \tag{5.202}$$

In Eq. (5.202), $\varpi_j^\alpha \otimes \varpi_{(\varrho_j)}^{\alpha,\,\beta-1}$ characterizes the rank permutation of $(x_j, x_{j\oplus1}, \ldots, x_{j\oplus(\alpha-1)})$ and the location permutation of $(x_{(\varrho_j \oplus m_1)}, x_{(\varrho_j \oplus m_2)}, \ldots, x_{(\varrho_j \oplus m_{\beta-1})})$, respectively. The $\ell^\alpha L^\beta$-filter weighs linearly the elements of $\mathbf{X}_{\alpha(\beta)}$ defined by Eq. (5.202) to provide an estimate of the desired response

$$\hat{s}_{\alpha(\beta)} = \mathbf{A}_{\alpha(\beta)}^T \mathbf{X}_{\alpha(\beta)} \quad \text{with} \quad \mathbf{A}_{\alpha(\beta)} = \left((\mathbf{a}_1^{\alpha,\,\beta})^T \mid (\mathbf{a}_2^{\alpha,\,\beta})^T \mid \cdots \mid (\mathbf{a}_N^{\alpha,\,\beta})^T \right)^T, \tag{5.203}$$

where $\mathbf{a}_l^{\alpha,\,\beta}$ is a coefficient vector of dimensions $P_N^\alpha P_{N-\alpha}^{\beta-1} \times 1$.

It can be shown that the location permutation indicator of the $\ell^{\alpha-1} L^\beta$-filter, $\varpi_{(\varrho_j)}^{\alpha-1,\,\beta-1}$, can be obtained from $\rho_j^{\alpha-1} \otimes \varpi_{(\varrho_j)}^{\alpha,\,\beta-1}$ by a linear transformation [40]

$$\varpi_{(\varrho_j)}^{\alpha-1,\,\beta-1} = \mathcal{Q} \left(\rho_j^{\alpha-1} \otimes \varpi_{(\varrho_j)}^{\alpha,\,\beta-1} \right), \tag{5.204}$$

where $\mathcal{Q}$ is a suitable $P_{N-\alpha+1}^{\beta-1} \times (N - \alpha + 1) P_{N-\alpha}^{\beta-1}$ matrix. The structure of the matrix $\mathcal{Q}$ is studied in [40]. It can also be shown that both $\mathbf{X}_{\alpha(\beta-1)}$ and $\mathbf{X}_{\alpha-1(\beta)}$ can be obtained by a linear transformation of $\mathbf{X}_{\alpha(\beta)}$.

5.7.3.2 $L^\alpha \ell^\beta$-filters

Similarly to $\ell^\alpha L^\beta$-filters we can define their duals, the $L^\alpha \ell^\beta$- filters. We construct the vector

$$
\mathbf{X}_{(\alpha)\beta} =
\begin{bmatrix}
x_{(1)} \left(\varpi_{(1)}^\alpha \otimes \varpi_{\ell_{(1)}}^{\alpha,\,\beta-1} \right) \\[6pt]
x_{(2)} \left(\varpi_{(2)}^\alpha \otimes \varpi_{\ell_{(2)}}^{\alpha,\,\beta-1} \right) \\
\vdots \\
x_{(N)} \left(\varpi_{(N)}^\alpha \otimes \varpi_{\ell_{(N)}}^{\alpha,\,\beta-1} \right)
\end{bmatrix},
\tag{5.205}
$$

where each order statistic $x_{(i)}$ in $\mathbf{x}_L$ is weighted, according to the location character- istics of α consecutive ordered samples $(x_{(i)}, x_{(i\oplus 1)}, \dots, x_{(i\oplus(\alpha-1))})$ as well as the additional rank information of the first β samples in $\mathbf{x}_{\ell_{(i)}}$, where

$$
\mathbf{x}_{\ell_{(i)}} = \left(x_{\ell_{(i)}}, x_{\ell_{(i)} \oplus 1}, \dots, x_{\ell_{(i)} \oplus (\beta-1)} \right)^T
\tag{5.206}
$$

and

$$
\varpi_{\ell_{(i)}}^{\alpha,\,\beta-1} = \rho_i^1 \otimes \rho_i^2 \otimes \cdots \rho_i^{\beta-1}.
\tag{5.207}
$$

The $L^\alpha \ell^\beta$-filter output is given by [40]

$$
\widehat{s}_{(\alpha)\beta} = \mathbf{A}_{(\alpha)\beta}^T \mathbf{X}_{(\alpha)\beta} \quad \text{where} \quad
\mathbf{A}_{(\alpha)\beta} = \left((\mathbf{a}_{(1)}^{\alpha,\,\beta})^T \mid (\mathbf{a}_{(2)}^{\alpha,\,\beta})^T \mid \cdots \mid (\mathbf{a}_{(N)}^{\alpha,\,\beta})^T \right)^T
\tag{5.208}
$$

in which $\mathbf{a}_{(i)}^{\alpha,\,\beta}$ is a $P_N^\alpha P_{N-\alpha}^{\beta-1} \times 1$ vector. Both $\mathbf{X}_{(\alpha-1)\beta}$ and $\mathbf{X}_{(\alpha)\beta-1}$ can be obtained by a linear transformation of $\mathbf{X}_{(\alpha)\beta}$. For the nonnegative integers α and β such that $\alpha + \beta = N$, all $L^\alpha \ell^\beta$- and all $\ell^\alpha L^\beta$- filters are equivalent to the $L^{N-1}\ell$- and ℓL^{N-1}- filters, respectively [40]. For $0 \le \alpha \le N-1$, $L^\alpha \ell$-filters are equivalent to ℓL^α- filters. Similarly, $L\ell^\alpha$-filters are equivalent to $\ell^\alpha L$-filters. Equivalence is interpreted as a one-to-one correspondence between the permutation observation vectors that are linearly weighted by the filter coefficients, such as $\mathbf{X}_{(a)1} \leftrightarrow \mathbf{X}_{1(a)}$, as well as a one-to-one mapping between the coefficient vectors, such as $\mathbf{A}_{(a)1} \leftrightarrow \mathbf{A}_{1(a)}$.

5.7.3.3 Lattice properties

For a given window size N, the number of $L^\alpha \ell^\beta$- filters is on the order of $\mathcal{O}(N^2)$, where each filter utilizes different type of rank and temporal ordering information. Let us assume a fixed observation vector of size N. Let $\mathcal{F}_{\ell L} = \{(\ell), (\ell L), \dots, (\ell L)^N\}$ be the set of all permutation filters, where $(\ell L)^k$ stands for the subset of filters $\{\ell^{k-1}L, \ell^{k-2}L^2, \dots, \ell L^{k-1}\}$. Its dual set is $\mathcal{F}_{L\ell} = \{L, (L\ell), \dots, (L\ell)^N\}$. We have already addressed the equivalence $(L\ell)^N \equiv (\ell L)^N$. It can be shown that $\mathcal{F}_{\ell L}$ constitutes a *complete lattice* [40]; that is, $\mathcal{F}_{\ell L}$ is a well-defined *partially ordered set* (poset) that has a least upper bound and a greater lower bound for any two elements of $\mathcal{F}_{\ell L}$.

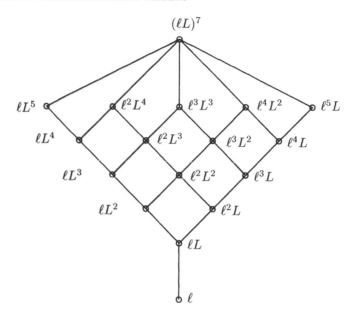

Fig. 5.13 The $\mathcal{F}_{\ell L}$ permutation filter lattice for $N = 7$. The least upper bound is $(\ell L)^7$ and the greatest lower bound is ℓ, which is the FIR filter (adapted from [40] by copyright permission from IEEE).

Indeed, once the rank-location permutation indicator of the $\ell^\alpha L^\beta$-filter is given, the rank-location permutation indicator of the $\ell^{k_1} L^{k_2}$ can be determined uniquely for $1 \leq k_1 \leq \alpha$ and $0 \leq k_2 \leq \beta$. Thus the rank-location ordering information from the $\ell^\alpha L^\beta$ to the $\ell^{k_1} L^{k_2}$ is always possible. However, the reverse is not possible.

The following properties have been proven [40]:

- The sets of $\ell^k L$-filters and ℓL^k-filters form complete lattices.

- For $\ell^{\alpha_1} L^{\beta_1}, \ell^{\alpha_2} L^{\beta_2} \in \mathcal{F}_{\ell L}$, $\ell^{\alpha_1} L^{\beta_1}$ covers $\ell^{\alpha_2} L^{\beta_2}$ if

$$(\alpha_2 = \alpha_1 + 1 \text{ and } \beta_2 = \beta_1) \text{ or } (\alpha_2 = \alpha_1 \text{ and } \beta_2 = \beta_1 + 1). \quad (5.209)$$

  Finding the supremum and infimum of two permutation filters reduces to simply finding the maximum and minimum of the respective orders of the given integers of the permutation filters. Accordingly $(\ell L)^N$ and ℓ filters are the top and bottom elements of the set $\mathcal{F}_{\ell L}$.

- The structure of $\mathcal{F}_{L\ell}$ can be obtained by replacing each filter in the lattice of $\mathcal{F}_{\ell L}$ by its dual filter. Hence it forms a complete lattice, whose top element is the $(L\ell)^N$-filter and the bottom element is the L-filter.

Figure 5.13 shows the diagram of the permutation filter lattice for $N = 7$. Lattices

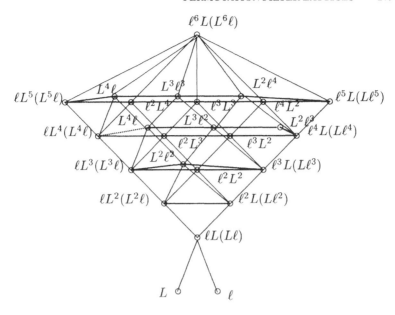

Fig. 5.14 The poset structure $\mathcal{F}_{\ell L} \cup \mathcal{F}_{L\ell}$ for $N = 7$ (adapted from [40] by copyright permission from IEEE).

$\mathcal{F}_{\ell L}$ and $\mathcal{F}_{L\ell}$ can be combined together. The resulting structure is depicted in Fig. 5.14. It does not form a lattice but only a poset, since there is no ordering relation between the linear (ℓ) and the L-filter. However, the structure that is above the $L\ell$ filter forms a complete lattice.

5.7.4 Optimal and Adaptive Design

It is simple to show that the optimal $\ell^\alpha L^\beta$-filter under the MSE criterion is given by

$$\mathbf{a}^o_{\alpha(\beta)} = \mathbf{R}^{-1}_{\alpha(\beta)}\mathbf{P}_{\alpha(\beta)}, \qquad (5.210)$$

where

$$\mathbf{P}_{\alpha(\beta)} = \mathrm{E}\left\{s(n)\mathbf{X}_{(\alpha)\beta}(n)\right\}, \qquad (5.211)$$

$$\mathbf{R}_{\alpha(\beta)} = \mathrm{E}\left\{\mathbf{X}_{\alpha(\beta)}(n)\mathbf{X}^T_{\alpha(\beta)}(n)\right\}. \qquad (5.212)$$

The solution given by Eq. (5.210) is unique when the correlation matrix in Eq. (5.212) is nonsingular. If the correlation matrix is singular, a solution can be found by using the singular value decomposition. It can be proved that the optimal $\ell^\alpha L^\beta$-filter, $\mathbf{A}^o_{\alpha(\beta)}$ specifies all other optimal filters $\mathbf{A}^o_{k_1(k_2)}$ for $1 \leq k_1 \leq \alpha$ and $0 \leq k_2 \leq \beta$.

$I_i := 0$;
for $m := 0$ to $\alpha + \beta - 2$
  $I_i := I_i \cdot (N - m) + (\gamma_i^m - 1)$
end
$I_i := I_i + 1$;

Fig. 5.15 Recursion for determining the nonzero element in ith subvector of $\mathbf{X}_{\alpha(\beta)}(n)$.

Obtain $\mathbf{x}_l(n)$, $\mathbf{x}_L(n)$ and $s(n)$;	Step 1
For $i := 1$ to N do	Step 2
$\quad$ find $\varrho_i(n)$ and $\ell_{(i)}(n)$;	
$\quad$ compute $I_i(n)$;	
Estimate the output and compute the estimation error	Step 3
$\quad \widehat{s}_{\alpha(\beta)}(n) = \sum_{i=1}^{N} \mathbf{a}_i^{\alpha, \beta}(I_i, n)\, x_i(n);$	
$\quad e(n) = s(n) - \widehat{s}_{\alpha(\beta)}(n);$	
Update the coefficient vector	Step 4
$\quad$ for $i := 1$ to N do	
$\quad\quad a_i^{\alpha, \beta}(I_i, n + 1) = a_i^{\alpha, \beta}(I_i, n) + \mu\, e(n)\, x_i(n);$	
Go to Step 1	Step 5

Fig. 5.16 LMS algorithm for the $\ell^\alpha L^\beta$-permutation filters (adapted from [40] by copyright permission from IEEE).

The complexity of permutation filters increases rapidly as $\alpha + \beta$ approaches N. Accordingly the computational cost for the inversion of $\mathbf{R}_{\alpha(\beta)}$ increases. Alternatively, the LMS algorithm can be employed in the filter design

$$\mathbf{A}_{\alpha(\beta)}(n + 1) = \mathbf{A}_{\alpha(\beta)}(n) + \mu\, e(n)\, \mathbf{X}_{\alpha(\beta)}(n), \qquad (5.213)$$

where $e(n) = s(n) - \widehat{s}_{\alpha(\beta)}(n)$ is the estimation error at n, and μ is the step-size. At any iteration n there are only N nonzero elements in $\mathbf{X}_{\alpha(\beta)}(n)$. Therefore the adaptation can be confined to these nonzero elements. Let I_i be the location of the nonzero sample x_i in $\varpi_i^\alpha \otimes \varpi_{(\varrho_i)}^{\alpha, \beta-1}$, the ith subvector of $\mathbf{X}_{\alpha(\beta)}(n)$. Let $\gamma_i^m \in \{1, 2, \ldots, N - m\}$ denote the location of the nonzero element in ρ_i^m for $m = 0, 1, \ldots, \alpha - 1$. The same parameter γ_i^m defines also the location of the nonzero element in $\xi_{(i)}^{m-\alpha+1}$ for $m = \alpha, \alpha + 1, \ldots, \alpha + \beta - 2$. Then I_i can be computed by the recursion described in Fig. 5.15 [40]. The steps of the algorithm are summarized in Fig. 5.16.

5.7.5 Application in Inverse Halftoning

Digital halftoning is the method for creating the illusion of continuous-tone pictures on displays capable of producing only binary pixels, such as laser printers, facsimiles, and plasma panels. Two are the most popular image halftoning techniques: ordered dithering and error diffusion. Inverse halftoning is needed whenever halftones need to be enlarged, reduced, or re-sampled, and when only the halftoned image is available. A class of inverse halftoning filters that utilizes the information provided by the rank-order permutation of the binary samples in an observation window is discussed in [41].

Let $\mathbf{x}_\ell$ denote the binary $N \times 1$ vector formed by rearranging the observation window in lexicographic ordering. Let $\mathbf{x}_L$ be the vector of observations sorted in ascending order of magnitude. Due to the binary nature of both $\mathbf{x}_\ell$ and $\mathbf{x}_L$, several methods can be used to resolve the ambiguities. Let us confine ourselves to the use of ranks 1 or 2 only as in [41]. That is, we consider all samples of equal values as having the same rank. More specifically, let us group the binary elements of $\mathbf{x}_l$ into two sets, $\mathcal{S}_0$ and $\mathcal{S}_1$, where $\mathcal{S}_0$ contains all zero-valued elements in $\mathbf{x}_\ell$, and $\mathcal{S}_1$ is the set of the remaining one-valued samples. We define the rank of each binary element x_j as

$$\varrho_j = \begin{cases} 1 & \text{if } x_j \in \mathcal{S}_0 \\ 2 & \text{if } x_j \in \mathcal{S}_1. \end{cases} \qquad (5.214)$$

Let $\boldsymbol{\rho}_j = (\rho_{j1}, \rho_{j2})^T$ be the rank indicator vector that quantifies the information that the sample x_j is "on" (i.e., black) or "off" (i.e., white). That is, for $l = 1, 2$ [41],

$$\rho_{jl} = \begin{cases} 1 & \text{if } x_j \in \mathcal{S}_{l-1} \\ 0 & \text{otherwise.} \end{cases} \qquad (5.215)$$

By definition, $\rho_{j\,\varrho_j} = 1$. The binary rank permutation indicator vector of order k is defined as

$$\boldsymbol{\varpi}_j^k = \begin{cases} \boldsymbol{\rho}_i \otimes \boldsymbol{\rho}_{j\oplus1} \otimes \cdots \otimes \boldsymbol{\rho}_{j\oplus(k-1)} & \text{if } k > 0 \\ 1 & \text{if } k = 0 \end{cases} \qquad (5.216)$$

for $j = 1, 2, \ldots, N$, where $\otimes$ denotes the Kronecker product and $\oplus$ represents the modulo N addition as in Eq. (5.191). The vector $\boldsymbol{\varpi}_j^k$ has length 2^k. For $k = 0$, it does not unveil any binary rank information, whereas for $k = N$ provides the binary rank information of all samples in $\mathbf{x}_l$. Only one element of $\boldsymbol{\varpi}_j^k$ is nonzero and takes the value 1. The location of this element in $\boldsymbol{\varpi}_j^k$ corresponds to one particular event of the joint rank permutation of the samples $(x_j, x_{j\oplus1}, \ldots, x_{j\oplus(k-1)})$. The parameter k determines the complexity and performance of the reconstruction algorithm and is a designer's choice. It can be shown [41] that

$$\boldsymbol{\varpi}_j^k = \boldsymbol{\varpi}_j^{k-1} \otimes \boldsymbol{\rho}_{j\oplus(k-1)}. \qquad (5.217)$$

That is, given $\boldsymbol{\varpi}_j^k$, the binary rank permutation vector of lower order can be obtained by a linear transformation

$$\boldsymbol{\varpi}_j^{k-1} = \mathcal{Q}_k \boldsymbol{\varpi}_j^k \quad \text{with} \quad \mathcal{Q}_k = \mathbf{I}_{2^{k-1}} \otimes (1, 1), \qquad (5.218)$$

where $\mathbf{I}_{2^{k-1}}$ is the $2^{k-1} \times 2^{k-1}$ identity matrix. The reconstruction algorithm for space-invariant inverse filtering is based on the $N \cdot 2^k$-dimensional vector

$$\mathbf{X}_k = \mathbf{x}_\ell \otimes \left((\varpi_1^k)^T \mid (\varpi_2^k)^T \mid \cdots \mid (\varpi_N^k)^T \right)^T. \qquad (5.219)$$

The output of the binary permutation reconstruction algorithm is written as

$$\widehat{s}(n) = \mathbf{A}_k^T \mathbf{X}_k \quad \text{with} \quad \mathbf{A}_k = \left((\mathbf{a}_1^k)^T \mid (\mathbf{a}_2^k)^T \mid \cdots \mid (\mathbf{a}_N^k)^T \right)^T, \qquad (5.220)$$

where $\mathbf{a}_j^k$ is a 2^k-dimensional vector. Equation (5.220) defines the filter output (estimate) $\widehat{s}(n)$ for the continuous tone $s(n)$. It can be easily seen that, for $k = 0$, no ranking information is used and the reconstruction reduces to an FIR filter, where only the spatial location is utilized. For $k = 1$, the spatial location and rank of x_j are used without further consideration of the neighboring samples. For $k = N$, the space-rank ordering of all samples are used.

The reconstruction filter weights $\mathbf{A}_k$ can be designed so as to minimize the error between the desired image $s(n)$ and the reconstructed image $\widehat{s}(n)$. If the MSE criterion is used, then, provided that $\mathbf{R}_k^{-1}$ is nonsingular, the optimal binary permutation filter is given by

$$\mathbf{A}_k = \mathbf{R}_k^{-1} \mathbf{p}_k, \qquad (5.221)$$

where $\mathbf{R}_k = \mathrm{E}\left\{ \mathbf{X}_k(n) \mathbf{X}_k^T(n) \right\}$ and $\mathbf{p}_k = \mathrm{E}\left\{ s(n) \mathbf{X}_k(n) \right\}$. The conditions for the nonsingularity of $\mathbf{R}_k$ repeat the arguments in [20]. It can be shown that [41]

$$\mathbf{X}_{k-l} = \mathrm{diag}(\underbrace{\mathcal{Q}_k, \mathcal{Q}_k, \ldots, \mathcal{Q}_k}_{N \text{ blocks}}) \mathbf{X}_k. \qquad (5.222)$$

Equation (5.222) implies that the correlation matrix $\mathbf{R}_{k-l}$, the crosscorrelation vector $\mathbf{p}_{k-l}$ and the optimal binary permutation filter $\mathbf{A}_{k-l}$ can be expressed in terms of $\mathbf{R}_k$, $\mathbf{p}_k$ and $\mathbf{A}_k$, respectively.

Using the LMS algorithm in raster scan, we obtain

$$\mathbf{A}_k(n+1) = \mathbf{A}_k(n) + \mu\, e(n)\, \mathbf{X}_k(n). \qquad (5.223)$$

Since there are only N nonzero elements in $\mathbf{X}_k$, the adaptation in Eq. (5.223) can be restricted to the weight vectors that correspond to these nonzero elements. Let I_j be the index of the nonzero coefficient of the jth subvector $\mathbf{a}_j^k$ to be updated that corresponds to the nonzero element of ϖ_j^k. It has been proved [41] that I_j is given by

$$I_j = \sum_{m=1}^k x_{j\oplus(m-1)}\, 2^{k-m} + 1. \qquad (5.224)$$

Accordingly the LMS updating equation given by Eq. (5.223) is rewritten as

$$\mathbf{a}_j^k(I_j; \ n+1) = \mathbf{a}_j^k(I_j; \ n) + \mu \, e(n) \, x_j(n), \quad j = 1, 2, \dots, N. \qquad (5.225)$$

The information provided by the dithering matrix can also be exploited in the algorithm to enforce constraints on $\widehat{s}(n)$. For details, the interested reader is referred to [41].

5.7.6 Extensions to Weighted-Order Statistics

Although in many signal processing applications a linear combination of the order statistics leads to highly effective filter structures, for processing signals with nonstationary mean levels and with abrupt changes, linear combinations lead to unavoidable blurring. Such processes are customary in image processing where important visual cues provided by the edges and fine details should be preserved.

Order statistic filters that constrain the output value to be identical to one of the input samples can be used in these cases. *Weighted-order statistics filters* (WOS) [14, 91, 90] fall into this category. Another suitable choice is the class of signal-adaptive filters reviewed in Section 5.12. Given the observation vector $\mathbf{x}_\ell$, a weight vector $\mathbf{a}$ and a threshold a_0, the output of a WOS filter is the estimate of the desired signal

$$\widehat{s} = a_0 \text{th largest element in } \{a_1 \diamond x_1, a_2 \diamond x_2, \dots, a_N \diamond x_N\}, \qquad (5.226)$$

where $a_i \diamond x_i$, $i = 1, 2, \dots, N$, denotes the duplication of x_i a_i times, that is,

$$a_i \diamond x_i = \underbrace{x_i, x_i, \dots, x_i}_{a_i \text{ times}}. \qquad (5.227)$$

A WOS filter corresponds to a threshold logic gate with positive weights and a positive threshold [91]. Stack filters generalize the concept of linear threshold gates inherent in WOS filters by allowing polynomial threshold logic gates. However, an N-long WOS filter is described by $N + 1$ parameters, a corresponding general stack filter is described by the 2^N elements of a truth table. For more details on stack filters the interested reader is referred to Chapter 2 of this book.

Stack and WOS filters are based on the threshold decomposition principle. That is, the filter operations are effectively performed independently on a set of thresholded signals, without the use of cross-level information [2]. WOS and stack filters can be applied for the removal of outliers. However, WOS and stack filters are not suitable for the restoration of images corrupted by interference with specific spectral content [2]. The class of selection permutation filters [5, 32, 6] can overcome the aforementioned limitations, but at the expense of a computationally intensive implementation. Another solution to the problem under discussion is to employ the class of *permutation weighted-order statistic filters* (PWOS) that were proposed in [2]. In this section the class of PWOS filters is reviewed. This class defines a filter lattice structure where the WOS filter defined by Eq. (5.226) is the greatest lower bound of the lattice.

The starting point in the design of PWOS filters is the rank permutation indicator ϖ_j^k associated with the input sample x_j that is defined in Eq. (5.193). Given the input vector $\mathbf{x}_\ell$ and the corresponding permutation indicators ϖ_j^k, $j = 1, 2, \ldots, N$, the P_N^k-dimensional coefficient vector $\mathbf{a}_j^k$ associated with the jth input sample is defined [2] as

$$\mathbf{a}_j^k = \left(a_{j(1)}^k, a_{j(2)}^k, \ldots, a_{j(P_N^k)}^k\right)^T, \tag{5.228}$$

where the elements are constrained to take only positive real values. By concatenating the above-defined vectors, we form the PWOS coefficient vector

$$\mathbf{A}^k = \left((\mathbf{a}_1^k)^T \mid (\mathbf{a}_2^k)^T \mid \cdots \mid (\mathbf{a}_N^k)^T\right)^T. \tag{5.229}$$

The output of the PWOS[k] filter is defined as [2]

$$\widehat{s}_k = a_0\text{th largest element in}$$
$$\left\{(\mathbf{a}_1^k)^T \varpi_1^k \diamond x_1, (\mathbf{a}_2^k)^T \varpi_2^k \diamond x_2, \ldots, (\mathbf{a}_N^k)^T \varpi_N^k \diamond x_N\right\}. \tag{5.230}$$

The PWOS filtering structure constitutes a space/time-varying WOS filter where the varying coefficients adjust to the rank and space/time ordering characteristics of the input samples. PWOS[0] is the ordinary WOS filter. In PWOS[$N - 1$] filter, the complete mapping from location to rank for all samples in the observation vector is used to assign the weights.

The PWOS[k] filter can be extended to the case where a permutation indicator not only specifies the ranks of consecutive space/time samples but also specifies the spatial/temporal location of an additional set of rank-ordered samples. Let PWOS[α, β] be the permutation filter that uses these rank/location indicators, where α and β refer to the number of consecutive space/time and order samples taken into account. The uncoupled rank-location permutation indicator vectors defined to be the Kronecker product of the rank permutation indicator ϖ_j^α - Eq. (5.193) - and the uncoupled location permutation indicator $\varpi_{(\varrho_j)}^{\alpha,\beta-1}$ [Eq. (5.201)],

$$\varpi_j^\alpha \otimes \varpi_{(\varrho_j)}^{\alpha,\beta-1} \tag{5.231}$$

are used to formulate the filter output of PWOS[α, β] filter as [2]

$$\widehat{s}_{\alpha(\beta)} = a_0\text{th largest element in}$$
$$\left\{(\mathbf{a}_1^{\alpha,\beta})^T(\varpi_1^\alpha \otimes \varpi_{(\varrho_1)}^{\alpha,\beta-1}) \diamond x_1 (\mathbf{a}_2^{\alpha,\beta})^T(\varpi_2^\alpha \otimes \varpi_{(\varrho_2)}^{\alpha,\beta-1}) \diamond x_2, \ldots,\right.$$
$$\left.(\mathbf{a}_N^{\alpha,\beta})^T(\varpi_N^\alpha \otimes \varpi_{(\varrho_N)}^{\alpha,\beta-1}) \diamond x_N\right\}. \tag{5.232}$$

It is seen that PWOS$[\alpha, 1]$ filters are identical to PWOS$[\alpha]$ filters. PWOS$[\alpha, \beta]$ filters where $\alpha + \beta = N$ use all the available information of the underlying permutation. Moreover all PWOS $[\alpha, \beta]$ having $\alpha + \beta = N$ are equivalent.

Let $\mathcal{F} = \{(\text{PWOS})^1, (\text{PWOS})^2, \ldots, (\text{PWOS})^N\}$ where $(\text{PWOS})^k$ stands for the subset of PWOS $[\alpha, \beta]$ filters with $\alpha + \beta = k$, $\alpha \geq 1$ and $\beta \geq 0$. $\mathcal{F}$ is proven to be a complete lattice [2]. The covering relation in the lattice is

$$\text{PWOS}[\alpha_1, \beta_1] \succ \text{PWOS}[\alpha_2, \beta_2] \quad \text{if}$$
$$(\alpha_1 \geq \alpha_2 \text{ and } \beta_1 > \beta_2) \text{ or } (\alpha_1 > \alpha_2 \text{ and } \beta_1 \geq \beta_2). \quad (5.233)$$

The infimum element is the WOS filter and the supremum element is the permutation filter.

The number of weights of PWOS filters increase rapidly for large window sizes or large values of $\alpha + \beta$. For example, a PWOS $[2, 2]$ of window size $N = 11$ has 10,890 coefficients [2]. Besides the increased memory requirements, the filter may also be difficult to be trained using adaptive algorithms. As a remedy, a class called *combination WOS filters* (CWOS) was proposed [2].

The output of the WOS, or PWOS, or CWOS filter can be described as

$$\hat{s}_k = a_0\text{th largest element in } \left\{ \bar{\mathbf{a}}_1^k \diamond x_1, \bar{\mathbf{a}}_2^k \diamond x_2, \ldots, \bar{\mathbf{a}}_N^k \diamond x_N \right\}, \quad (5.234)$$

where

$$\bar{\mathbf{A}}^k = \left((\bar{\mathbf{a}}_1^k)^T \mid (\bar{\mathbf{a}}_2^k)^T \mid \cdots \mid (\bar{\mathbf{a}}_N^k)^T \right)^T = \mathcal{P}^k \mathbf{A}^k, \quad (5.235)$$

and $\mathcal{P}^k$ is a matrix that has the structure in the case of a PWOS $[k]$,

$$\mathcal{P}^k = \begin{bmatrix} (\varpi_1^k)^T & \mathbf{0}^T & \cdots & \mathbf{0}^T \\ \mathbf{0}^T & (\varpi_2^k)^T & & \mathbf{0}^T \\ \vdots & & \ddots & \vdots \\ \mathbf{0}^T & \mathbf{0}^T & \cdots & (\varpi_N^k)^T \end{bmatrix}. \quad (5.236)$$

The algorithm developed in [89] can be used to derive the coefficients of $\bar{\mathbf{A}}^k$ that minimize the mean absolute error. Let $\tilde{\mathbf{A}}^k = \left(a_0, (\bar{\mathbf{A}}^k)^T \right)^T$ be the extended coefficient vector. This algorithm is briefly described below.

Let us decompose the observation vector $\mathbf{x}_\ell(n)$ into a sequence of binary vectors

$$\mathbf{x}_\ell^m(n) = (x_1^m(n), x_2^m(n), \ldots, x_N^m(n))^T, \quad x_j^m = \mathcal{U}(x_j - m); \quad j = 1, 2, \ldots, N, \quad (5.237)$$

where $0 \leq m < \infty$ is the threshold level and $\mathcal{U}(\cdot)$ is the unit step function. The original value x_j is obtained by

$$x_j(n) = \int_0^\infty x_j^m(n) \, dm. \quad (5.238)$$

Let $\widetilde{\mathbf{x}}_\ell^m(n) = \left(-1, (\mathbf{x}_\ell^m(n))^T\right)^T$ be the extended observation vector at the mth binary level. By threshold decomposition, the estimate of the desired signal is

$$\widehat{s}_k(n) = \int_0^\infty \mathcal{U}\left((\widetilde{\mathbf{A}}^k(n))^T \widetilde{\mathbf{x}}_\ell^m(n)\right) dm. \qquad (5.239)$$

The error between the desired signal and the estimate, $e(n) = s(n) - \widehat{s}_k(n)$, is given by

$$\int_0^\infty \left(s^m(n) - \mathcal{U}((\widetilde{\mathbf{A}}^k(n))^T \widetilde{\mathbf{x}}_\ell^m(n))\right) dm. \qquad (5.240)$$

The mean absolute error (MAE) is obtained by

$$\mathrm{E}\{|e(n)|\} = \mathrm{E}\left\{\left|\int_0^\infty \left(s^m(n) - \mathcal{U}((\widetilde{\mathbf{A}}^k(n))^T \widetilde{\mathbf{x}}_\ell^m(n))\right) dm \right|\right\}. \qquad (5.241)$$

By making use of the stacking property, we can deduce that the quantity inside the absolute value operator is monotonic [2], accordingly

$$\mathrm{E}\{|e(n)|\} = \mathrm{E}\left\{\int_0^\infty \left| \left(s^m(n) - \mathcal{U}((\widetilde{\mathbf{A}}^k(n))^T \widetilde{\mathbf{x}}_\ell^m(n))\right)\right| dm\right\}. \qquad (5.242)$$

At each binary level the absolute error is equal to the squared error; Therefore

$$\mathrm{E}\{|e(n)|\} = \mathrm{E}\left\{\int_0^\infty \left(\left(s^m(n) - \mathcal{U}((\widetilde{\mathbf{A}}^k(n))^T \widetilde{\mathbf{x}}_\ell^m(n))\right)\right)^2 dm\right\}. \qquad (5.243)$$

The optimal coefficient vector $(\widetilde{\mathbf{A}}^k)^o(n)$ that minimizes the MAE, given by Eq. (5.243), subject to the constraint that all coefficients should be nonnegative, can be obtained adaptively as follows:

$$\widetilde{\mathbf{a}}_j(n+1) = \mathbf{\Pi}\left[\widetilde{\mathbf{a}}_j(n) + \mu\, e(n)\, \mathcal{U}(x_j(n) - \widehat{s}(n))\right], \quad j = 1, 2, \ldots, N, \qquad (5.244)$$

$$\widetilde{\mathbf{a}}_0(n+1) = \mathbf{\Pi}\left[\widetilde{\mathbf{a}}_0(n) - \mu\, e(n)\right], \qquad (5.245)$$

where $\mathbf{\Pi}[\cdot]$ is the projection operator

$$\mathbf{\Pi}[\alpha] = \begin{cases} \alpha & \text{if } \alpha \geq 0 \\ 0 & \text{otherwise.} \end{cases} \qquad (5.246)$$

For a more detailed discussion on adaptive weighted median filtering, the interested reader is referred to [90].

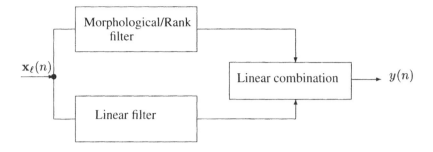

Fig. 5.17   Structure of MRL filter.

5.8   MORPHOLOGICAL-RANK-LINEAR FILTERS

Increasing discrete morphological systems and rank or stack filters are closely related, because they can be represented as maxima of morphological erosions [55]. A general approach for the design of morphological/rank filters is via a gradient-based adaptive optimization [78]. Such an approach can be applied to both increasing and nonincreasing systems and to both binary and real-valued signals. In Sections 5.5 and 5.7 we argued that, in many cases, it is useful to mix linear and nonlinear subfilters in the same hybrid filter. *Morphological-rank-linear filters* (MRL) is a general class of nonlinear systems that contains the morphological, rank, and linear filters as special cases [66]. Their structure is depicted in Fig. 5.17. In this section we review their adaptive optimal design [66]. Let $\mathcal{G}_{(i)}(\mathbf{x}_\ell)$ be the function that selects the ith order statistic from $\mathbf{x}_\ell$. The MRL filter is defined as the shift-invariant system whose output is given by

$$y(n) = \gamma \, \mathcal{G}_{(i)}(\mathbf{x}_\ell(n) + \mathbf{a}) + (1 - \gamma) \, \mathbf{b}^T \mathbf{x}_\ell(n), \tag{5.247}$$

where $\mathbf{a}, \mathbf{b} \in \mathbb{R}^N$ and $\gamma \in \mathbb{R}$. Thus, the MRL filter is a linear combination of a morphological/rank filter and a linear filter. The vector $\mathbf{b}$ corresponds to the coefficients of the linear filter and the vector $\mathbf{a}$ represents the coefficients of the morphological/rank filter. $\mathbf{a}$ is called the *structuring element*, because the morphological filter becomes the morphological erosion and dilation by a flat structuring function equal to $\mp a$ within its support for $i = 1$ and $i = N$, respectively. Besides the vectors $\mathbf{a}$ and $\mathbf{b}$, the rank i and the mixing parameter γ are also trainable parameters in the filter design. For $\gamma \in [0, 1]$, the MRL filter becomes a convex combination of its components.

Let $\varphi(\alpha), \alpha \in \mathbb{R}$, be the unit sample function defined as

$$\varphi(\alpha) = \left\{ \begin{array}{ll} 1 & \text{if } \alpha = 0 \\ 0 & \text{otherwise.} \end{array} \right. \tag{5.248}$$

The application of the unit sample function to a vector $\boldsymbol{\alpha} \in \mathbb{R}^N$ is performed componentwise. Based on Eq. (5.248), the ith location indicator vector defined in

Eq. (5.11) can be expressed as

$$\boldsymbol{\xi}_{(i)}(\mathbf{x}_\ell) = \frac{\varphi(x_{(i)}\mathbf{1}_N - \mathbf{x}_\ell)}{\mathbf{1}_N^T\varphi(x_{(i)}\mathbf{1}_N - \mathbf{x}_\ell)}. \tag{5.249}$$

The location indicator vector marks the locations in $\mathbf{x}_\ell$ where the ith order statistic occurs. It is a function of $\mathbf{x}_\ell$ and it is now defined to have a unit area, namely $\mathbf{1}_N^T\boldsymbol{\xi}_{(i)}(\mathbf{x}_\ell) = 1$.

For i fixed, if $\boldsymbol{\xi}_{(i)}(\mathbf{x}_\ell)$ is constant in a neighborhood of some $\mathbf{x}_{\ell,0}$, then $\mathcal{G}_{(i)}(\mathbf{x}_\ell)$ is differentiable at $\mathbf{x}_{\ell,0}$ and [66]

$$\frac{\partial\mathcal{G}_{(i)}(\mathbf{x}_\ell)}{\partial\mathbf{x}_\ell}\Big|_{\mathbf{x}_\ell=\mathbf{x}_{\ell,0}} = \boldsymbol{\xi}_{(i)}(\mathbf{x}_{\ell,0}). \tag{5.250}$$

At points where the function $\partial\mathcal{G}_{(i)}(\mathbf{x}_\ell)$ is not differentiable, one may assign the vector $\boldsymbol{\xi}_{(i)}(\mathbf{x}_\ell)$ as an one-sided value of the discontinuous derivative. To avoid abrupt changes, the unit sample function, $\varphi(\alpha)$, can be replaced by smoothed functions, such as

$$\varphi_\sigma(\alpha) = \exp\left(-\frac{\alpha^2}{2\sigma^2}\right) \quad \text{or} \quad \varphi_\sigma(\alpha) = \text{sech}^2\left(\frac{\alpha}{\sigma}\right). \tag{5.251}$$

Moreover, instead of working with an integer rank parameter i, we can work with a real variable ρ implicitly defined via the following scaling function [66]:

$$i \equiv \lceil\frac{N-1}{1+\exp(-\rho)} + 0.5\rceil. \tag{5.252}$$

Accordingly the filter coefficient vector to be specified is

$$\mathbf{A} = \left(\mathbf{a}^T \mid \rho \mid \mathbf{b}^T \mid \gamma\right)^T. \tag{5.253}$$

The usual approach to adaptively adjust the vector $\mathbf{A}$ is to define a cost function $\varepsilon(\mathbf{A})$, estimate its gradient, and update $\mathbf{A}$ by the method of steepest descent:

$$\mathbf{A}(n+1) = \mathbf{A}(n) - \frac{\mu}{2}\nabla\,\varepsilon(\mathbf{A})\,|_{\mathbf{A}=\mathbf{A}(n)}. \tag{5.254}$$

In using as cost function the average squared error

$$\varepsilon(\mathbf{A}(n)) = \frac{1}{M}\sum_{k=n-M+1}^{n} e^2(k) \tag{5.255}$$

where $M = 1, 2, \ldots$ is a memory parameter, we obtain the *averaged-LMS* algorithm

$$\mathbf{A}(n+1) = \mathbf{A}(n) + \frac{\mu}{M}\sum_{k=n-M+1}^{n} e(k)\frac{\partial y(k)}{\partial\mathbf{A}}\,|_{\mathbf{A}=\mathbf{A}(n)}. \tag{5.256}$$

It can be shown that [66]

$$\frac{\partial y}{\partial \mathbf{a}} = \gamma \frac{\varphi_\sigma(\mathcal{G}_{(i)}(\mathbf{x}_\ell + \mathbf{a})\mathbf{1}_N - \mathbf{x}_\ell - \mathbf{a})}{\mathbf{1}_N^T \varphi_\sigma(\mathcal{G}_{(i)}(\mathbf{x}_\ell + \mathbf{a})\mathbf{1}_N - \mathbf{x}_\ell - \mathbf{a})}, \tag{5.257}$$

$$\frac{\partial y}{\partial \rho} = \gamma \left(1 - \frac{1}{N} \mathbf{1}_N^T \varphi_\sigma(\mathcal{G}_{(i)}(\mathbf{x}_\ell + \mathbf{a})\mathbf{1}_N - \mathbf{x}_\ell - \mathbf{a}) \right), \tag{5.258}$$

$$\frac{\partial y}{\partial \mathbf{b}} = (1 - \gamma) \mathbf{x}_\ell, \tag{5.259}$$

$$\frac{\partial y}{\partial \gamma} = \mathcal{G}_{(i)}(\mathbf{x}_\ell + \mathbf{a}) - \mathbf{b}^T \mathbf{x}_\ell, \tag{5.260}$$

where the partial derivative in Eq. (5.258) is a design choice (i.e., it has not been derived analytically). A detailed study of the convergence properties of the algorithm defined by Eq. (5.256) can be found in [66].

5.9 $\ell + L$-FILTERS

Another hybrid filter class that is closely related to $L\ell$- and MRL filters is the linear combination of an ℓ- and L-filter proposed in [92], that is,

$$\hat{s}(n) = \mathbf{a}^T \begin{bmatrix} \mathbf{x}_\ell(n) \\ \mathbf{x}_L(n) \end{bmatrix}, \tag{5.261}$$

where $\mathbf{a}$ is a $2N$-dimensional coefficient vector. Let us denote this structure $\ell + L$-filter. The optimal design of $\ell + L$-filter was demonstrated for the unbiased estimation of a constant signal s that has been corrupted by zero-mean noise $v(n)$ having a symmetric pdf about zero [92]. In the following, a brief description of the optimal design is reviewed.

The unbiased estimation of the mean value by the filter output of Eq. (5.261) should satisfy the following constraints:

$$\mathbf{a}^T \begin{bmatrix} \mathbf{0} \\ \mathrm{E}\{\mathbf{v}_L\} \end{bmatrix} = 0, \tag{5.262}$$

$$\mathbf{a}^T \mathbf{1}_{2N} = 1, \tag{5.263}$$

where $\mathrm{E}\{\mathbf{v}_L\}$ is the mean vector of the ordered noise samples. The conditions do not depend on s, hence s will not appear on the estimator. The $\ell + L$-filter that minimizes the MSE subject to Eqs. (5.262) and (5.263) satisfies

$$\mathbf{\Upsilon}_{2N} \, \mathbf{a} = -\frac{\psi_1}{2} \mathbf{1}_{2N} - \frac{\psi_2}{2} \begin{bmatrix} \mathbf{0} \\ \mathrm{E}\{\mathbf{v}_L\} \end{bmatrix}, \tag{5.264}$$

where ψ_1 and ψ_2 are Lagrange multipliers and

$$\Upsilon_{2N} = \begin{bmatrix} \mathrm{E}\{\mathbf{v}\mathbf{v}^T\} & \mathrm{E}\{\mathbf{v}\mathbf{v}_L^T\} \\ \mathrm{E}\{\mathbf{v}_L\mathbf{v}^T\} & \mathrm{E}\{\mathbf{v}_L\mathbf{v}_L^T\} \end{bmatrix}. \tag{5.265}$$

Υ is not invertible, because its rank is $2N - 1$. Moreover, the vector $(\mathbf{1}_N^T \mid -\mathbf{1}_N^T)^T$ spans its kernel. The vectors $\mathbf{1}_{2N}$ and $(\mathbf{0}_N^T \mid \mathrm{E}\{\mathbf{v}_L^T\})^T$ are orthogonal to $(\mathbf{1}_N^T \mid -\mathbf{1}_N^T)^T$; therefore they are in the range of Υ. Accordingly the solutions of Eq. (5.264) define a one-dimensional affine space. We seek solutions of the form $\mathbf{a} = (0 \mid \mathbf{a}'^T)^T$ that cancel one dimension in the estimator. If Υ is partitioned as

$$\Upsilon_{2N} = \begin{bmatrix} v_{11} & v^T \\ v & \Upsilon_{2N-1} \end{bmatrix}. \tag{5.266}$$

then the optimal $\mathbf{a}'^o$ is given by

$$\mathbf{a}'^o = \frac{\Upsilon_{2N-1}^{-1}\mathbf{1}_{2N-1}}{\mathbf{1}_{2N-1}^T\,\Upsilon_{2N-1}^{-1}\mathbf{1}_{2N-1}}. \tag{5.267}$$

Therefore

$$\widehat{s} = (\mathbf{a}'^o)^T \begin{bmatrix} \mathbf{x}_\ell' \\ \mathbf{x}_L \end{bmatrix}, \tag{5.268}$$

where $\mathbf{x}_\ell' = (x_2, x_3, \ldots, x_N)^T$. Any vector of the form

$$\begin{bmatrix} 0 \\ \mathbf{a}'^o \end{bmatrix} + k \begin{bmatrix} \mathbf{1}_N \\ -\mathbf{1}_N \end{bmatrix}, \quad k \in \mathbb{R} \tag{5.269}$$

is also a solution of Eq. (5.264).

5.10   AFFINE ORDER STATISTIC FILTERS

The weighted-order statistic (WOS) filters incorporate both temporal and rank order information by first weighting the input observations according to their spatial/temporal order and then by employing a selection operation based on rank order. Their output is confined to be one of the input samples. Accordingly their performance in near Gaussian environments is poor. WOS filters usually admit only positive weights, which implies lowpass filter characteristics. Several alternatives have been proposed to overcome the aforementioned drawbacks, such as the FIR-WOS hybrid filter [90], the $L\ell$-filters [64, 27], the permutation filters [5], the permutation lattices [40]. However, for $L\ell$- and permutation filters, the complexity increases with the window size, a fact that limits those filters to operate only with relatively small windows.

The class of *affine order statistic filters* [23] consists of two subclasses:

- The subclass of *weighted-order statistic (WOS) affine filters* whose reference point is a data-dependent function of the order statistics.

- The subclass of *FIR affine filters* whose reference point is a linear FIR filter.

A tunable nonlinear function, called the *affinity function*, is used in both cases to modify the values, and hence the influence of observations in the estimation process. The affinity function is defined as a nonlinear mapping of the observations onto $[0, 1]$:

$$\mathcal{D}^{\tilde{x},\gamma} : x_j \mapsto \mathcal{D}_j^{\tilde{x},\gamma} \in [0, 1], \quad j = 1, 2, \ldots, N, \tag{5.270}$$

where $\tilde{x} \in (-\infty, \infty)$ and $\gamma \in [0, \infty)$. The real-valued number $\mathcal{D}_j^{\tilde{x},\gamma}$ is a metric of the proximity of x_j to the reference point $\tilde{x}$ as measured by $\mathcal{D}^{\tilde{x},\gamma}$ [23]. If $\tilde{x}$ corresponds to one of the order statistics, $\mathcal{D}_j^{\tilde{x},\gamma}$ coincides with the fuzzy location-rank membership function described in Section 5.2, namely

$$\mathcal{D}_j^{\tilde{x},\gamma} = m_{\tilde{\Re}}(x_j, \tilde{x}). \tag{5.271}$$

The Gaussian membership function defined by Eq. (5.34) was adopted in [23].

5.10.1 WOS Affine FIR Filters

Let $\{c_1, c_2, \ldots, c_N\}$ be a set of nonnegative coefficients and let us define the reference point as

$$\tilde{x} = k\text{th largest element in } \{c_1 \diamond x_1, c_2 \diamond x_2, \ldots, c_N \diamond x_N\}. \tag{5.272}$$

The output of a WOS affine FIR filter is defined as [23]

$$\hat{s}_\gamma = \frac{\sum_{j=1}^N b_j \, \mathcal{D}_j^{\hat{x},\gamma} \, x_j}{\sum_{j=1}^N |b_j| \, \mathcal{D}_j^{\hat{x},\gamma}}, \tag{5.273}$$

where b_j are the filter coefficients. Therefore the WOS affine FIR filter weighs each observation according to its reliability as well as its natural order. Since a Gaussian affinity function has been adopted, the values $\mathcal{D}_j^{\tilde{x},\gamma}$ depend on the distance of the observation x_j from the reference point that is the WOS estimate given by Eq. (5.272). For $\gamma \to \infty$, the affinity function is constant on its entire domain, and the estimator defined by Eq. (5.273) weighs all observations according to their spatial/temporal order. On the contrary, for $\gamma \to 0$, the affinity function shrinks to a δ impulse function at $\tilde{x}$, and the estimate is equal to the WOS filter output $\tilde{x}$. A common choice for $\tilde{x}$ is the median of input observations as in the MTM filter [51].

5.10.2 FIR Affine L-Filters

The FIR affine L-filters can be considered a duals of WOS affine FIR filters. Let $\mathbf{c}$ be an N-dimensional coefficient vector. We select the reference point to be the output of an FIR filter $\widetilde{x} = \mathbf{c}^T \mathbf{x}_\ell$. Then the FIR affine L-filter output is defined by

$$\hat{s}_\gamma = \frac{\sum_{i=1}^N a_i \, \mathcal{D}_{(i)}^{\widetilde{x},\gamma} \, x_{(i)}}{\sum_{i-1}^N |a_i| \, \mathcal{D}_{(i)}^{\widetilde{x},\gamma}}, \tag{5.274}$$

where $\mathcal{D}_{(i)}^{\widetilde{x},\gamma}$ measures the affinity between the ith order statistic and the reference point. For $\gamma \to \infty$, the FIR affine L-filter reduces to an L-filter, and for $\gamma \to 0$, it reduces to an FIR filter with coefficient vector $\mathbf{c}$.

5.10.3 Adaptive Design

Let us confine ourselves to the case of median affine FIR filters, where $\widetilde{x} = x_{(\nu)}$. An adaptive design of the median affine FIR filter was developed in [23]. Under the MSE criterion, we attempt to find the median affine FIR filter defined by the parameter vector $(\gamma \mid \mathbf{b}^T)^T$ that minimizes

$$\varepsilon(\gamma, \mathbf{b}) = \mathrm{E}\left\{(s(n) - \hat{s}(n))^2\right\}. \tag{5.275}$$

Let us first consider the optimization with respect to γ for a fixed coefficient vector $\mathbf{b}$. It can be shown that

$$\frac{\partial \varepsilon}{\partial \gamma} = -2\mathrm{E}\left\{\frac{e(n)}{\gamma^2} \frac{\sum_{j=1}^N |b_j| \mathcal{D}_j^{x_{(\nu)},\gamma}(\mathrm{sgn}(b_j)\, x_j - \hat{s})(x_j - x_{(\nu)})^2}{\sum_{j=1}^N |b_j| \, \mathcal{D}_j^{x_{(\nu)},\gamma}}\right\}. \tag{5.276}$$

Accordingly the steepest descent algorithm for γ is

$$\gamma(n+1) = \gamma(n) - \frac{\mu_\gamma}{2} \frac{\partial \varepsilon}{\partial \gamma} \Big|_{\gamma = \gamma(n)}. \tag{5.277}$$

By dropping out the expectation operator in Eq. (5.276) and using the resulting instantaneous estimate in Eq. (5.277), an LMS algorithm for updating γ is obtained. The optimization with respect to the filter weights b_j yields the following LMS descent updating equation [23]:

$$\begin{aligned} b_j(n+1) &= b_j(n) + \mu_b \, e(n)\mathcal{D}_j^{x_{(\nu)}(n),\gamma(n)} \sum_{k=1}^N b_k(n)\, \mathcal{D}_k^{x_{(\nu)}(n),\gamma(n)} \\ &\quad \cdot (\mathrm{sgn}\,(b_k(n))x_j(n) - \tanh(b_j(n))\, x_k(n)). \end{aligned} \tag{5.278}$$

5.11   WEIGHTED MYRIAD FILTERS

Median filters and their generalizations based on order statistics reviewed in this chapter are derived to be optimal for the Laplacian noise distribution [4]. Another distribution that accurately models the impulsive noise process is the α-*stable distribution* [58]. The family of α-stable distributions has a parameter α ($0 < \alpha \leq 2$), called the *characteristic exponent*, that controls the heaviness of their tails. The smaller α is, the heavier tailed distribution is obtained. For $\alpha = 2$, the Gaussian distribution results, while for $\alpha = 1$, the *Cauchy distribution* is obtained.

Weighted myriad filters have been proposed as a class of robust nonlinear filters based on α-stable distributions [28]. Consider N independent i.i.d. observations that form the vector $\mathbf{x}_\ell = (x_1, x_2, \ldots, x_N)^T$ drawn from a Cauchy distribution with location parameter β and scaling factor $K > 0$:

$$f(x; \beta) = \left(\frac{K}{\pi}\right) \frac{1}{K^2 + (x - \beta)^2}. \tag{5.279}$$

The sample myriad filter, defined by [28, 36]

$$\widehat{\beta}_K = \text{myriad}(K; x_1, x_2, \ldots, x_N) = \arg\min_\beta \sum_{j=1}^{N} \log[K^2 + (x_j - \beta)^2], \tag{5.280}$$

is the maximum likelihood estimate of the location parameter β for the Cauchy distribution. Clearly, Eq. (5.280) defines an M-estimator [70]. The sample myriad includes the sample mean as a limit for $K \to \infty$.

For a given $K > 0$ and an N-dimensional coefficient vector $\mathbf{a}$, the weighted myriad filter is defined by [36]

$$\widehat{\beta}_K(\mathbf{a}, \mathbf{x}_\ell) = \arg\min_\beta \mathcal{H}_K(\beta, \mathbf{a}, \mathbf{x}_\ell), \tag{5.281}$$

where

$$\mathcal{H}_K(\beta, \mathbf{a}, \mathbf{x}_\ell) \stackrel{\text{def}}{=} \prod_{j=1}^{N} [K^2 + a_j (x_j - \beta)^2]. \tag{5.282}$$

The formulation of the weighted myriad as a maximum likelihood estimate constrains the weights a_j to be nonnegative. For nonnegative weights and $K > 0$, $\mathcal{H}_k(\beta)$ is positive for all β and goes to ∞ as $\beta \to \pm\infty$. $\widehat{\beta}_K$ occurs at one of the local minima of $\mathcal{H}(\beta)$. It can be shown that $\widehat{\beta}_K$ satisfies [36]

$$\sum_{j=1}^{N} \frac{a_j(\widehat{\beta}_K - x_j)}{K^2 + a_j(\widehat{\beta}_K - x_j)^2} = 0. \tag{5.283}$$

It can also be shown that $\mathcal{H}_K(\beta)$ has L local minima and $(L-1)$ local maxima, where $1 \leq L \leq N$. For nonnegative weights it can be proved that all the local extrema occur within the interval $[x_{(1)}, x_{(N)}]$. Therefore $x_{(1)} \leq \widehat{\beta}_K \leq x_{(N)}$. To compute the weighted myriad, one has to find the roots of the first derivative of $\mathcal{H}_K(\beta)$ with respect to β, to choose the ones that are local minima of $\mathcal{H}_K(\beta)$, and to test all the local minima to find the global minimum [36]. Accordingly a direct computation of the weighted myriad is a nontrivial and prohibitively expensive task. Kalluri and Arce [37] proved that the mapping

$$T(\beta) = \frac{\sum_{j=1}^{N} h_j(\beta) x_j}{\sum_{j=1}^{N} h_i(\beta)}, \tag{5.284}$$

where

$$h_j(\beta) = \frac{1}{S_j} \varphi(\frac{x_j - \beta}{S_j}), \quad S_j = \frac{K}{\sqrt{a_j}}; \quad j = 1, 2, \ldots, N, \tag{5.285}$$

and

$$\varphi(v) = \frac{1}{v} \frac{d\rho(v)}{dv} \quad \text{with} \quad \rho(v) = \log(1 + v^2), \tag{5.286}$$

has the roots of the first derivative of $\mathcal{H}_K(\beta)$ with respect to β as fixed points and shown that the weighted myriad is one of them. They also proposed that the fixed points of $T(\beta)$ can be found by the following recursion:

$$\beta_{m+1} = \beta_m + T(\beta_m). \tag{5.287}$$

As K gets larger the number of local minima of $\mathcal{H}(\beta)$ decreases. Letting $K \to \infty$ in Eq. (5.283) and keeping the weights finite, we obtain

$$\widehat{\beta}_\infty = \frac{\mathbf{a}^T \mathbf{x}_\ell}{\mathbf{1}_N^T \mathbf{a}}. \tag{5.288}$$

For $K = 0$, the objective function $\mathcal{H}_0(\beta)$ is zero, whenever β is one of the input samples. In this case there are N local minima, one at each input sample. Any of the input samples could be the output. The so-called weighted mode-myriad filter is defined by [28]

$$\widehat{\beta}_0 = \lim_{K \to 0} \widehat{\beta}_K(\mathbf{a}, \mathbf{x}_\ell). \tag{5.289}$$

The output of the weighted mode-myriad filter is the most repeated input sample, if unique. When the most repeated sample is not unique, the filter output is [36]

$$\widehat{\beta}_0 = \arg\min_{x_j \in \mathcal{X}} \prod_{\substack{l=1 \\ l \neq j}}^{N} a_j (x_l - x_j)^2, \tag{5.290}$$

where $\mathcal{X}$ is the set of the most repeated values among the input samples.

Given an input vector $\mathbf{x}_\ell$, a weight vector $\mathbf{a}$, and the parameter K, let us denote the output of the weighted myriad filter as $y_K(\mathbf{a}, \mathbf{x}_\ell)$. Assuming that a desired signal

$s(n)$ is also available, we aim at determining the optimal weight vector $\mathbf{a}$ under the MAE criterion

$$\varepsilon'(\mathbf{a}, K) = \mathrm{E}\left\{|y_K(\mathbf{a}, \mathbf{x}_\ell) - s(n)|\right\} \tag{5.291}$$

subject to the constraints

$$a_j \geq 0, \quad j = 1, 2, \ldots, N. \tag{5.292}$$

The objective function defined by Eq. (5.291) is nonconvex. Accordingly it may have multiple local minima. Using the signed error LMS algorithm, we can obtain the following algorithm to update the filter weights [36]:

$$a_j(n+1) =$$

$$\Pi\left[a_j(n) + \mu\,\mathrm{sgn}(e(n))\frac{\dfrac{y(n) - x_j(n)}{\left[1 + \dfrac{a_j(n)}{K^2}(y(n) - x_j(n))^2\right]^2}}{K^2\left\{\eta + \displaystyle\sum_{l=1}^{N}\dfrac{a_l(n)}{K^2}\dfrac{1 - \dfrac{a_l(n)}{K^2}(y(n) - x_l(n))^2}{\left[1 + \dfrac{a_l(n)}{K^2}(y(n) - x_l(n))^2\right]^2}\right\}}\right], \tag{5.293}$$

where $\Pi[\alpha]$ is the projection operator defined by Eq. (5.246) and η is a parameter introduced to avoid a zero denominator.

5.12 SIGNAL-ADAPTIVE ORDER STATISTIC FILTERS

Both adaptive L- and $L\ell$-filters depend on the availability of a noise-free image to be used as a reference image. It has been demonstrated that in principle the location-invariant LMS L-filter can be modified so that it minimizes the overall output filter, thus alleviating the need for having a reference image. When a reference image is not available, another possibility is to use a *signal-adaptive* filter. The smoothing properties of signal-adaptive filters change at each image pixel according to a local SNR measure that is used to adapt the filter window size as well. In this section signal-dependent adaptive L-filter structures and the signal-adaptive median (SAM) are reviewed.

5.12.1 Signal-Dependent Adaptive L-filters

A signal-dependent adaptive filter structure is developed aiming at a different treatment of the image pixels close to the edges from those that belong to homogeneous regions. The signal-dependent adaptive L-filter structure adjusts its smoothing prop-

erties at each point according to the local image content in order to achieve edge
preservation as well as maximum noise suppression in homogeneous regions. It consists of two LMS adaptive L-filters whose outputs $y_L(k)$ and $y_H(k)$ are combined to
give the final response as follows:

$$y(n) = y_L(n) + \beta(n)\{y_H(n) - y_L(n)\} = \beta(n)y_H(n) + [1 - \beta(n)]\,y_L(n), \quad (5.294)$$

where $\beta(n)$ is a signal-dependent weighting factor. $\beta(n)$ can be chosen so as it minimizes the mean-squared error between the filter output $y(n)$ given by Eq. (5.294) and
the desired response $s(n)$ [10]. Let $\mathbf{a}_H(n)$ and $\mathbf{a}_L(n)$ denote the L-filter coefficient
vectors driven by the high frequency and the low-frequency data, respectively. It can
easily be shown that $\mathrm{E}\left\{(s(n) - y(n))^2\right\}$ is minimized for

$$\beta(n) = \frac{(\mathbf{R}(n)\mathbf{a}_L(n) - \mathbf{p}(n))^T (\mathbf{a}_H(n) - \mathbf{a}_L(n))}{(\mathbf{a}_H(n) - \mathbf{a}_L(n))^T \mathbf{R}(n) (\mathbf{a}_H(n) - \mathbf{a}_L(n))}. \quad (5.295)$$

Equation (5.295) implies that both $\mathbf{a}_H(n)$ and $\mathbf{a}_L(n)$ are of the same dimensions. It
can be interpreted as follows. In homogeneous regions, we should have

$$\mathbf{R}(n)\mathbf{a}_L(n) = \mathbf{p}(n), \quad (5.296)$$

that is, $\mathbf{a}_L$ is the optimal L-filter that minimizes the MSE between the filter output
and the desired response. Therefore $\beta(n)$ equals 0. Similarly close to the edges we
should have

$$\mathbf{R}(n)\mathbf{a}_H(n) = \mathbf{p}(n). \quad (5.297)$$

On substituting Eq. (5.297) into Eq. (5.295), we see that $\beta(n)$ equals 1. In order to
avoid the estimation of the correlation matrix $\mathbf{R}(n)$ and the crosscorrelation vector
$\mathbf{p}(n)$, we replace the optimal signal-dependent weighting factor of Eq. (5.295) by
another one that shares the same favorable properties, namely by the *local signal-to-
noise ratio* measure

$$\beta(n) = 1 - \frac{\sigma_v^2}{\widehat{\sigma}_x^2(n)} \quad (5.298)$$

where σ_v^2 is the noise variance and $\widehat{\sigma}_x^2(n)$ is the variance of the noisy input observations. The adaptive L-filters $\mathbf{a}_L(n)$ and $\mathbf{a}_H(n)$ may use different window sizes. In
such a case the coefficient $\beta(n)$ given by Eq. (5.298) can be used as a signal-dependent
switch between the two LMS adaptive L-filters, that is,

$$y(n) = \begin{cases} y_H(n) & \text{if } \beta(n) > \beta_\tau \\ y_L(n) & \text{otherwise,} \end{cases} \quad (5.299)$$

where $0 < \beta_\tau < 1$ is a threshold that determines a trade-off between noise suppression and edge preservation. In the reported experiments, β_τ is chosen to be either 0.6
or 0.75. As far as the threshold β_τ is concerned, a same parameter is also found in
signal-adaptive median filters [10], and a selection of its proper value does not pose
any difficulty.

Table 5.10   Noise reduction and mean absolute error reduction (in dB) achieved by the signal-dependent normalized LMS L-filter structures in the restoration of "Lenna" corrupted by mixed impulsive and additive Gaussian noise

Noise Type	Filter parameters	NR	MAER
Mixed impulsive and additive Gaussian	L, H: 3×3; β_τ =0.75	-9.024	-9.552
	L: 5×5, H: 3×3; β_τ=0.75	-13.224	-13.928

Subsequently, an application is described where the signal-dependent filter is employed. The coefficients $\mathbf{a}_L$ and $\mathbf{a}_H$ that were derived during the operation of the algorithm throughout the training image are averaged and applied subsequently to filter the pixel blocks that fall into the homogeneous regions or close to the image edges. Two structures are considered. The first one uses two 3×3 adaptive L-filters that were trained by different regions of the corrupted input image. The second structure employs a larger window in homogeneous regions than that close to the edges. In both structures the pixels that belong to the homogeneous image regions are used to adapt the coefficients of the $\mathbf{a}_L$ adaptive L-filter, while those that are close to the image edges are used to adapt the coefficients of the $\mathbf{a}_H$ adaptive L-filter. Any of the adaptive L-filters that were described in Section 5.3 (e.g., the location-invariant LMS L-filter, the normalized LMS L-filter, or the modified LMS L-filter with nonhomogeneous step-sizes) can be included in the signal-dependent structure. In the experiment reported in this section, we used the normalized LMS L-filter. Table 5.10 summarizes the noise reduction and the mean absolute error reduction achieved by the signal-dependent adaptive normalized LMS L-filter structures in the restoration of "Lenna" depicted in Fig. 5.3(a) which was corrupted by mixed impulsive and additive Gaussian noise. From inspection of Table 5.10 it is seen that the signal-dependent adaptive L-filter structure provides better results when the window of the normalized LMS L-filter used in the homogeneous image regions has larger dimensions (e.g., 5×5) than the one trained by using pixels close to the image edges. A comparison of the entries in Tables 5.10 and 5.1 shows that the signal-dependent adaptive L-filter with different window sizes in homogeneous and edge regions is the best filter for both noise types. The superior performance is attributed to the larger window size in homogeneous regions. In this case the local SNR measure given by Eq. (5.298) is computed twice by calculating the local variance of the noisy input observations for both window sizes. An image pixel is declared as an edge pixel if either $\beta_{3 \times 3}(n)$ or $\beta_{5 \times 5}(n)$ exceeds the threshold. For completeness, it should be noted that when the image corruption is not too severe, the impulse detection and removal mechanism described in [10] can be included in the filter structure. This way pixels that pass the test, defined by Eq. (5.299), will not be misread as edges when they are actually impulses. The output of the signal-dependent normalized LMS L-filter structure that employs different window sizes in homogeneous regions and close to the edges, is depicted in Fig. 5.18.

Fig. 5.18 Output of the signal-dependent adaptive L-filter structure that employs two normalized LMS L-filters of dimensions 5 ×5 and 3 × 3 to smooth the image data in homogeneous regions and close to the edges, respectively; (reprinted by copyright permission from IEEE).

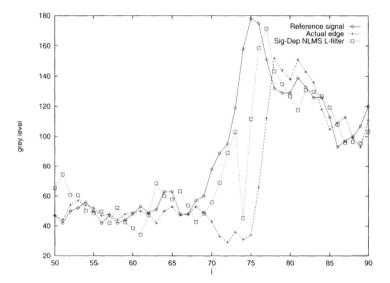

Fig. 5.19 Performance of the signal-dependent LMS L-filter structure in tracking a time-varying edge (reprinted by copyright permission from IEEE).

Next we investigate the ability of the proposed adaptive L-filters to handle nonstationarity in image data. We study the ability of the signal-dependent adaptive L-filter in tracking a time-varying edge as shown in Fig. 5.19, where we focus on a part of the 128th row in the 2nd and 9th frames of "Trevor White." Variant motion has caused different edges to appear at the two time instants. Although the signal-dependent normalized LMS L-filter has been trained using as reference an edge different from the one that appears in the noisy input image, it tends to track the latter one thus demonstrating its ability to track slow time-variations in the input signal characteristics.

5.12.2   Morphological Signal-Adaptive Median Filter

A novel extension of the SAM filter is proposed in [83], called *morphological signal-adaptive median* (MSAM) filter. Two modifications are introduced in the SAM filter [10] in order to alleviate its disadvantages/limitations: (1) the impulse detection mechanism of SAM is enhanced so that it detects not only impulses of a constant amplitude but randomly valued impulses as well; (2) unlike SAM which employs isotropic filter windows of dimensions 3×3 to 11×11, MSAM implements an anisotropic window adaptation based on binary morphological erosions/dilations with predefined structuring elements. The MSAM filter performs well against the contaminated Gaussian or impulsive noise corruption. It does not require the a priori knowledge of a noise-free image, but only of certain noise characteristics that can easily be estimated from the noisy image. It adapts its behavior based on a local SNR measure, thus achieving edge preservation and noise smoothing in homogeneous image regions.

We let σ_v^2 denote the variance of the image noise that is known or has been estimated from the noisy image beforehand. In the case of impulsive noise, we let q_{v+} and q_{v-} denote the probability of occurrence of positive impulses (i.e., having intensity $s_{\max} = 255$), and negative impulses (i.e., having intensity $s_{\min} = 0$), respectively. The noisy image pixel values $x(n)$ are then determined by the model

$$x(n) = \begin{cases} s_{\min} & \text{with probability } q_{v-} \\ s_{\max} & \text{with probability } q_{v+} \\ x(n) & \text{with probability } 1 - (q_{v-} + q_{v+}), \end{cases} \quad (5.300)$$

where $x(n)$ is an image pixel corrupted possibly by additive white or signal-dependent noise. The output of the MSAM filter is expressed as in the standard SAM filter [10], namely

$$y(n) = \widetilde{x}_{(\nu)}(n) + \beta(n)[x(n) - \widetilde{x}_{(\nu)}(n)]. \quad (5.301)$$

Here $\widetilde{x}_{(\nu)}(n)$ is the modified median, that is, the median of the pixels that remain after the removal of impulses from the local window. In Eq. (5.301), $\beta(n)$ is a weighting coefficient that is used to adapt the window size according to whether a flat region or an edge has been met. In homogeneous regions total noise suppression is achieved, because a large window is employed with $\beta(n)$ being close to 0. Edges are also well preserved because in this case a small window size is used with $\beta(n)$ is close to 1. The window size increase/decrease procedure is explained below. Two major modifications in the standard SAM filter [10] are introduced by the MSAM filter:

1. MSAM employs the morphological operations of dilation and erosion with certain predefined structuring elements (SEs), in order to vary anisotropically the window size with respect to the local image content. This modification has been motivated by the work reported in [82].

2. MSAM employs two impulse detectors: one for constant impulses (either positive or negative) and another one for randomly valued impulses. Impulse detection is done only in the initial window of dimensions 3×3.

In the following, the steps of the algorithm are presented.

Step 1: Constant value impulse detection. The method proposed in [10] is employed. The filter performs detection of constant value impulses in an initial window of dimensions 3×3 by using a signal-dependent threshold $\tau_{v-}(n)$ for negative impulses given by

$$\tau_{v-}(n) = c\left[s_{\min} - \tilde{x}_{(\nu)}(n)\right] \leq 0 \tag{5.302}$$

and another one for positive impulses defined by

$$\tau_{v+}(n) = c\left[s_{\max} - \tilde{x}_{(\nu)}(n)\right] \geq 0, \tag{5.303}$$

where $c = 5/6$. If $\left[x(n) - \tilde{x}_{(\nu)}(n)\right] < \tau_{v-}$, then $x(n)$ is declared a negative impulse. On the contrary, if $\left[x(n) - \tilde{x}_{(\nu)}(n)\right] > \tau_{v+}$, then $x(n)$ is detected as a positive impulse.

Step 2: Randomly valued impulse detection. Motivated by the randomly valued impulse detection mechanisms developed in [53, 80], two additional thresholds are incorporated in the standard SAM filter. They are defined as

$$\theta_1(n) \quad = \quad x_{\max} - x_{\min_2}, \tag{5.304}$$
$$\theta_2(n) \quad = \quad x_{\max_2} - x_{\min}, \tag{5.305}$$

where $x_{\min}$ is the minimum pixel value, $x_{\min_2}$ is the second minimum pixel value, $x_{\max}$ is the maximum pixel value, and $x_{\max_2}$ is the second maximum pixel value in the initial window. If the absolute difference $|x(n) - \tilde{x}_{(\nu)}(n)|$ is greater than any of the thresholds θ_1 or θ_2, then $x(n)$ is a very small or very large value in terms of its neighboring pixels and possibly is a randomly valued impulse. Random-value impulse detection is performed only if the current pixel has not previously been detected as a constant impulse and only when $\beta_{3\times 3}(n) > \beta_\tau$. This implies that either an edge region or a possible randomly valued impulse in the initial 3×3 window has been detected in the subsequent steps 3 and 4. If the current pixel is an impulse, either a constant or a randomly valued one, it is excluded from the estimation of the median at the current and at any future window centered at n. That is, it is not considered in the estimation of the modified median employed in Eq. (5.301).

Step 3: Calculation of the weighting coefficient $\beta(n)$. The coefficient $\beta(n)$ is given by the expression [10]

$$\beta(n) = \begin{cases} 0 & \text{if } \gamma_1 \sigma_v^2 \geq \hat{\sigma}_x^2 \\ \left[1 - \gamma_1 \dfrac{\sigma_v^2}{\hat{\sigma}_x^2}\right]^{\gamma_2} & \text{otherwise.} \end{cases} \tag{5.306}$$

In Eq. (5.306), $\hat{\sigma}_x^2$ denotes the image variance that is estimated from the local "windowed" histogram [10] from which the current pixel is excluded if it is detected as an impulse. Parameters γ_1 and γ_2 are appropriately chosen in the range $[0, 1]$. The parameter γ_1 controls the threshold on the local signal to noise ratio up to which the high-frequency components are entirely suppressed. The parameter γ_2 controls the noise suppression close to edges.

Step 4: Decision whether the current pixel belongs to a homogeneous region or to an edge. The weighting factor $\beta(n)$ calculated in step 3 is compared to a predefined threshold β_τ. If it is smaller than β_τ, then the current pixel is assumed to belong to a homogeneous region. Otherwise, the current pixel belongs to an edge. The threshold β_τ lies in the interval $[0, 1]$. Its selection is accomplished in accordance with the degree of corruption and the nature of the noise. For highly corrupted images, its value is lower than 0.5. If the image is corrupted by pure Gaussian noise of relatively medium variance, the threshold lies in the range $[0.65, 0.85]$. A reliable method for the choice of the threshold β_τ is described in [10]. Methods employing local statistics are reported in [81].

Step 5: Novel window adaptation procedure. The proposed MSAM further differs from the SAM filter in the window adaptation procedure used. SAM employs isotropic filter windows of dimensions 3×3 up to 11×11. In contrast to SAM, an anisotropic window adaptation procedure is proposed based on the mathematical morphology operations of binary erosion/dilation with predefined structuring elements (SEs). Four structuring elements are employed: B_1, B_2, B_3, and B_4 and their symmetric counterparts, B_1^s, B_2^s, B_3^s, and B_4^s. These structuring sets are illustrated in Fig. 5.20(a). They are divided into even-angle SEs (B_1, B_2, B_1^s, B_2^s) and odd-angle SEs (B_3, B_4, B_3^s, B_4^s). The window size increase is performed by a dilation operation, $W \oplus B_i$, where W denotes the current filter window. The "direction" of increase depends on the choice of B_i. The result of the window growing $W \oplus B_i$ for an original 3×3 window size W is shown in Fig. 5.20(a). The thin dots belong to the original window W while the bold dots denote the new pixels that have been appended to W to form a new (larger) window. In a similar fashion the window size decrease is performed by an erosion operation $W \ominus B_i$. This is demonstrated in Fig. 5.20(b) for B_1, B_1^s, B_2 and B_2^s. Note that only the SEs indicated in 5.20(b) are used for reducing the window size. The procedure of the window adaptation begins with a 3×3 square window and checks whether the central pixel belongs to an edge. The algorithm is as follows:

1. If it does not belong to an edge,

    (a) An attempt is made to increase the window size by using the odd-angle SEs.

        (i) If an edge is hit (i.e., $\beta(n) > \beta_\tau$), the current odd-angle SE is excluded and the even-angle SEs that compose the SE under consideration are tested for possible window increase. For example, if

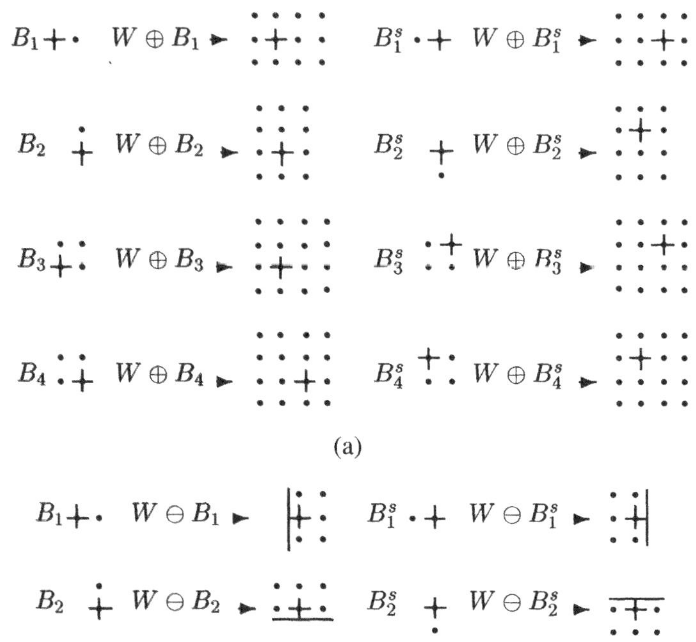

(a)

(b)

Fig. 5.20 (a) Window size increase through dilation; (b) window size decrease through erosion.

> B_4 is excluded, B_1 and B_2 are tested for possible use in window increase. This means that the sides of the mask are also separately checked expecting that an edge is possibly met at one side only, thus allowing window increase at the other sides. In doing so, maximal window increase is achieved.
>
> (ii) If an edge is not met, then the current odd-angle SE is used to increase the window size. The corresponding even-angle SEs are then excluded.

(b) In the next step the odd-angle SEs that were not excluded in the previous step, are tested again. In our example, B_3, B_3^s, and B_4^s remain to be tested. If it is known from a previous step that a window side hits an edge, this side is not considered again.

(c) The procedure continues until all the odd-angle and all the even-angle SEs are excluded or until at least one side reaches a predefined maximal size (e.g., 11 pixels).

2. If the pixel belongs to an edge, the goal is to expand the mask in the neighboring regions that are homogeneous. That is, the current pixel is labeled a border

pixel, and the window increase is done toward the side of the edge where the pixel belongs to. To do so, the opposite side of the edge must be found and the increase of the filter window toward that direction must be prohibited. This is done as follows: The average pixel value on each of the four sides of a window of dimensions 3×3 is derived and the absolute difference between the average side pixel values and the current pixel is calculated. The side that corresponds to the greater difference is removed. The difference is a measure of deviation of the side pixels from the current one. The side that deviates the most is possibly the side that should be removed. The decrease of the initial window size is achieved by an erosion with one of the SEs B_1, B_1^s, B_2, and B_2^s. Subsequently the window increases toward the remaining sides in the way described above by using appropriate SEs. For example, if an erosion with B_1 is performed, B_1, B_2, B_2^s, B_3, and B_4^s are used to increase further the window.

Finally, if the current pixel is detected as an impulse, the factor $\beta(n)$ is set to 0. After the "optimal" window size has been determined with respect to the local image content, the filter output is estimated by Eq. (5.301). The modified median $\tilde{x}_{(\nu)}(n)$ in Eq. (5.301) is calculated only twice: at the initial window for impulse detection and at the final window for the filter output estimation. In contrast the classical SAM filter requires median estimation at every step of the isotropic window adaptation procedure in case that impulsive noise is present. A detailed analysis of the computational complexity of the MSAM against that of SAM filter can be found in [83].

The noise-free images "Airfield" and "Bridge" have been corrupted by adding white i.i.d. noise whose pdf is a Gaussian mixture given by

$$v \sim (1 - \epsilon)\mathcal{N}(0, \sigma) + \epsilon \mathcal{N}\left(0, \frac{\sigma}{\epsilon}\right), \tag{5.307}$$

with mean value $E\{v\}$ being close to 0 and variance being $\sigma_v^2 = \sigma^2(1 - \epsilon + 1/\epsilon)$. The contamination factor ϵ, along with the initial standard deviation σ, determines the degree of corruption. The result of this noise corruption process is a mixture of Gaussian and impulsive noise (contaminated Gaussian noise) of varying characteristics according to the value of both ϵ and σ. Thus the proposed filter performance can be evaluated on a wide range of different combinations of noise distributions and corruption levels. For more detail, the reader is referred to [25]. The case of pure impulsive, Laplacian and uniformly distributed noise is additionally investigated in [83].

Four objective criteria have been evaluated for each pair of noisy and filtered images: the SNR, the PSNR, the MAE, and the MSE, defined by

$$\widehat{\text{SNR}} = 10 \log_{10} \frac{\widehat{\sigma}_s^2}{\widehat{\sigma}_e^2},$$

$$\widehat{\text{PSNR}} = 10 \log_{10} \frac{255^2}{\widehat{\sigma}_e^2},$$

(a)   (b)

Fig. 5.21   Noise-free images. (a) Airfield and (b) Bridge.

$$\widehat{\text{MAE}} = \frac{1}{N_1 N_2} \sum_{n_1=1}^{N_1} \sum_{n_2=1}^{N_2} |e(n_1, n_2)|,$$

$$\widehat{\text{MSE}} = \frac{1}{N_1 N_2} \sum_{n_1=1}^{N_1} \sum_{n_2=1}^{N_2} e^2(n_1, n_2), \qquad (5.308)$$

where $e(n_1, n_2) = y(n_1, n_2) - s(n_1, n_2)$ is the output noise, $s(n_1, n_2)$ is the noise-free image, and $y(n_1, n_2)$ denotes the filtered image. In Eq. (5.308), $\widehat{\sigma}_s^2$ is the variance of the noise-free image and $\widehat{\sigma}_e^2$ is the variance of the output noise.

Two different noise-free images are chosen, namely the images "Airfield" and "Bridge." Initially both types were corrupted by adding white i.i.d. noise whose pdf is a Gaussian mixture given by Eq. (5.307). We will compare the performances of the LMS L- and $L\ell$-filters to those of the MSAM and SAM filters. For the adaptive LMS order statistic filters, a preprocessing phase (i.e., a training phase) is employed to derive the filter coefficients at convergence. Toward this end, a corrupted version of image "Bridge" is employed in the training phase. The value $\mu = 10^{-7}$ is used in the filter coefficient updating equations during the training. A corrupted version of image "Airfield" by the same kind of noise is used as a test image. The coefficients determined at the end of the training phase are applied to the test image for noise smoothing. A window size of 3×3 is used in both training and testing phases. Images "Airfield" and "Bridge", both of dimensions 512×512, are shown in Fig. 5.21(a) and Fig. 5.21(b). We consider the case where the test image is corrupted by noise having a contaminated Gaussian distribution given by Eq. (5.307) with values $\epsilon = 0.1$ and $\sigma^2 \simeq 586.5$. This leads to a high corruption of the original image by mixed impulsive and Gaussian noise. The corrupted test image is called "Airm0s3l01" and is shown in Fig. 5.22(a). The training image obtained by corrupting the original image "Bridge" with the same noise is depicted in Fig. 5.22(b). Fig. 5.22(c) through (e) shows the outputs of the adaptive location invariant L-, adaptive $L\ell$-, and morphological SAM filters. For comparison purposes, the output of the SAM filter of dimensions 3×3 is shown in Fig. 5.22(f).

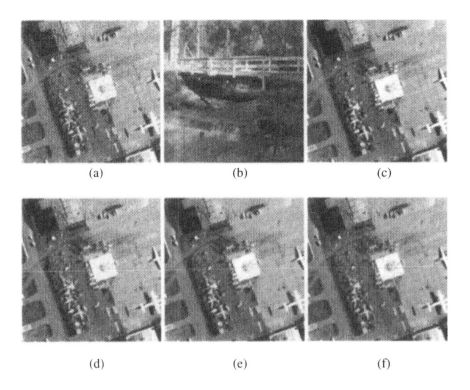

Fig. 5.22   (a) Noisy test image Airm0s3l01. (b) Noisy training image Brim0s3l01. Filtered Airm0s3l01 by (c) the adaptive location invariant L-filter; (d) the adaptive $L\ell$-filter; (e) the morphological SAM filter; and (f) the SAM filter.

The quantitative figures of merit for the various filters employed in the study are tabulated in Table 5.11 after the noisy versions of images "Airfield" and "Bridge" are filtered. The respective figures of merit for the SAM and the median filter of dimensions 3×3 are included in Table 5.11 for comparison. Additional results for pure Gaussian and pure impulsive noise can be found in [83]. In the following, we outline the major conclusions derived from the experimental results reported in [83].

1. For low SNR and very impulsive images, the morphological SAM filter attains the best performance. The $L\ell$-filter is the second best.

2. For images corrupted by Gaussian noise, the adaptive $L\ell$ filter attains the best performance. The MSAM filter is found to be the second best filter.

3. For pure impulsive noise, MSAM outperforms the other filters in terms of SNR, PSNR and MSE values whereas SAM attains the smallest MAE.

4. In all cases, the location invariant L-filter is not far behind. However, this filter has the advantages of a simple structure, limited memory requirements, and

Table 5.11   **Figures of merit when a corrupted version of the image "Airfield" is filtered. The best results are shown in bold.**

Filter	$\widehat{SNR}$	$\widehat{PSNR}$	$\widehat{MAE}$	$\widehat{MSE}$
Initial	3.034	15.608	26.467	1788.423
MSAM	**11.721**	**24.295**	**11.005**	**242.177**
$L\ell$	11.650	24.224	12.159	254.632
L	11.553	24.127	11.840	254.127
SAM	9.272	21.846	13.157	425.840
Median	11.188	23.762	12.273	273.558

small computational complexity compared to the $L\ell$-filter as well as the SAM and MSAM filters.

It is worth noting that all filters under study outperform the median filter. The inspection of Fig. 5.22(a) through Fig. 5.22(f) reveals that the MSAM-filtered image is perceived to have better visual quality than the other filtered images. Edges are preserved and not blurred, as is evident in the other filters under study. Mixed impulsive and Gaussian noise is smoothed very well, a fact that is not true for the other filters.

5.13   CONCLUSIONS

In this chapter we describe adaptive order statistic filters that stem from three fundamental models, namely the L-filter, the $L\ell$-filter, and the ranked-order filters. Two design methodologies are demonstrated. The first methodology is based on the LMS algorithm, and finds the filter coefficients that minimize an objective function through an iterative procedure. Commonly used objective functions are either the MSE or the MAE. The minimization of the objective function can be either unconstrained or subject to constraints. Both optimization problems were treated in the chapter. The second methodology employs a local signal-to-noise ratio measure for two purposes: first, to treat differentially the image pixels that belong to homogeneous regions and those close to image edges and second, to vary the filter window size, thus offering maximal noise suppression in homogeneous regions. The resulting filters are commonly known as signal-adaptive filters. Adaptive LMS order statistic filters rely on the availability of a reference signal to be used in the optimization procedure. Only the location-invariant filter structure can deliver filtered images that do not depend heavily on the reference images used in the training procedure. Signal-adaptive filters do not need a reference image in their design as well and constitute another alternative when a reference image is not available.

REFERENCES

1. G. Arce and M. Tian. Order statistic filter banks. *IEEE Trans. Image Processing*, 5(6): 827-837, June 1996.

2. G.R. Arce, T.A. Hall, and K.E. Barner. Permutation weighted order statistic filter lattices. *IEEE Trans. Image Processing*, 4(8): 1070-1083, August 1995.

3. J. Astola, P. Haavisto, and Y. Neuvo. Vector median filters. *Proc. IEEE*, 78(4): 678-689, April 1990.

4. J. Astola and Y. Neuvo. Optimal median type filters for exponential noise distributions. *Signal Processing*, 17: 95-104, 1989.

5. K.E. Barner and G.R. Arce. Permutation filters: A class of nonlinear filters based on set permutations. *IEEE Trans. Signal Processing*, 42(4): 782-798, April 1994.

6. K.E. Barner and G.R. Arce. Design of permutation order statistic filters through group colorings. *IEEE Trans. Circuits and Systems–Part II: Analog and Digital Signal Processing*, 44(7): 531-548, July 1997.

7. K.E. Barner, A. Flaig, and G.R. Arce. Fuzzy time-rank relations and order statistics. *IEEE Signal Processing Letters*, 5(10): 252-255, October 1998.

8. V. Barnett. The ordering of multivariate data. *J. Royal Statistical Society A*, 139, Part 3: 318-354, 1976.

9. M. Bellanger. *Adaptive Digital Filters and Signal Analysis*. Dekker, New York, 1987.

10. R. Bernstein. Adaptive nonlinear filters for simultaneous removal of different kinds of noise in images. *IEEE Trans. Circuits and Systems*, 34(11): 1275-1291, November 1987.

11. N.J. Bershad. On the optimum gain parameter in LMS adaptation. *IEEE Trans. Acoustics, Speech, and Signal Processing*, 35(7): 1065-1068, July 1987.

12. A.C. Bovik, T.S. Huang, and D.C. Munson, Jr. A generalization of median filtering using linear combinations of order statistics. *IEEE Trans. Acoustics, Speech, and Signal Processing*, 31(6): 1342-1350, December 1983.

13. J.C. Brailean, R.P. Kleihorst, S. Efstratiadis, A.K. Katsaggelos, and R.L. Lagendijk. Noise reduction for dynamic images sequences: A review. *Proc. IEEE*, 83(9): 1272-1292, September 1995.

14. D.K. Brownrigg. The weighted median filter. *Communications of the ACM*, 27: 807-818, August 1984.

15. Q. Cai, W. Yang, Z. Liu, and D. Gu. Recursion adaptive order statistics filters for signal restoration. In *Proc. IEEE Int. Symp. Circuits and Systems*, San Diego, CA, 1992, pp. 129-132.

16. L. Castedo, C.Y. Tseng, and L.J.Griffiths. An adjustable constraint approach for robust adaptive beamforming. In J. Vandewalle, R. Boite, M. Moonen, and J. Oosterlinck, eds., *Signal Processing VI: Theories and Applications*. Elsevier, Amsterdam, 1992, pp. 1121-1124.

17. S. Chen and G.R. Arce. Microstatistic LMS filters. *IEEE Trans. Signal Processing*, 41(3): 1021-1034, March 1993.

18. L.O. Chua and T. Roska. The CNN paradigm. *IEEE Trans. Circuits and Systems-I*, 40(3): 147-156, March 1993.

19. P.M. Clarkson and G.A. Williamson. Constrained adaptive order statistic filters for minimum variance signal estimation. In *Proc. 1992 IEEE Int. Conf. Acoustics, Speech, and Signal Processing*, Vol. IV, San Francisco, 1992, pp. 253-256.

20. H. Cramer. *Mathematical Methods of Statistics*. Princeton University Press, Princeton, 1963.

21. P.S.R. Diniz, M.L.R. deCampos, and A. Antoniou. Analysis of LMS-Newton adaptive filtering algorithms with variable convergence factor. *IEEE Trans. Signal Processing*, 43(3): 617-627, March 1995.

22. A. Feuer and E. Weinstein. Convergence analysis of LMS filters with uncorrelated Gaussian data. *IEEE Trans. Acoustics, Speech, and Signal Processing*, 33(1): 222-230, February 1985.

23. A. Flaig, G.R. Arce, and K.E. Barner. Affine order statistic filters: "Medianization" of linear filters. *IEEE Trans. Signal Processing*, 46(8): 2101-2112, August 1998.

24. O.L. Frost, III. An algorithm for linearly constrained adaptive array processing. *Proc. IEEE*, 60: 926-935, 1972.

25. M. Gabbouj and I. Tabus. TUT noisy image database. Technical Report ISBN 951-722-281-5, Signal Processing Lab., Tampere Univ. of Technology, Tampere, Finland, December 1994.

26. M. Gabrani, C. Kotropoulos, and I. Pitas. Cellular LMS *L*-filters for noise suppression in still images and image sequences. In *Proc. IEEE Int. Conf. Image Processing*, Austin, Texas, 1994, pp. 358-362.

27. P.P. Gandhi and S.A. Kassam. Design and performance of combination filters for signal restoration. *IEEE Trans. Signal Processing*, 39(7): 1524-1540, July 1991.

28. J.G. Gonzalez and G.R. Arce. Weighted myriad filters: A robust filtering framework derived from α-stable distributions. In *Proc. IEEE Int. Conf. Acoustics, Speech, and Signal Processing*, Vol. 5, Atlanta, GA, 1996, pp. 2833-2836.

29. L.J. Griffiths and C.W. Jim. An alternative approach to linearly constrained adaptive beamforming. *IEEE Trans. Antennas and Propagation*, 30(1): 27-34, January 1982.

30. M.M. Hadhoud and D. W. Thomas. The two-dimensional adaptive LMS (TDLMS) algorithm. *IEEE Trans. Circuits and Systems*, 35(5): 485-494, May 1988.

31. R.C. Hardie and G.R. Arce. Ranking in R^p and its use in multivariate image estimation. *IEEE Trans. Circuits and Systems for Video Technology*, 1(2): 197-209, June 1991.

32. R.C. Hardie and K.E. Barner. Extended permutation filters and their application to edge enhancement. *IEEE Trans. Image Processing*, 5(6): 855-867, June 1996.

33. S. Haykin. *Adaptive Filter Theory*. Prentice Hall, Englewood Cliffs, NJ, 1986.

34. L.L. Horowitz and K.D. Senne. Performance advantage of complex LMS for controlling narrow-band adaptive arrays. *IEEE Trans. Acoustics, Speech, and Signal Processing*, 29(3):722-736, June 1981.

35. A.K. Jain. *Fundamentals of Digital Image Processing*. Prentice Hall, Englewood Cliffs, NJ, 1989.

36. S. Kalluri and G.R. Arce. Adaptive weighted myriad filter algorithms for robust signal processing in α-stable noise environments. *IEEE Trans. Signal Processing*, 46(2): 322-334, February 1998.

37. S. Kalluri and G.R. Arce. Fast algorithms for weighted myriad computation by fixed-point search. *IEEE Trans. Signal Processing*, 48(1): 159-171, January 2000.

38. N. Kalouptsidis and S. Theodoridis, eds. *Adaptive System Identification and Signal Processing Algorithms*. Prentice Hall, London, 1993.

39. S.A. Kassam and M. Aburdene. Multivariate median filters and their extensions. In *Proc. 1991 IEEE Int. Symp. Circuits and Systems*, 1991, pp. 85-88.

40. Y.-T. Kim and G.R. Arce. Permutation filter lattices: A general order statistic filtering framework. *IEEE Trans. Signal Processing*, 42(9): 2227-2241, September 1994.

41. Y.T. Kim, G.R. Arce, and N. Grabowski. Inverse halftoning using binary permutation filters. *IEEE Trans. Image Processing*, 4(9): 1296-1311, September 1995.

42. V. Koivunen, N. Himayat, and S.A. Kassam. Nonlinear filtering techniques for multivariate images-Design and robustness characterization. *Signal Processing*, 57(1): 81-91, February 1997.

43. C. Kotropoulos and I. Pitas. Constrained adaptive LMS *L*-filters. *Signal Processing*, 26(3): 335-358, 1992.

44. C. Kotropoulos and I. Pitas. Optimum nonlinear signal detection and estimation in the presence of ultrasonic speckle. *Ultrasonic Imaging*, 14(3): 249-275, July 1992.

45. C. Kotropoulos and I. Pitas. Adaptive nonlinear filters for digital signal/image processing. In C. Leondes, ed., *Control and Dynamic Systems*, Vol. 67. Academic Press, San Diego, CA, 1994, pp. 263-318.

46. C. Kotropoulos and I. Pitas. Multichannel *L*-filters based on marginal data ordering. *IEEE Trans. Signal Processing*, 42(10): 2581-2595, October 1994.

47. C. Kotropoulos and I. Pitas. Adaptive LMS *L*-filters for noise suppression in images. *IEEE Trans. Image Processing*, 5(12): 1596-1609, December 1996.

48. C. Kotropoulos and I. Pitas. Adaptive multichannel marginal *L* filters. *Optical Engineering*, 38(4): 688-704, April 1999.

49. C. Kotropoulos, S. Tsekeridou, M. Gabrani, and I. Pitas. Adaptive LMS order statistic filters with variable step-sizes. In *Proc. Int. Conf. Telecommunications*, Vol. I, Halkidiki, Greece, 1998, pp. 465-469.

50. J.S. Lee. Digital image enhancement and filtering by use of local statistics. *IEEE Trans. Pattern Analysis and Machine Intelligence*, 2:165-168, March 1980.

51. Y.H. Lee and S.A. Kassam. Generalized median filtering and related nonlinear filtering techniques. *IEEE Trans. Acoustics, Speech, and Signal Processing*, 33(3): 672-683, June 1985.

52. J.N. Lin, X. Nie, and R. Unbehauen. Two-dimensional LMS adaptive filter incorporating a local mean estimator for image processing. *IEEE Trans. Circuits and Systems–Part II*, 40(7): 417-428, July 1993.

53. S.K. Mitra M. Lightstone, E. Abreu and K. Arakawa. A new filtering approach for the removal of impulse noise from highly corrupted images. *IEEE Trans. Image Processing*, 5(6): 1012-1025, June 1996.

54. C.L. Mallows. Some theory of nonlinear smoothers. *Annals of Statistics*, 8(4): 695-715, 1980.

55. P. Maragos and R.W. Schafer. Morphological filters–Part II: Their relations to median, order-statistic, and stack filters. *IEEE Trans. Acoustics, Speech, and Signal Processing*, 35: 1170-1184, August 1987.

56. W.B. Mikhael and S.M. Ghosh. Two-dimensional variable step-size sequential adaptive gradient algorithms with applications. *IEEE Trans. Circuits and Systems*, 38(12): 1577-1580, December 1991.

57. L. Naaman and A.C. Bovik. Least-squares order statistic filters for signal restoration. *IEEE Trans. Circuits and Systems*, 38(3): 244-257, March 1991.

58. C.L. Nikias and M. Shao. *Signal Processing with Alpha-Stable Distributions and Applications*. Wiley, New York, 1995.

59. N. Nikolaidis and I. Pitas. Multichannel L-filters based on reduced ordering. *IEEE Trans. Circuits and Systems for Video Technology*, 6(5): 470-482, October 1996.

60. F. Palmieri. *Nonlinear Filtering for Robust Signal Processing*. Ph.D. thesis. University of Delaware, August 1987.

61. F. Palmieri. A backpropagation algorithm for multilayer hybrid order statistic filters. In *Proc. 1989 IEEE Int. Conf. Acoustics, Speech, and Signal Processing*, Glasgow, 1989, pp. 1179-1182.

62. F. Palmieri and C.G. Boncelet Jr. Optimal MSE linear combination of order statistics for restoration of markov processes. In *Proc. 20th Ann. Conf. Information Sciences*, Princeton, NJ, 1986.

63. F. Palmieri and C.G. Boncelet Jr. A class of adaptive nonlinear filters. In *Proc. 1988 IEEE Int. Conf. Acoustics, Speech, and Signal Processing*, New York, 1988, pp. 1483-1486.

64. F. Palmieri and C.G. Boncelet Jr. $L\ell$-filters–A new class or order statistic filters. *IEEE Trans. Acoustics, Speech, and Signal Processing*, 37(5): 691-701, May 1989.

65. A. Papoulis. *Probabilities, Random Variables and Stochastic Processes*. 3rd ed., McGraw-Hill, New York, 1991.

66. L.F.C. Pessoa and P. Maragos. MRL-filters: A general class of nonlinear systems and their optimal design for image processing. *IEEE Trans. Image Processing*, 7(7): 966-978, July 1998.

67. I. Pitas. Marginal order statistics in color image filtering. *Optical Engineering*, 29(5): 495-503, May 1990.

68. I. Pitas and P. Tsakalides. Multivariate ordering in color image restoration. *IEEE Trans. Circuits and Systems for Video Technology*, 1(3): 247-259, September 1991.

69. I. Pitas and A.N. Venetsanopoulos. LMS and RLS adaptive L-filters. In *Proc. 1990 IEEE Int. Conf. Acoustics, Speech, and Signal Processing*, Albuquerque, 1990, pp. 1389-1392.

70. I. Pitas and A.N. Venetsanopoulos. *Nonlinear Digital Filters: Principles and Applications*. Kluwer Academic, Norwell, MA, 1990.

71. I. Pitas and A.N. Venetsanopoulos. Adaptive filters based on order statistics. *IEEE Trans. Signal Processing*, 39(2): 518-522, February 1991.

72. I. Pitas and A.N. Venetsanopoulos. Order statistics in digital image processing. *Proc. IEEE*, 80(12): 1893-1921, December 1992.

73. I. Pitas and S. Vougioukas. LMS order statistic filter adaptation by backpropagation. *Signal Processing*, 25: 319-335, 1991.

74. K.N. Plataniotis, D. Androutsos, S. Vinayagamoorthy, and A.N. Venetsanopoulos. Color image processing using adaptive multichannel filters. *IEEE Trans. Image Processing*, 6(7): 933-949, July 1997.

75. C.S. Regazzoni and A. Teschioni. A new approach to vector median filtering based on space filling curves. *IEEE Trans. Image Processing*, 6(7): 1025-1037, July 1997.

76. S. Roy. On *L*-filter design using the gradient search algorithm. In *Proc. SPIE Nonlinear Image Processing II*, Vol. 1451, 1991, pp. 254-256.

77. M.J. Rude and L.J. Griffiths. A linearly constrained adaptive algorithm for constant modulus signal processing. In L. Torres, E. Masgrau, and M.A. Lagunas, eds., *Signal Processing V: Theories and Applications*. Elsevier, Amsterdam, 1990, pp. 237-240.

78. P. Salembier. Adaptive rank order based filters. *Signal Processing*, 27: 1-25, April 1992.

79. D.T.M. Slock. On the convergence behavior of the LMS and the normalized LMS algorithms. *IEEE Trans. Signal Processing*, 41(3): 2811-2825, September 1993.

80. R. Sucher. A recursive nonlinear filter for the removal of impulse noise. In *1995 IEEE Int. Conf. Image Processing*, 1995, pp. 183-186.

81. X.Z. Sun and A.N. Venetsanopoulos. Adaptive schemes for noise filtering and edge detection by use of local statistics. *IEEE Trans. Circuits and Systems*, 35(1): 57-69, January 1988.

82. I.D. Svalbe. The geometry of basis sets for morphologic closing. *IEEE Trans. Pattern Analysis and Machine Intelligence*, 13(12): 1214-1224, December 1991.

83. S. Tsekeridou, C. Kotropoulos, and I. Pitas. Adaptive order statistic filters for the removal of noise from corrupted images. *Optical Engineering*, 37(10): 2798-2816, October 1998.

84. C.Y. Tseng and L.J. Griffiths. A unified approach to the design of linear constraints in minimum variance adaptive beamformers. *IEEE Trans. Antennas and Propagation*, 40(12): 1533-1542, December 1992.

85. J. Vandewalle and T. Roska. Cellular neural networks. *Int. J. Circuit Theory and Applications*, 20, September-October 1992.

86. T. Viero, K. Öistämö, and Y. Neuvo. Three dimensional median-related filters for color image sequence filtering. *IEEE Trans. Circuits and Systems for Video Technology*, 4(2): 129-142, April 1994.

87. B. Widrow and S.D. Stearns. *Adaptive Signal Processing*. Prentice Hall, Englewood Cliffs, NJ, 1985.

88. G.A. Williamson and P.M. Clarkson. On signal recovery with adaptive order statistic filters. *IEEE Trans. Signal Processing*, 40(10): 2622-2626, October 1992.

89. L. Yin and Y. Neuvo. Fast adaptation and performance characteristics of FIR-WOS hybrid filters. *IEEE Trans. Signal Processing*, 42(7): 1610-1628, July 1994.

90. L. Yin, R. Yang, M. Gabbouj, and Y. Neuvo. Weighted median filters: A tutorial. *IEEE Trans. Circuits and Systems–Part II: Analog and Digital Signal Processing*, 43(3): 157-192, March 1996.

91. O. Yli-Harja, J. Astola, and Y. Neuvo. Analysis of properties of stack and weighted median filters. *IEEE Trans. Acoustics, Speech, and Signal Processing*, 39(2): 395-410, February 1991.

92. S. Zozor, E. Moisan, and P. Amblard. Revisiting the estimation of the mean using order statistics. *Signal Processing*, 68: 155-173, 1998.

6 Video Segmentation Based on Multiple Features for Interactive and Automatic Multimedia Applications

R. CASTAGNO[†], A. CAVALLARO, F. ZILIANI, and T. EBRAHIMI

Swiss Federal Institute of Technology, Lausanne, Switzerland

6.1 INTRODUCTION

Video segmentation is a basic tool for a new generation of video processing applications. The work presented in the literature in recent years shows a growing interest in content-based manipulation of video information. Many of the emerging multimedia applications, such as audiovisual databases, educational and training systems, virtual reality, entertainment, as well as other specific advanced communication applications (remote surveillance and expertise, tele-medicine, etc.), rely on the possibility of manipulating, distributing, indexing and accessing audiovisual information based on its content.

Future multimedia applications will therefore require a new model for the coding and the description of information, in which reality is viewed as composed by independent audiovisual objects integrated to create an *audiovisual scene*. In this framework, new standards are being developed for content-based *coding* (MPEG-4) and *description* (MPEG-7) of audiovisual information. In parallel, a demand for efficient ways of extracting and characterizing relevant objects in video sequences has emerged. This chapter describes a flexible solution to this problem.

A key feature of the system is the distinction between the two levels of segmentation: the region and the object segmentation. *Regions* are homogeneous areas of the

[†]Currently with Nokia Mobile Phones, Tampere, Finland

images that are extracted automatically by the computer. Semantically meaningful *objects* are obtained by grouping regions automatically or by means of user interaction, according to the specific application. This splitting relieves the computer of ill-posed semantic problems, and allows a higher level of flexibility in the use of the results.

The scheme presented in this chapter is a flexible tool that addresses the segmentation of video sequences into regions that are homogeneous in color, texture, motion (or combination thereof), as well as the tracking of these regions along time. In addition the system provides information about the characteristics of the extracted regions: this information, close to the concept of "metadata," can be exploited by higher-level tools in order to produce content description or to implement simple content-based decision schemes.

As mentioned above, the regions produced by the proposed algorithm do not have semantic content per se; they constitute the main bricks on which a semantic decomposition of the visual scene can be built. We provide two examples of how this goal can be achieved in two sections dedicated to applications. The first addresses the interactive manipulation of objects in a video sequence. The user has the possibility, by means of a simple graphic user interface, of grouping together a number of regions that constitute a semantic object with respect to a specific application. The proposed tool is then capable of extracting and tracking the selected regions over time, thus allowing the manipulation of the entire object, such as for video editing.

The second application demonstrates, instead, the capability of a completely automatic operation, when the specific application is determined in advance. In traffic surveillance, it is interesting to concentrate the attention of the system on the areas of the scene that have changed with respect to a reference situation, such as an empty highway. The objective of the system is then to analyze the changed areas, segment them into regions and track them over time. Moreover data such as the size, the position, and the speed of the regions will allow an improved level of scene understanding. For instance, it may be possible to identify larger objects such as trucks, evaluate the speed of vehicles, and spot potentially dangerous situations such as a car standing still on the edge of the road.

The chapter is organized as follows. In Section 6.2 we briefly discuss the automatic and the interactive approaches to video segmentation. Section 6.3 analyzes the implications associated to the use of multiple image features for the segmentation of video sequences. In Section 6.4 the clustering approach used in the algorithm is presented: based on the traditional fuzzy C-means algorithm, we propose an extension that takes into due account the spatial adjacency of pixels and the relative level of reliability of the image features used in the process. In Section 6.5 we discuss the problem of temporal tracking of the extracted regions over time, thus completing the description of the region level segmentation tool. Sections 6.6 and 6.7 are dedicated to two examples of applications of the proposed segmentation tool, the former in interactive mode for content-based video manipulation, the latter for automatic traffic surveillance.

6.2 AUTOMATIC AND SUPERVISED SEGMENTATION: A BRIEF OVERVIEW

6.2.1 A Historical Perspective

Segmentation has been for a long time a central problem in image processing, and in the past it has been regarded as mainly related to the *analysis* of images. Numerous algorithms have been proposed for the segmentation of still images, in the context of various applications, such as the analysis of satellite [1] and medical images[2], or the control of manufacturing processes [3]. In these schemes the final application of the segmentation was often well known and determined a priori, and such knowledge could be integrated in the process itself. User supervision has been introduced initially as a generic help to the algorithm, and eventually maintained when needed for very specific applications (e.g., medical image analysis).

When the necessary computer power became available, the segmentation problem was addressed also in the field of video, in addition to the still image applications. Probably as a follow-up of the results obtained in the field of still image segmentation, the attention of researchers was mainly devoted to the development of automatic methods. For instance, this was clearly reflected in the mandate of the working group that dealt with segmentation in the framework of MPEG-4. The corresponding Core Experiment N2 was named "Comparison of *Automatic* Segmentation Techniques." Supervision was regarded as a negative characteristic of proposed methods.

It can be argued that considering segmentation as a task achieved by means of completely automatic methods is equivalent to stating the problem under the most difficult conditions. In this case, in fact, the task overlaps with the thorny fields of scene understanding and artificial intelligence.

However, if the application allows a certain degree of user interaction, this can be used to improve the analysis results, and reduce the complexity of the problem at the same time. The idea is to start by using as far as possible the best available automatic tools. Whenever the application permits, additional knowledge provided interactively by the user can be incorporated in order to improve the performance. User interaction should not be seen as a substitute for mediocre tools but rather as a complement to overcome difficult cases.

In addition to this, interaction should represent a basic feature of analysis tools that are dedicated to multimedia applications. The very nature of multimedia goes beyond the simple merging of video, audio, and data. Specially in this context, the possibility of interacting with the different forms of information can represent an added value of the method rather than a limitation. In his book *Being Digital* [4], Nicholas Negroponte states: "Interaction is implicit in all multimedia. If the intended experiences were passive, then closed-captioned television and subtitled movies would fit the definition of video, audio and data combined". In the following, some forms of interactivity in segmentation tools are reviewed. A more complete discussion on these issues is presented in [5] and [6].

6.2.2 *Automatic* Segmentation?

As mentioned above, fully automatic video segmentation has been a widely addressed problem in the recent literature. However, a common weak point of many
automatic segmentation methods proposed in the literature is the dependence of their
performance on numerous parameters, which need to be fixed by the user, until an
acceptable result is obtained. This fine-tuning process can be considered to a great
extent as a form of hidden interaction.

In many cases the reaction of the system to the changes in the parameters is not
easily predictable, which makes the tuning process a delicate and time-consuming
one. Furthermore the parameters to be manipulated are often related to the deep
structure of the algorithm, which makes it difficult to envisage user-friendly interfaces
for the necessary tuning procedure.

There is one notable exception to the consideration above: when the targeted
application is well specified and known a priori, it is possible to tune in advance the
parameters of the algorithm. If the system eventually works in controlled conditions,
an automatic operation can be achieved as a trade-off for flexibility.

An example of such an application is surveillance where conditions such as
the lighting, the position of the cameras, the nature of the objects to extract, or a
"reference" condition of the scene (e.g., a road without traffic) are known. In this
case a tool such as the one presented in this contribution can be integrated with
the additional information obtained by this knowledge, and achieve completely an
automatic operation mode. Such application is presented in Section 6.7.

6.2.3 Initial User Interaction

The user interacts with the system by performing an appropriate initialization, by
specifying the initial analysis constraints, or even by refining some preliminary
automatic segmentation results (e.g., by correcting an initial segmentation of the first
frame). Based on this initial guidance, the system is then capable to carry on the rest
of the process in an automatic mode.

Examples that this form of initial interaction might include are as follows:

- Identification and classification of a subset of significant points, to be used as
  training set for the follow-up of the segmentation process [7, 8, 9, 10].

- Drawing a rough outline of the object(s) to be extracted during the segmentation. Such an approach could be envisaged in a video-conference system
  where the user could identify by means of a mouse click, or by touch of screen,
  the foreground and the background of the scene.

- Determination of the number of regions or objects to be identified in the scene.

- Improvement of a tentative segmentation of the first frame, for instance, by
  merging clusters or refining the boundaries.

6.2.4   User Refinement

The user is able to interact during the progress of the segmentation process, or even with the final results, in order to refine the results, rather than influencing the process as in the initial interaction.

Examples of user refinement are as follows:

- Adjusting the results by improving the visual quality (e.g., by improving the appearance of borders).

- Merging and/or splitting regions so as to obtain semantically significant objects.

- Taking care of events that change significantly the scene content (e.g., scene cuts, appearance or disappearance of objects).

- Dealing with occlusions by reconstructing the temporal continuity of objects that have temporarily disappeared or have been occluded by others.

6.2.5   Interactive Segmentation Schemes: Some Recent Developments

An interesting example of a partially supervised segmentation scheme is presented in [10] by Chalom and Bove. The method takes into account a set of image features, and a multi-modal statistical model of regions is developed based on a limited amount of training data selected by the user. For each picture element, a vector of features (including motion, texture, color and texture information) is extracted. The method aims at modeling parametrically the probability density function (PDF) of each feature vector. It is assumed that the distribution of a particular feature can be approximated by a sum of uni-modal or multi-modal Gaussian PDF's. Since the distribution is not known a priori, the user supplies a set of training points, from which the system approximates the PDF. The remaining points are then classified to one of the regions specified by the user.

The algorithm proposed by Salembier et al. in [11] mainly aims at improving the coding efficiency by means of a region based scheme. However, an accurate tracking of the resulting regions allows the user to select groups of regions and manipulate them separately, thus achieving object based functionalities.

Another example of how interaction can be used in order to improve the quality of a segmentation scheme is given in [7]. The method addresses in particular the segmentation of still magnetic resonance images (MRI). Although referred to still images, this example shows that not only does the user's interaction ease the task of the algorithm, but it also allows to obtain results that better fit the specific application in which the algorithm is used. The algorithm is based on a fuzzy C-means. Section 6.4 presents a general description of the algorithm, which is used also in our approach. The authors exploit here the fact that the application is well determined a priori: a few *high-quality* labeled data are selected from each class as an initialization for the method. The knowledge of the user plays here an important role in such a delicate field.

Recently a special issue of the *IEEE Transactions on Circuits and Systems for Video Technology* has been dedicated to object-based video coding and description [12]. The interested reader will find there a broader survey of recent proposals in this field.

In the following sections we introduce in more detail our proposal. Our scheme shares with Chalom's method [10] the multiple-feature analysis but proposes a different approach to user interaction based on a distinction between the concepts of region and object. We also adopt a different scheme for the actual clustering of the picture elements.

6.3   REGION EXTRACTION BASED ON MULTIPLE FEATURES

6.3.1   Regions and Objects

The concepts of *region* and *object* have been commonly used in the literature related to segmentation and computer vision in general. In most cases, the two terms are used as synonyms. However, in this contribution we propose a distinction between the two, which represents a key characteristic of the proposed approach to video segmentation.

We define a *region* as an area of the frame that is homogeneous according to given quantitative criteria, such as the gray level, the color, the texture, the motion, or a combination of them, in the most general case. It is important to stress that at this stage we do not require an area to have any intrinsic semantic meaning in order to be classified as region.

Our definition of *object* is in full accordance to the concept of video object as it is defined in the framework of MPEG-4, "an entity in a scene that a user is allowed to access (seek, browse), and manipulate (cut and paste)." Unlike the above-mentioned regions, objects are strongly characterized by their semantic content, so they can easily lack global coherence in color, texture, and movement. According to these definitions, and without loosing generality in most real-world situations, we assume that an object is constituted by one or more regions. The grouping of regions into objects is dependent on a semantic interpretation of the scene, which can in turn depend on the specific application.

An example of how regions and objects are related is given in Figs. 6.1 to 6.5. Figure 6.1 shows a frame of sequence "Akiyo," and Fig. 6.2 shows a segmentation at the region level. For sake of simplicity, we have selected a case in which the coherence within each region is based on gray level and color.

Figures 6.3 to 6.5 show three examples of how the same segmentation at the region level can yield different segmentations at the object level, depending on the specific application. Figure 6.3 could correspond to an editing application, in which the whole person needs to be extracted and manipulated. In Fig. 6.4, the object is constituted of a single region: in this case we can imagine video-conferencing applications, in which the user wants the highest possible quality for the object of

Fig. 6.1   A frame of sequence "Akiyo."

Fig. 6.2   A segmentation into regions of the same frame.

Fig. 6.3   The object of interest is Akiyo.

Fig. 6.4   The object of interest in a video-conference application is the face of Akiyo.

Fig. 6.5   The object of interest is a sensible area in the background.

interest, still tolerating lower quality or temporary degradation in other regions. The last example, Fig. 6.5, refers to a situation that is likely to occur in news production, where the producer is able to create the video content by merging information from different sources.

6.3.2   The Semantic Step

These definition and the examples allow us to introduce here the deeper motivation for the distinction between regions and objects, which largely affects the structure of the proposed approach. From the previous discussion, semantics emerge as the key to this distinction. In order to obtain the regions, no semantic knowledge is required from the system, but a *semantic step* is needed in order to group the regions into objects. Our knowledge of the human body and the particular application that we might have in mind tell us that a pink jacket, a white shirt, a face, and the hair constitute the object "person." However, the problem of making the same decision appears definitely ill-posed to a computer.

These considerations, together with the analysis of other proposals to which we made previous references, suggested the idea of inserting the user interaction exactly

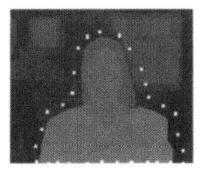

Fig. 6.6 The user draws an approximate contour around the object of interest.

Fig. 6.7 The corresponding regions are grouped into the desired object.

at this step. Unlike in [10] and [7], we do not suggest anything to the computer about how to build its "favorite regions," but we choose to add our understanding of the scene on top of *its* interpretation, stepping up from regions to objects. In Figs. 6.6 and 6.7 a simple graphic user interface based on the generation of "snakes" is demonstrated: the system proposes a segmentation into regions, and the user selects to group some of them into a semantically meaningful object.

It should, however, be pointed out that for specific applications the grouping of regions can also be performed in an automatic mode. For instance, the system could aggregate all regions that move from frame to frame, obtaining results similar to those of motion-based segmentation methods. Another possibility could be that of selecting regions that show facial characteristics (the presence of the eyes, etc.), by coupling the scheme to a face detection system.

In the case of traffic surveillance, it is possible to exploit the a priori knowledge that "everything that changes with respect to a reference frame is potentially interesting." By integrating a scene change detection algorithm in the proposed region segmentation method, it is possible to achieve a completely automatic operation mode for this specific application. Such application is presented in detail in Section 6.7.

The following sections deal with the specific method that we have chosen in order to achieve segmentation at the region level. It is important to stress that the extraction and the tracking of regions take place in an automatic mode.

6.3.3 Use of Multiple Features

Once we have accepted that the semantics will be taken care of by the user interaction, and having eliminated therefore a part of the ill-posedness of the segmentation problem, we aim at automatically extracting the regions by exploiting the coherence among different features. In this aspect the starting point of our method is similar to that of Chalom and Bove proposed in [14, 10]. Each frame in the sequence is analyzed by a program that extracts a vector of feature values for each pixel.

To avoid the *curse of dimensionality* problem [15], it is important to choose a significant set of features. In our method we choose one feature describing each of the following categories: color, motion, texture, and position.

As far as *color* information is concerned, several color spaces may be used as features (RGB, YUV, HLS, Lab, Luv, and others). The RGB color space is certainly the simplest, since it is often directly generated by the acquisition system. However, the color representation it provides does not separate the luminance information from the chrominance information. The YUV color space is a simple linear transformation of the RGB color space that performs such a separation, and it is used in many practical applications. Luv and Lab color spaces, aim at describing colors as they are perceived by human beings. They are nonlinear transformations of the RGB components and thus more complex to define than the YUV. In our technique we chose the YUV color space, as it appeared as the best compromise, within our scope, between representation quality and complexity. However, the generality of the segmentation method can be extended, according to the application, to the Lab or Luv.

Regarding the *motion category*, the horizontal and vertical components of the optical flow are chosen as features.

The *texture* information is simply represented by the variance of the image luminance in a small neighborhood surrounding the pixel.

The *position* is taken into account by the absolute x and y coordinates of each image pixel.

Each of these features can be appropriately preprocessed by a filter chosen from a repertoire (median, morphological, lowpass, etc.).

The joint use of different kinds of information is a common feature of many segmentation schemes. In particular, several methods have been proposed that perform spatiotemporal segmentation. Typical schemes consist of obtaining a segmentation based on spatial (color/gray level) information and iteratively improving it by means of the motion information, by projecting the segmentation from frame to frame and using motion information for region merging [11, 16, 17]. Graphs are often used to represent the mutual adjacency of regions and guide the merging process.

In this approach we propose, instead, that all the available information be used in parallel to exploit the existing correlation among all the features without intermediate decision steps based on partial information. Our objective is to simplify the scheme, avoiding iterative passages from one domain to the other (spatial-temporal).

The use of data that differ in range and importance poses the twofold problem of accounting for their different ranges of variation as well as for the different importance that has to be attributed to the different kinds of data. The following sections present the proposed solutions to this problem, with the ultimate goal of appropriately defining a distance in the feature vector space.

6.3.4 Scaling

The features that we propose to use in our segmentation scheme belong to four groups: color, motion, position, and texture. Each is characterized by range of possible values. Color information typically ranges from 0 to 255. Motion spans a more limited interval (e.g., $[-10, \ldots, +10]$ pixels/frame). The x and y coordinates are limited by the size of the image. Texture information shows the biggest variations. In order to process this information in parallel, it is necessary to introduce some form of normalization that allows us to define a distance that is easily measurable. A common solution is known as Mahalanobis distance, which is, for example, adopted in [18], [19], and [14]. The value of the jth feature at pixel k is normalized with respect to the standard deviation of that feature over the entire image:

$$\overline{f_j} = \frac{\sum_{k=0}^{N-1} f_{kj}}{N}, \tag{6.1}$$

$$\sigma_j^2 = \frac{\sum_{k=0}^{N-1} (f_{kj} - \overline{f_j})^2}{N}, \tag{6.2}$$

$$\widehat{f_{kj}} = \frac{f_{kj}}{\sigma_j}, \tag{6.3}$$

where N is the total number of pixels in the image. It should be observed here that the expression of Eq. (6.2) is striclty valid under the assumption that the features are uncorrelated. A more correct formulation would require the evaluation and the inversion of the covariance matrix. For simplicity, the use of Eq. (6.2) is a common choice in the literature. The use of the Mahalanobis distance induces therefore a distance between two vectors $\mathbf{f}_m$ and $\mathbf{f}_n$ in the feature vector space that can be defined as

$$d_{mn}^2 = d(\mathbf{f}_m, \mathbf{f}_n) = \|\widehat{\mathbf{f}_m} - \widehat{\mathbf{f}_n}\|_2 = \sum_{j=0}^{F-1} \frac{(f_{mj} - f_{nj})^2}{\sigma_j^2}, \tag{6.4}$$

where F is the total number of features.

6.3.5 A Priori Weighting

6.3.5.1 Generalities The normalization proposed above represents only a preliminary step for a meaningful definition of distance in the vector space. The features that have been used in the proposed method differ not only in their range of variation but also in their level of reliability and, in general, the weight they should be given in the segmentation process.

In the proposed approach we have chosen to adopt two criteria for the evaluation of the reliability of the different features. The first criterion, the a priori reliability, is based on the robustness of the optical flow estimation, and is used to establish the relative weight of motion information with respect to color information. The

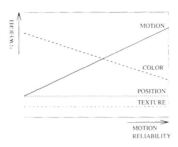

Fig. 6.8   The relative weight of the different type of features as a function of the reliability of motion information.

a posteriori reliability is evaluated based on the significance that each feature has had in the characterization of a region in previous segmentation stages. These two estimates are integrated in the segmentation process and allow the system to adapt to the characteristics of the scene without the need of time-consuming parameter setting often required by other segmentation schemes.

The a priori reliability estimation is introduced in Section 6.3.5.2. The a posteriori criterion can also be considered as a refinement of the reliability estimate, and will be introduced, according to the logical flow of the algorithm, in Section 6.4.5.

6.3.5.2   Estimation of feature reliability   Motion estimation is known to be quite accurate, and therefore reliable, in textured areas. In uniform areas, instead, the motion fields are usually quite noisy. Finally, on edges, the motion information tends to be inaccurate, since gradient-based estimation methods typically impose smoothness constraints on edges, which results in an estimation error when objects with different motions overlap.

On the other hand, the spatial information (gray level and color) is reliable in uniform areas and on edges, while it tends to yield oversegmentation in textured areas. The considerations above suggest devising a method that allocates relative weight to features according to their local degree of reliability.

In particular, we need to evaluate an index of reliability for the motion information, characterized by a higher value in textured areas, and a lower value in uniform areas and along edges. This index of reliability is then used to weight the spatial and the motion information according to the qualitative graph of Fig. 6.8. It can be observed that as the reliability of the motion information increases, the motion information has more weight in the segmentation process, while the color information is less important. The opposite happens is areas (e.g., uniform zones) where the motion information is less reliable, and color represents a more useful criterion of homogeneity. Texture and position information are assumed to maintain the same level of reliability over the entire image.

In the following, we present two alternative methods than can be used for the evaluation of such index. The first is based on qualitative consideration and is

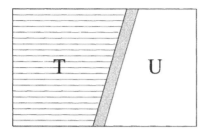

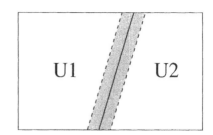

Fig. 6.9 Border between a textured (**T**) and a uniform **U** area. In the gray area, high local variance values are assigned to pixels in the uniform zone.

Fig. 6.10 Border between two uniform areas (**U1** and **U2**). In the gray area, high local variance values are assigned to pixels in the uniform zones on both sides of the border.

independent of the chosen motion estimation scheme: we will briefly introduce it for sake of generality. The second is based on information directly obtained from a specific optical flow estimation method, and has been used for the simulations presented in this chapter.

Fuzzy rule approach   The evaluation of the index is based on the evidence that "the more the area is textured, the more the motion estimation is reliable and should be weighted accordingly." We simply measured the standard deviation of the gray level over a 3×3 window and used the obtained value as an index of reliability in order to weight the spatial and the motion information according to the graph in Fig. 6.8.

It should be observed that this simple evaluation of the standard deviation yields two kinds of errors:

1. In Fig. 6.9 we show the case in which a textured area **T** is adjacent to a uniform area **U**: when the measurement window is positioned across the two zones, it detects high variance, which results in an area of wrong estimation in the uniform area (in gray).

2. Similarly, as shown in Fig. 6.10, on the edge between two uniform areas **U1** and **U2**, the window detects high variance on both sides of the border, even if there is no texture.

As an example, Fig. 6.11 shows a frame of sequence "table tennis," and Fig. 6.12 an image, in which the gray level represents the standard deviation calculated over a 3×3 running window. The two errors mentioned above are clearly visible around the hand and around the arm, respectively. It can be observed that the error is of similar nature in both cases; namely areas without texture are assigned high variance values. We reduce this problem by performing a morphological erosion of the high variance areas by means of a min filter of appropriate size. The result of such filtering is shown in Fig. 6.13, where the artifacts are significantly reduced. This information is adopted as an index of reliability for the motion information. The

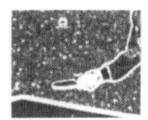

Fig. 6.11 An original frame of sequence "table tennis."

Fig. 6.12 Evaluation of the amount of texture in the frame of Fig. 6.11.

Fig. 6.13 Reliability of motion information in the frame of Fig. 6.11 after the min-filtering.

hand, which is highly textured, will be segmented mostly based on its motion, thus reducing the occurrence of oversegmentation. The sharp edge between the arm and the background, instead, is correctly *not* interpreted as a textured area.

Despite its simplicity this criterion is, on one hand, quite effective and has the advantage of being generic, since it does not depend on the method used to estimate the motion. On the other hand, several motion estimation algorithms provide an index of reliability of the estimate, which can be directly integrated in our method. We present this second solution in the next paragraph.

Confidence measure derived from the optical flow estimation method   The algorithm proposed in this chapter uses the motion information that can be provided by different motion estimation algorithms. A valuable comparison of different techniques is presented in [20]. In our tests we obtained satisfactory results by applying the method proposed by Lucas and Kanade in [21] and modified according to the work of Simoncelli, Adelson, and Heeger [22]. Furthermore this optical flow estimation algorithm provides an objective measurement of the local level of reliability of the motion information. Shi and Tomasi adopted this criterion of reliability in order to evaluate the texture properties of pictures areas, and they achieved improved tracking performances [23].

We will not go into a detailed description of the method but will just report here the results of the discussion in [20]. The reliability of the estimates for a given pixel **k** can be evaluated using the eigenvalues $\lambda_1 \geq \lambda_2$ of the matrix:

$$M = A^T W^2 A = \begin{bmatrix} \sum_{\mathbf{h} \in \eta_k} W^2(\mathbf{h}) I_x^2(\mathbf{h}) & \sum_{\mathbf{h} \in \eta_k} W^2(\mathbf{h}) I_x(\mathbf{h}) I_y(\mathbf{h}) \\ \sum_{\mathbf{h} \in \eta_k} W^2(\mathbf{h}) I_y(\mathbf{h}) I_x(\mathbf{h}) & \sum_{\mathbf{h} \in \eta_k} W^2(\mathbf{h}) I_y^2(\mathbf{h}) \end{bmatrix},$$

(6.5)

where the summations are intended over a small spatial neighborhood η_k of the pixel. $W(\mathbf{h})$ is a window function that gives more influence to pixels in the center of the neighborhood, and I_x and I_y are the spatial gradients of the gray levels in directions x and y respectively.

Fig. 6.14 A frame of sequence "mobile and calendar."

Fig. 6.15 The index of reliability obtained from the eigenvalues of matrix M in Eq. (6.5).

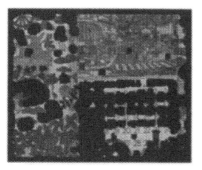

Fig. 6.16 Areas in the original gray level image have reliable motion estimation and will be segmented mainly according to motion information.

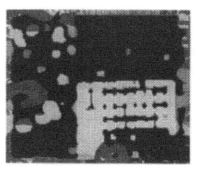

Fig. 6.17 Areas in the original gray level image have unreliable motion estimation and will be segmented mainly according to color information.

The product of the two eigenvalues λ_1 and λ_2 is used as a confidence measure. Furthermore, the method proposed in [21, 22] sets a condition $\lambda_2 \geq \tau$ on the smallest eigenvalue for a velocity to be evaluated. If the condition is not met, no velocity value is assigned to the pixel, and the confidence measure set to 0. We assumed the value $\tau = 1$ in our simulations. Since this process involves the evaluation of gradients over a neighborhood of pixels, as in Eq. (6.5), we applied to the results of the estimation the same kind of morphological post–processing that has been presented in Section 6.3.5.2. This index of reliability replaces the standard deviation (Fig. 6.8) in order to establish relative weights for the features.

Results of this analysis are shown in Figs. 6.14 to 6.17. Figure 6.14 shows an original frame of sequence *mobile and calendar*, and Fig. 6.15 shows the values of the measure of confidence for the motion information: a lighter gray level indicates higher reliability. The two following test images Fig. 6.16 and Fig. 6.17 have been produced in order to better explain the mechanism of the estimation of reliability. In the first, the pixels with the original gray level are those where motion information has a reliability above the average while the others are colored in black. The second shows the dual situation. It can be seen that textured areas are correctly identified.

6.4   CLUSTERING BY CONSTRAINED FUZZY C-MEANS

6.4.1   Introduction

Once we have selected the features needed for the segmentation process and we have defined an appropriate distance in the vector space, a clustering is needed in order to segment the image into regions.

In image clustering techniques, the original image is analyzed so as to extract a set of features for each pixel, as described in the previous sections. Regions in the scenes are then assumed to constitute *clusters* in the feature space, which can be seen as "clouds" of points in an F-dimensional feature space. Clustering techniques aim at grouping the points in the feature space by associating them to *centroids* or *class centers*, which are supposed to best represent each of the clusters.

Objective function methods are based on the choice of a criterion (or objective function) that measures the "desirability" of clustering candidates for each of the c clusters. Typically local extrema of the function are defined as optimal clusterings. The clustering process corresponds therefore to the minimization of a functional. For example, if one chooses as the similarity measure the Euclidean distance in the F-dimensional data space ($\mathbf{f}_k \in \Re^F$), and as a measure of cluster quality the overall within-group sum of squared errors (WGSS) between the $\mathbf{f}_k$ and the class centers $\boldsymbol{\mu}_i$, the objective function will be the sum of squared errors. The WGSS functional is the most extensively used clustering criterion, and it has been adopted in the well-known ISODATA scheme [6, 24].

The way in which we establish the membership of each pixel to the clusters influences the performance of the clustering method itself. In Section 6.4.2 we will briefly discuss two such criteria, namely the *hard c-partitions* and the *fuzzy partitions*.

The ideal case is where the different clusters are well separated one from another, so that it would be easy to determine the membership of one pixel to one and only one class. This way, the distribution of pixels can be efficiently represented by a hard c-partition.

However, in more practical situations, and specially in the case of multidimensional feature space, the borders between the classes are not crisp, and fuzzy c-partitions offer a more flexible way of representing them. This characteristic represents also an advantage in terms of convergence of the clustering algorithms [25, 26]. It can be demonstrated [24, 6] that the continuity of the membership function (and thus its differentiability) is at the base of the convergence of the fuzzy C-means algorithm at least to a local minimum.

After a short presentation of the properties of hard and fuzzy partitions, in the following of this section we will introduce the clustering method that has been chosen in the proposed approach, known as fuzzy C-means (FCM). The fuzzy C-means algorithm can be considered as a fuzzy generalization of the hard C-means (HCM) algorithm. The reader interested in an introduction to HCM is referred to classical sources such as [24] or to [6, 27].

6.4.2   Hard and Fuzzy Partitions

As discussed in the preceding section, the objective of clustering techniques is to determine appropriate *partitions* of the feature space. We can distinguish two families of partitions: hard partitions Ω_{hard} and fuzzy partitions Ω_{fuzzy}. In order to present the difference between the two, we will adopt a matrix representation of the partitions.

Hard c-partition space   Let $\Re^F$ be an F-dimensional space, and let $\Phi = \{\mathbf{f}_1, \ldots, \mathbf{f}_N\}$ be a finite set, with $\mathbf{f}_k \in \Re^F$. Let $\mathcal{M}_{cN}$ be the set of $(c \times N)$ matrices, with $2 \le c < N$. The space Ω_{hard} of hard c-partitions for Φ is the set:

$$\Omega_{hard} = \left\{ U \in \mathcal{M}_{cN} \mid u_{ik} \in \{0, 1\} \, \forall i, k; \; \sum_{i=1}^{c} u_{ik} = 1 \right.$$
$$\left. \forall k; \; 0 < \sum_{k=1}^{N} u_{ik} < N \, \forall i \right\}, \tag{6.6}$$

where u_{ik} is defined as *membership value* of element $\mathbf{f}_k$ to class i. The condition, $\sum_{i=1}^{c} u_{ik} = 1$, says that each object $\mathbf{f}_k$ belongs to one and only one class. The second condition $0 < \sum_{k=1}^{N} u_{ik} < N \forall i$ says that no class is empty, nor do any of the classes correspond to the whole space Φ. Ω_{hard} represents the space of the solutions of the partition problem, which needs to be explored in order to obtain the segmentation. In order to get some sense of the complexity of the problem, it is worth remembering that the cardinality of the space is

$$\mid \Omega_{hard} \mid = \frac{1}{c!} \left[\sum_{i=1}^{c} \binom{c}{i} (-1)^{c-i} \, i^N \right] \tag{6.7}$$

If $c = 10$ and $N = 25$, there are roughly 10^{18} hard c-partitions of the 25 points in 10 classes. It goes without saying that an appropriate strategy is needed to attack the problem without resorting to an exhaustive search.

Fuzzy c-partition space   Let $\Re^F$ be an F-dimensional space, and let $\Phi = \{\mathbf{f}_1, \ldots, \mathbf{f}_N\}$ be a finite set, with $\mathbf{f}_k \in \Re^F$. Let $\mathcal{M}_{cN}$ be the set of $(c \times N)$ matrices, with $2 \le c < N$. The space Ω_{fuzzy} of fuzzy c-partitions for Φ is the set:

$$\Omega_{fuzzy} = \{ U \in \mathcal{M}_{cN} \mid u_{ik} \in [0, 1] \subset \Re \, \forall i, k;$$

	Germany	Belgium	Italy	France	Switzerland	Netherlands
German	1	0	0	0	1	0
French	0	0	0	1	0	0
Italian	0	0	1	0	0	0
Dutch	0	1	0	0	0	1

	Germany	Belgium	Italy	France	Switzerland	Netherlands
German	1	0	0	0	0.65	0
French	0	0.43	0	1	0.25	0
Italian	0	0	1	0	0.10	0
Dutch	0	0.57	0	0	0	1

$$\sum_{i=1}^{c} u_{ik} = 1 \ \forall k; \ 0 < \sum_{k=1}^{N} u_{ik} < N \ \forall i\} \tag{6.8}$$

This definition is identical to Eq. (6.7), except for the fact the u_{ik} can be a real number in the interval $[0, \ldots, 1]$ rather than an integer belonging to the set $\{0, 1\}$. The matrix U is called *membership matrix*, and its elements u_{ik} account for the *membership value* of element $\mathbf{f}_k$ of the set $\mathbf{\Phi}$ to class i.

In the following, a simple example shows the difference between hard and fuzzy partition. Let us assume that our goal is to cluster some European countries according to the spoken languages. In this case the set $\mathbf{\Phi}$ is represented by the countries:

$$\mathbf{\Phi} = \{\text{Germany, Belgium, Italy, France, Switzerland, Netherlands}, \ldots\},$$

and the dimension F of the feature space is 1 (the language). Let define $c = 4$ classes corresponding to 4 languages (Dutch, French, German, Italian). In case a hard partition of the space is produced, the result would be a truth table: This partition would reflect the reality in a very partial way, specially as far as Switzerland and Belgium are concerned. In using a fuzzy partition of the set, our resulting table would be more realistic[1]: This discussion could be extended to a "multidimensional" case by taking into account the religious groups and the linguistic groups at the same time. Then the rigidity imposed by hard partitions would be more evident, since, for example, Switzerland would be simply classified as "a German speaking protestant country," without taking into account the rich diversity of the its people in reality.

---

[1]Small minorities such as the German in Belgium and the Reto-rumantsch in Switzerland are not taken into account.

6.4.3 The Standard Fuzzy C-Means Algorithm

In the previous section we have discussed the characteristics of hard and fuzzy partitions, reaching the conclusion that fuzzy partitions can often represent a more powerful way of representing complex multidimensional cluster configurations, where
borders between classes are not crisp and fuzziness plays an important role. The
fuzzy C-means (FCM) algorithm represents a generalization of methods based on
the within-groups sum of square (WGSS) errors functions, in order to generate hard-
partitions of multidimensional sets such as the hard C-means.

This extension was initially reported by Dunn in [28], and generalized into a
family of fuzzy clustering algorithms based on a least-squares error criterion [24].
Given $\Phi = \{f_1, \ldots, f_N\} \subset \Re^F$, which represents our data set and the desired number
of classes c, $2 \leq c \leq N$, the algorithm aims at finding a fuzzy partition U of the data
set containing N elements [24, 6]:

$$U \mid u_{ik} \in [0,1] \forall i,k; \sum_{i=1}^{c} u_{ik} = 1 \forall k; 0 < \sum_{k=1}^{N} u_{ik} < N \forall i, \qquad (6.9)$$

where u_{ik} represents the membership value of feature vector f_k to the class i. The
algorithm aims at evaluating the partition that minimizes the functional expressed by

$$J_{FCM}(U, \mu) = \sum_{k=1}^{N} \sum_{i=1}^{c} u_{ik}^m (d_{ik})^2, \qquad (6.10)$$

where $\mu = \{\mu_1, \ldots, \mu_c\}$, is the set of the centroids corresponding to each of the
classes and $m \in [1, \infty)$ is a weighting exponent that controls the amount of fuzziness.
d_{ik} is the distance between the ith centroid μ_i and the feature vector corresponding
to the kth pixel, f_k. The FCM algorithm is based on the fuzzy C-means theorem,
reported in the following. The reader is referred to [24, 6] for a demonstration.

The fuzzy C-means theorem   Let $\| \cdot \|$ be an inner product induced norm in $\Re^F$, and
let $m \in [1, \ldots, \infty)$ be the *fuzzy exponent*. We define the two sets S_k and $\bar{S}_k$ as

$$\begin{aligned} S_k &= \{i \mid 1 \leq i \leq c; \ d_{ik} = \| f_k - \mu_i \| = 0\}, \\ \bar{S}_k &= \{1, \ldots, c\} - S_k. \end{aligned}$$

(U, μ) may be globally minimal for J_{FCM} only if:

$$S_k = \emptyset \Rightarrow u_{ik} = \frac{1}{\sum_{s=1}^{c} (d_{ik}/d_{sk})^{2/(m-1)}}, \qquad (6.11)$$

$$S_k \neq \emptyset \Rightarrow u_{ik} = 0 \ \forall i \in \bar{S}_k \text{ and } \sum_{i \in S_k} u_{ik} = 1, \qquad (6.12)$$

$$\mu_i = \frac{\sum_{k=1}^{N}(u_{ik})^m \mathbf{f}_k}{\sum_{k=1}^{N}(u_{ik})^m}, \quad 1 \leq i \leq c. \tag{6.13}$$

The fuzzy C-means theorem allows the definition of the fuzzy C-means clustering algorithm as Picard iterations through necessary conditions:

1. Fix c so that $2 \leq c \leq N$ and $m, 1 \leq m < \infty$. Initialize $U^{(0)} \in \Omega_{fuzzy}$. Let an inner-product norm be defined over $\Re^F$. At each step $l = 0, 1, \ldots$:

2. Evaluate the c fuzzy cluster centers (or *centroids*) $\{\mu_1^{(l)}, \ldots, \mu_c^{(l)}\}$ according to Eq. (6.13) and using $U^{(l)}$ as membership matrix.

3. Update the membership matrix according to Eqs. (6.11) and (6.12) using the centroids $\{\mu_1^{(l)}, \ldots, \mu_c^{(l)}\}$ evaluated at step 2, obtaining $U^{(l+1)}$.

4. Compare $U^{(l+1)}$ and $U^{(l)}$ according to an appropriate criterion, such as a matrix norm $\|\cdot\|$. If $\|U^{(l+1)} - U^{(l)}\| \leq \epsilon$ then STOP; otherwise, return to step 2.

The fuzzy C-means algorithm iterates, evaluating at each step new centroids and a new fuzzy partition, until stability is reached. For further details about the implementation of the fuzzy C-means algorithm, as well as for a discussion about its convergence properties, the reader is referred to [24] and [19, 18].

In the evaluation of the distances d_{ik}, the different features are weighted according to the criterion introduced in Section 6.3.5. In particular, for each pixel k we evaluate a vector $\mathbf{w}_k = (w_{k1} \ldots w_{kF})^T$ in which w_{kj} represents the relative weight of the jth feature in pixel k, evaluated according to the estimate of reliability presented in Section 6.3.5.2. After the introduction of the weighting factor, Eq. (6.4) becomes

$$d_{ik}^2 = \sum_{j=1}^{F} w_{kj} \frac{(f_{kj} - \mu_{ij})^2}{\sigma_j^2}. \tag{6.14}$$

In order to demonstrate the effectiveness of the proposed weighting procedure, the FCM algorithm has been applied in intramode (i.e., without tracking) on some frames of sequences "foreman," "mobile and calendar," and "table tennis." The three original frames are shown in Fig. 6.18.

In order to provide a means of comparison, Figs. 6.19 to 6.21 were obtained by imposing fixed weights to the different groups of features on the whole frame. Table 6.1 shows the weightings that were used to obtain the three series of images of Figs. 6.19 to 6.21.

In particular, the segmentation masks of Fig. 6.19 have been obtained by attributing to the color features a predominant weight compared to motion information. The segmentation masks of Fig. 6.20 have been obtained by attributing the same relative weight to color features and to motion features. Finally, Fig. 6.21 shows the results obtained by giving higher weight to the motion features compared to the color information.

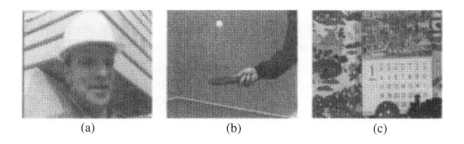

(a)   (b)   (c)

Fig. 6.18 (a) An original frame of sequence "foreman;" (b) An original frame of sequence "table tennis;" (c) An original frame of sequence "mobile and calendar."

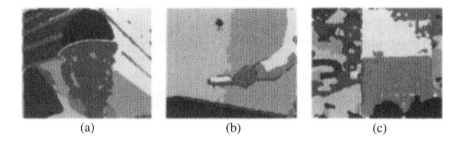

(a)   (b)   (c)

Fig. 6.19 Three segmentation masks obtained by attributing a higher weight to color information on the whole image. (a) Sequence "foreman;" (b) Sequence "table tennis;" (c) Sequence "mobile and calendar."

Table 6.1   Fixed weighting options adopted in the three sets of test images depicted in Figs. 6.19, 6.20, and 6.21

	Color	Motion	Position	Texture
Fig. 6.19	60%	20%	10%	10%
Fig. 6.20	40%	40%	10%	10%
Fig. 6.21	20%	60%	10%	10%

The segmentation masks of Fig. 6.19, obtained by a mostly color-based segmentation, show a better subjective quality. In particular, the borders of the regions are quite well defined, and uniform areas are well segmented. However, a more careful analysis shows that some image areas with similar color appearance, but different motion, are grouped in the same region. This is the case, for example, in the upper right part of the calendar image of Fig.6.19(c): the calendar is moving upward while the background is slowly sliding to the right. Since the motion information has very

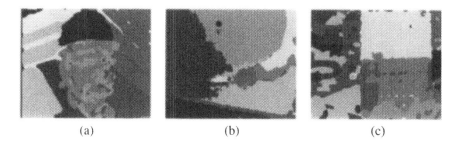

(a)   (b)   (c)

Fig. 6.20   Three segmentation masks obtained by attributing the same weight to color and motion information. (a)a) Sequence "foreman;" (b) Sequence "table tennis;" (c) Sequence "mobile and calendar."

(a)   (b)   (c)

Fig. 6.21   Three segmentation masks obtained by attributing a higher weight to motion information. (a) Sequence "foreman;" (b) Sequence "table tennis;" (c) Sequence "mobile and calendar."

little weight in the segmentation of Fig. 6.19(c), the border between the calendar and the background has not been detected and a spurious region is created.

When motion information is taken into account, the distinction between the two differently moving areas of Fig. 6.18 is correctly detected, as can be seen in Fig. 6.21(c). On the other hand, by examining the images of Fig. 6.21, it can be easily observed that a generalized use of motion information as the predominant segmentation criterion over the entire image is not efficient. The definition of the borders is worse than in Fig. 6.19, and noisy motion fields result in visually unpleasant and unexploitable segmentation.

In the following, the results obtained with the reliability-based (a priori) weighting strategy are presented. Figure 6.22 shows the weight attributed to motion information after the reliability analysis performed on the images in Fig. 6.18. The corresponding test images for color information are shown in Fig. 6.23. In both series of images, a lighter gray level indicates a higher weight attributed to motion and color information

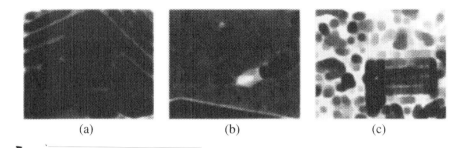

Fig. 6.22 The local distribution of the weight of motion features based on their a priori reliability. A lighter gray indicates a higher weight. (a) Sequence "foreman;" (b) Sequence "table tennis;" (c) Sequence "mobile and calendar."

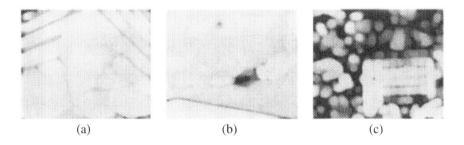

Fig. 6.23 The local distribution of the weight of color features based on their a priori reliability. A lighter gray indicates a higher weight. (a) Sequence "foreman;" (b) Sequence "table tennis;" (c) Sequence "mobile and calendar."

respectively[2]. The local attribution of weights to the texture and position features is the same as for the previous examples: 10% each.

Figure 6.24 show the results obtained after the first series of FCM iterations when the reliability-based weighting strategy described in Section 6.3.5.2 is adopted. These images show that the a priori evaluation of feature reliability allows us to improve significantly the results of the first stage of the segmentation scheme. The use of a weighting strategy based on the a priori estimation of reliability allows us to obtain results that match those of the previous examples in the areas where each of the two strategies (prevalence of color information or motion information) produces the best results.

In particular, it can be seen that the image area relative to the face of the man in sequence "foreman" appears to have a very unreliable estimation of the optical flow, Fig. 6.22(a), which is also confirmed by the results obtained in Fig. 6.21(a),

[2]In order to improve the quality of the display on paper, the gray levels in these images have undergone a gamma correction.

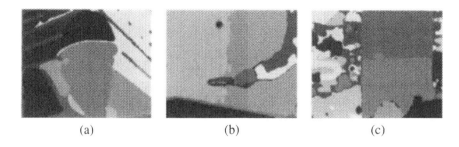

(a)   (b)   (c)

Fig. 6.24 Three segmentation masks obtained after one series of FCM iterations by weighting the features according to their a priori reliability: (a) Sequence "foreman;" (b) Sequence "table tennis;" (c) Sequence "mobile and calendar."

where mainly motion information is used for its segmentation. For this reason, color information is attributed a predominant importance, as shown in Fig. 6.23(a), and the segmentation is based almost exclusively on color, yielding the good results shown in Fig. 6.24(a).

As far as the sequence "table tennis" is concerned, the situation is quite similar. The only area where the motion is estimated to be highly reliable is that of the hand, on the remaining part of the image, color information is mostly used for the segmentation.

The results for sequence "mobile and calendar" demonstrate the importance of the use of motion information in areas where it is reliable. The upper half of the calendar has a different motion from the background to its left and its right, but a very similar color and texture. The a priori reliability of motion information is high in these areas, and the algorithm attributes to it a higher weight, as shown in Fig. 6.22(c). As a result the two regions are correctly separated, as can be seen in Fig. 6.24(c). In the same image it is possible to appreciate the fact that the border between the two regions is well defined and more accurate than in Fig. 6.21(c). This is a positive side effect of the morphological filtering performed as a postprocessing to the evaluation of the eigenvalues λ_1 and λ_2. The motion information is used exactly inside the areas where it is reliable (e.g., the upper part of the calendar), but the definition of the borders, where the motion information is much less reliable, is instead based on the color.

6.4.4   Spatial Constraint

Notice that the functional of Eq. (6.10) does not take into account the spatial adjacency of the data elements. This may result in a lack of spatial coherence of the resulting clusters. In order to reduce this problem, Schroeter proposed in [18, 19] the introduction of a spatial constraint that biases the algorithm so as to encourage adjacent pixels to be assigned to the same class. The proposed modification, called constrained fuzzy C-means (CFCM) aims at the minimization of the objective function:

$$J_{CFCM}(U, \boldsymbol{\mu}) = \sum_{k=1}^{N} \left(\sum_{i=1}^{c} u_{ik}^{m}(d_{ik})^2 + \sum_{h \in \eta_k} \beta \parallel \mathbf{u}_h - \mathbf{u}_k \parallel^2 \right), \qquad (6.15)$$

where $\mathbf{u}_k = (u_{1k}, \ldots, u_{ck})^T$, η_k is a neighborhood of pixel k, and β is a parameter that regulates the strength of the spatial constraint. In our experiments the neighborhood of pixel k is chosen to be four-connected. As is the case for the FCM algorithm, the fuzzy partition of the data set can be obtained by iteratively updating the centroids and the membership value of each vector to the classes. In our approach, the standard FCM algorithm is used to obtain an initial segmentation, and the spatial constraint is introduced in a second round of iterations aimed at refining the result.

Once the iterations have stabilized, the algorithm is stopped, and the fuzzy partition is hardened. The latter means that each pixel is assigned to the class for which it shows the highest membership value.

6.4.5 A Posteriori Weighting

6.4.5.1 Evaluation of significance   The goal of this analysis is to assess the significance of features for the segmentation process. In other words, the process aims at establishing which features are the most important in the creation of each cluster during the first series of FCM iterations.

For each cluster resulting from the first series of FCM iterations $\Phi_1, \ldots, \Phi_c$, and for each feature, the average value of the jth feature f_j over cluster Φ_i is evaluated:

$$\bar{f}_{ij} = \frac{\sum_{k \in \Phi_i} f_{kj}}{N_{\Phi_i}}, \qquad (6.16)$$

where N_{Φ_i} is the number of pixels that belong to cluster Φ_i. The variance of the jth feature over the ith cluster Φ_i is then given by

$$\sigma_{ij}^2 = \frac{\sum_{k \in \Phi_i} \left(f_{kj} - \bar{f}_{ij} \right)^2}{N_{\Phi_i}}. \qquad (6.17)$$

In order to establish the level of significance of the jth feature for the creation of cluster Φ_i, the quantity σ_{ij}^2 needs to be compared with the value σ_j^2 of the variance of the feature j over the entire image (Eq. (6.2)). In particular, a high value of the ratio σ_j^2/σ_{ij}^2 indicates that the distribution of the values of the jth feature in cluster Φ_i is more concentrated than it is over the entire image, therefore suggesting that feature j has been significant for the creation of that cluster. On the other hand, a lower value of the ratio indicates that the distribution of the feature values over the cluster is mainly due to noise, and therefore the values of the jth feature have not been particularly significant for the creation of cluster Φ_i.

6.4.5.2   *Assignment of the a posteriori weights* The significance of the generic feature presented above is that it shows how little purpose it has in the overall scheme if used in absolute terms. This information proves more valuable when used in relative terms, that is, by comparing the significance of the different features in the characterization of a specific cluster. Since the features belonging to the four groups (color, position, texture, and motion) are extracted and processed using different methods, any comparison may be meaningful only if performed among features belonging to the same group. This statement has been confirmed by simulations results.

In order to implement the weighting strategy related to the a posteriori reliability estimation, the ratios σ_j/σ_{ij}, $i = 1, \ldots, c$, are evaluated for all the features belonging to a group of features Γ (color, texture, position, or motion). As for the a priori reliability estimates in Eq.(6.14), the expression of the distance in the feature space is further modified by inclusion of the a posteriori weighting term.

The *significance weight* $w_{ij}^{(s)}$ is attributed to feature j in cluster Φ_i according to

$$w_{ij}^{(s)} = \frac{\sigma_j^2/\sigma_{ij}^2}{\sum_{\gamma \in \Gamma} \sigma_\gamma^2/\sigma_{i\gamma}^2}, \quad j \in \Gamma. \tag{6.18}$$

In practice, the importance attributed to this feature is redistributed (in relative terms) within each group of features, thus favoring those that proved more significant in the previous step of the segmentation process. The structure of Eq.(6.18) ensures the necessary normalization of the weights to yield a total of 1. The distance measure in the feature space of Eqs.(6.4) and (6.14) is therefore further modified to include the new weights of Eq. (6.18). The complete expression of the distance of Eq. (6.14) is therefore modified to yield

$$d_{ik}^2 = d^2(\mathbf{f}_k, \mu_i) = \sum_{j=1}^{F} w_{kj}^{(r)} w_{ij}^{(s)} \frac{(f_{kj} - \mu_{ij})^2}{\sigma_j^2}. \tag{6.19}$$

By repeating the demonstration of the FCM theorem (Sec. 6.4.3) and taking into account the new definition of the distance, we can demonstrated that the expressions of the membership values u_{ik} in Eq. (6.11) remain unchanged, provided that the distance is evaluated according to Eq. (6.19). As for the expression of the centroid values, Eq. (6.13) is slightly modified, yielding

$$\mu_{ij} = \frac{\sum_{k=1}^{N} (u_{ik})^m w_{kj}^{(r)} f_{kj}}{\sum_{k=1}^{N} (u_{ik})^m w_{kj}^{(r)}}, \quad 1 \leq i \leq c, \ 1 \leq j \leq f. \tag{6.20}$$

To demonstrate how this mechanism works, the following example is provided, with reference to the frame of sequence "table tennis." The output of the first series of FCM iterations was shown in Fig. 6.24(b). Figure 6.25 shows the gray level representation of the horizontal and vertical components of the motion vectors for

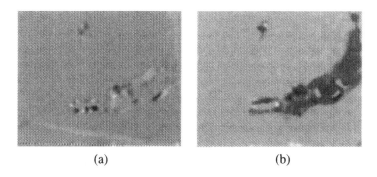

(a) (b)

Fig. 6.25 Temporal features for frame 4 of sequence "table tennis."(a) Horizontal component; (b) vertical component.

frame 4 of sequence "table tennis" after the median filtering. In this gray level representation, the 0 (no motion) is represented by the gray of level 128. Lighter gray levels represent the movements to the right and upward, while darker gray is relative to the movement to the left or downward, respectively.

In particular, let us concentrate our attention on the region corresponding to the hand of the player. The dominant motion is the vertical one. This is confirmed by the aspect of the two separate test images corresponding to the horizontal and the vertical components of the motion, Figs. 6.25(a) and 6.25(b), respectively. The horizontal components appear quite noisy all over the frame, whereas the vertical component clearly identifies the moving areas. This impression is confirmed by the numerical output of the segmentation process and the subsequent significance analysis: The variances over the whole frame are $\sigma_h^2 = 2.4$ and $\sigma_v^2 = 11.9$ for the horizontal and the vertical component, respectively. The variances of the same features over the hand region are, instead, $\sigma_{hand,h}^2 = 18.3$ and $\sigma_{hand,v}^2 = 16.8$. The ratios σ_j^2/σ_{ij}^2 for the two features are, respectively, 0.13 and 0.7. Applying Eq. (6.18), we obtain the values of $w_{hand,h}^{(s)} = 0.15$ and $w_{hand,v}^{(s)} = 0.85$. This repartition of the weight within the motion features accounts well for the different level of significance visible in Figs. 6.25(a) and 6.25(b).

Foregoing results are easily confirmed by the visual appearance of the sequence. The fact that the hand has a predominant motion is evident to the observer. However, the evaluation of the significance of the features and the estimation of the a posteriori reliability plays a determinant role also in less evident cases. In fact the values of the a posteriori weights give us a hint about the criteria that the algorithm has used while segmenting a frame. In other words, it tells us *why* a cluster is a cluster, and why certain pixels have been grouped together. In the example of the hand in "table tennis" the answer is visually evident; in other situations such is not the case. Nevertheless, this clue about the internal functioning of the algorithm is valuable when it comes to extending the segmentation process over time. Intuitively the exploitation of the a posteriori weights is a usefull tracking strategy.

After the evaluation of the weights $w_{ij}^{(s)}$, a second iteration of the FCM algorithm is performed, using the modified expression of distance from Eq. (6.19).

6.5 TEMPORAL TRACKING

6.5.1 Introduction

The algorithmic steps described so far account for the segmentation of single frames in a video sequence. However, it is important that the segmentation obtained for a single frame is coherent with that obtained for temporally adjacent frames. In the case of the algorithm proposed in this chapter, the objective is to track the homogeneous regions along time. If one or more regions have been associated (interactively or by using some additional knowledge) so as to constitute a semantically meaningful object, the tracking at the region level will induce a tracking at the object level, thus allowing the extraction of the actual spatiotemporal visual objects.

The issue of tracking in video segmentation is a challenging one, and it has been addressed widely and deeply in the literature. A survey of the possible solution goes beyond the scope of this work. The interested reader is referred to [6] for a short survey and richer bibliographical reference on the subject.

The tracking mechanism chosen for different segmentation schemes is closely related to the clustering scheme adopted. One needs to identify the information that best represents the clustering in one frame, and then use it to ensure that coherence with the clustering results in temporally adjacent frames. In the case of the FCM scheme adopted in the proposed method, clustering information is essentially "stored" in the vectorial values of the centroids, and in the membership matrix $U = \{u_{ik} \in [0, 1]\}$ that associates to each pixel k its level of membership to the ith class, u_{ik}. The iterative structure of the FCM algorithm requires an initialization step, and this step appears as the most natural phase in which the segmentation results from previous frames can be "injected" in the process relative to the current frame.

Theoretical considerations, supported by simulation results, suggest that the information provided by the membership matrix is used in the tracking phase. A thorough discussion of the rational behind this choice can be found in [6].

6.5.2 Tracking by Initialization of the Membership Matrix

The membership matrix is a representation of the segmentation that retains much more information about the segmentation than the simple set of centroids. Moreover it is possible to modify the matrix so as to retain information on modifications in the pixel assignment (e.g., due to postprocessing) that took place outside the FCM or CFCM iterations.

The mechanism that allows the evaluation of the projected segmentation mask and the initialization of the membership matrix is shown in Fig. 6.26. Unlike the centroids, the membership matrix is related to the physical position of the pixels

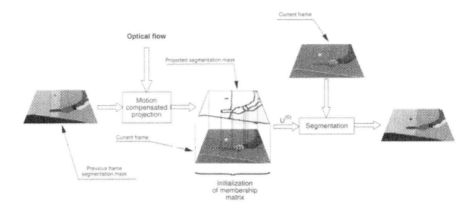

Fig. 6.26 The steps of the initialization of the membership matrix based on the projection of the segmentation mask from a previous frame.

belonging to the different classes, and the motion that occurred between the previous and the current frames needs to be accounted for. This preliminary step is carried on by projecting the segmentation mask of the previous frame onto the current one by taking into due account the information provided by the optical flow estimation.

Let $\mathcal{S}^{(n-1)}$ be the segmentation mask obtained for frame $(n-1)$. We define a projected segmentation mask on frame n by *warping* the $\mathcal{S}^{(n-1)}$ so as to obtain the predicted motion-compensated mask $\hat{\mathcal{S}}^{(n)}$. For the kth pixel, located at coordinates (k_x, k_y), $\hat{\mathcal{S}}^{(n)}$ is evaluated according to

$$\hat{\mathcal{S}}^{(n)}(k_x, k_y) = \mathcal{S}^{(n-1)}\left(x - v_x(k_x, k_y), y - v_y(k_x, k_y)\right), \qquad (6.21)$$

where $v_x(k_x, k_y)$ and $v_y(k_x, k_y)$ are the horizontal and the vertical component of the motion vector associated to pixel k in the optical flow of frame n, respectively. Based on the projected segmentation mask $\hat{\mathcal{S}}^{(n)}$, the initial membership matrix $U^{(0)}$ (Section 6.4.3) for the kth pixel of frame n, located at coordinates (k_x, k_y), is defined as

$$u_{ik}^{(0)} = \begin{cases} 1 & \text{if } \hat{\mathcal{S}}^{(n)}(k_x, k_y) = i \\ 0 & \text{otherwise.} \end{cases} \qquad (6.22)$$

After this initialization the FCM algorithm proceeds as described in Section 6.4.3: at the first step, new centroids are evaluated according to the newly initialized matrix U, and the iterations continue. Already at this first iteration, the centroids are well adapted to the characteristic of the new frame, including the possible occurrence of illumination changes, which is a problem when the initialization is done directly on centroids.

This tracking at the region level eventually induces a tracking at the object level: regions that are attributed to the same object by the initial user interaction are grouped together in the following frames, thus reconstructing the successive temporal instantiations of the visual objects (VOP in the MPEG-4 terminology).

6.6   OBJECT-BASED VIDEO MANIPULATION

The scheme proposed in this dissertation allows the *automatic* extraction of a number of homogeneous regions from a video sequence and their tracking along time. The segmentation at the region level should in general be considered as an intermediate result, since the real value of a segmentation method resides in its capability of providing input for higher level processing so as to achieve semantically meaningful results. When it comes to grouping these regions in order to obtain *objects*, it has been shown in Figs. 6.1, 6.2, 6.3, and 6.4 that the same segmentation in regions can then yield different segmentation at the *object* level, depending on the application. When the criterion for the creation of the objects is semantic, human interaction remains unchallenged as far as flexibility, understanding, accuracy is concerned. The goal of the next two sections is to provide some examples of how the segmentation into regions provided by this algorithm can successfully be used in content-based multimedia applications.

The first application is closely related to the initial motivation of this work, namely the emerging object-based video manipulation schemes for interactive multimedia applications. In this case the regions obtained are grouped together by means of user interaction so as to obtain semantically meaningful spatiotemporal objects. The visual objects constitute the basic element of the MPEG-4 coding standard for the coding of visual information in multimedia environment. Moreover the possibility of addressing single objects in video sequences opens interesting perspectives in the field of video editing, archiving, virtual reality and hypertextual data representation.

In order to influence the creation of the audiovisual objects starting from a segmentation into regions, we propose to use a *graphic user interface* (GUI) similarly to what is done in popular image processing software packages, such as Adobe Photoshop. The user could draw (e.g., by means of a mouse) an approximate contour of the object of interest in the initial frame (as shown in Fig. 6.27, phase 1). The same initial contour is simultaneously matched with the underlying region segmentation (phase 2). By means of a snake-based algorithm (Section 6.3.2) the contour is fitted to the objects of interest. In the specific case, the snake convergence is driven directly by the underlying segmentation at the region level (Fig. 6.27, phase 3).

The selection of the object results in the selection of a certain number of underlying regions, which will be "flagged" as belonging to the object of interest. At each following frame, the selected regions are extracted so that the object of interest is tracked along time (Fig. 6.28).

MPEG-4 is the first content-based audiovisual coding standard where the information is understood as composed by objects, which are coded independently, and can be therefore accessed and manipulated separately. In particular, MPEG-4 provides the tool for the coding of the different audiovisual objects that constitute an audiovisual scene. It should be, however, observed that the upcoming standard will not specify the methods to be used for the creation of the audiovisual objects. A description of the main characteristics of the standard can be found, among others sources, in [29].

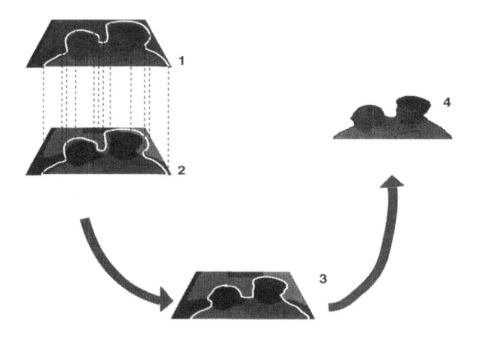

Fig. 6.27 An example of a simple GUI for the grouping of extracted regions into semantic objects.

Fig. 6.28 Extraction of the visual object selected in Fig. 6.27 from sequence "mother and daughter."

A very simple example of object-level manipulation is shown in Fig. 6.29: the man on the left extracted from sequence "hall monitor" has been re-inserted in the sequence with a delay of 20 frames. This functionality is the video equivalent of the *cut and paste* operation in text and image processing. Despite its simplicity this small "trick" demonstrates the opportunities that object-based manipulation can open in fields such as video editing and virtual reality.

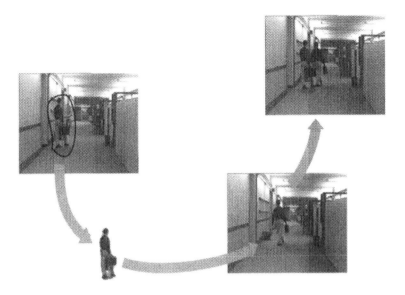

Fig. 6.29 A simple example of object-based editing: Video cut and paste. (The original sequence "hall monitor" is courtesy of Texas Instruments.)

6.7 AUTOMATIC TRAFFIC SURVEILLANCE

6.7.1 Introduction

The segmentation tool proposed in Section 6.4 allows the user to interact with the image analysis process and to guide the final result, as shown in Section 6.6. This interaction is necessary when there is no a priori information on the image data or on the properties of the final objects that need to be extracted.

This section is instead dedicated to automatic segmentation. We intend to provide an example of how the segmentation tool presented in this chapter can be efficiently adapted so as to constitute the basis for a completely automatic traffic surveillance system. This adaptation is made by introducing in the analysis process some additional information, specific to the application. In the case presented in this section, a scene change detection algorithm is applied to the frames of the sequence, before the actual segmentation takes place. In this way it is possible to analyze and extract significant information from the areas of the scene that have changed with respect to a reference frame, which are the most significant in view of traffic surveillance.

In principle, it can be stated that in such applications "everything that changes with respect to a reference frame is potentially interesting." By integrating this rational with the tool presented in Section 6.4, the system is capable of automatically detecting and segmenting objects of interest.

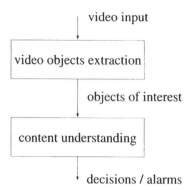

video input

video objects extraction

objects of interest

content understanding

decisions / alarms

Fig. 6.30 Block diagram of an advanced video surveillance system. The video input is processed in order to provide to a content understanding module only the objects of interest that are moving in the scene.

The evolution of advanced video surveillance systems has lead from the so-called first generation CCTV systems (1960–1980) to the second generation PC based systems (1980-today) [30]. This favored the introduction of automatic digital image processing techniques to assist an human operator in video surveillance tasks, thus reducing the need for continuous human attention. The user intervention can therefore be limited to higher level tasks. The user interprets critical situations, makes decisions in doubt cases, and chooses the most appropriate action when an alarm is generated.

As discussed in [31], an automatic tool for video surveillance should achieve the extraction of the information of interest from the video input, and it should provide reliable and filtcrcd data to a content understanding module. The basic structure of the system is depicted in Fig. 6.30. Here, the initial visual data is filtered so as to pass to the content understanding step a reduced amount of data that provides the most significant information on the scene content.

In this framework the use of segmentation techniques in video surveillance systems represents an important preliminary step. For instance, in a traffic monitoring system, it might be interesting to segment out the single vehicles, and evaluates some basic characteristics, such as the position and the speed. Based on these data, it is possible to determine whether a vehicle is stopped along the road, or whether it is violating speed limits.

We propose an approach to video surveillance based on the concept of *focus of attention*. The attention of the system is driven by a change detection mask, defined by the statistical approach that is described in Section 6.7.3. This provides a binary mask that allows us to delimit the areas of interest in the scene. In Section 6.7.5 we explain how these areas are segmented into spatiotemporal homogeneous regions, which are then tracked along the sequence as described in Section 6.7.6. This procedure provides a trajectory for each region that may help to detect automatically the alarm.

6.7.2   Scene Change Detection Techniques

Despite many differences that characterize each surveillance scenario, image motion, as defined in [32], represents a crucial information for most of them. In particular, image motion detection is relevant in surveillance applications where the viewing system is static and one of the goals is to locate mobile objects in the observed scene. The result of a motion detection analysis is a binary mask that indicates the presence or absence of motion for each pixel of the image. Since moving objects generate changes in the image intensity, motion detection is highly related to temporal change detection. However, this relationship is not unique. On the one hand, temporal changes of image intensity can be generated by noise or other external causes like illumination drifts. On the other hand, moving objects generate difficult-to-predict perturbations in the luminance temporal changes. An example is the area referred to as uncovered background, which does not belong to a moving object, but it is generally detected as temporally changed. Another critical area is the overlap of the same object in two successive frames. This area is hard to be detected as changed, when the object is not sufficiently textured.

The last two problems are less critical when the temporal changes are computed between the current image of the sequence and a reference frame that represents only the scene background [33, 34]. In many surveillance applications, this simplified scenario is generally feasible, since it is often possible to choose the reference image when there are no foreground objects in the scene. Another approach consists in reconstructing a background image by integrating the background information from following frames of the sequences [35]. The use of a fixed reference image has an interesting advantage for surveillance applications. With this approach it is possible to detect objects in the scene even if they suddenly stop moving. It is also possible to detect objects that have been removed from the scene. These are important features in most surveillance applications. However, this approach highlights shadows and reflections that often show the same changing properties as the moving objects. The spatial accuracy of the results are therefore degraded. In order to reduce this effect, it is possible to combine the results obtained using the reference frame and the consecutive frames.

Earlier change detection techniques perform a threshold operation on a simple difference image in order to simply extract the moving objects. In Fig. 6.31 we show an example of what we obtain by computing the difference between two images. We observe a high difference (white pixels) in those areas interested by motion. In those areas where motion is not present, the difference represents the noise in the two images. This difference is generally low, but not zero. An ad hoc thresholding strategy is often required to discriminate between pixels changed due to noise or to a real change [36]. Another threshold related to the minimum acceptable dimension of the changed regions is sometimes needed. This is another parameter that has to be

Reference frame   Current frame   Absolute difference

Fig. 6.31 The left and center images represent the same scene with a temporal distance of few seconds. The absolute difference of their luminance intensities is displayed in the right image. Note that it shows several pixels whose values are different from zeros (white pixels) also in areas where no real motion is present. (The highway scenes are available in Postscript/Adobe Portable Document Format via anonymous FTP from ftp://ftp.wiley.com/public/sci_tech_med/image_video/.)

tuned according to the particular scene represented by the sequence. The thresholds that are needed to extract the changed areas have to be tuned manually according to the sequence characteristics and often they need an update along the sequence itself. This main drawback limits this approach for automatic applications.

A method to determine the values of the parameters required by the change detection has been proposed by Teschioni in [37]. The technique aims at the best parameter selection by evaluating the robustness performances of the low-level image processing module (object detector) providing results to the module responsible for the description of the image content (high-level). The optimization of the high level module is performed automatically by searching the best parameters set of the low-level one. However, this optimization step comes after an interactive phase where the user selects the ideal output of the image processing module.

Another problem related to motion detection techniques arises if the contrast of moving objects is not sufficiently high compared to the camera noise. In this case there might not exist a unique threshold able to get rid of noise and preserve, at the same time, the motion information, and detected objects have not precise edges.

To face the problem of false and imprecise object edges due to reflections and shadows, different techniques have been presented in literature. A motion detection based on the difference between consecutive frames and a fuzzy engine based on heuristic rules is used in [38]. To discriminate false moving objects, Anzalone presented a texture change detection [39]. This enforces the reliability of motion detection with respect to reflections. However, the use of a morphological operator in post-processing requires parameter tuning. In [40] the case of an outdoor scenario is considered. The technique extracts the information of occlusion, velocity and length of vehicles and updates periodically the background. A major drawback is that this method is not general and requires hypothesis on the vehicles motion and camera's view.

To overcome these problems an approach based on a statistical decision rule is adopted. The method we apply was proposed by Aach [41] and has already found several applications in computer vision [16, 17]. It is presented in the next section.

6.7.3   Statistical Change Detection

Instead of simply thresholding the difference image, the statistical change detection approach compares the statistical behavior of a small neighborhood, η_k, of each pixel position k in the squared difference image with a model $\mathcal{M}$ of the noise that is expected to affect the squared difference image. A significance test is used to compare the statistics in η_k with the model $\mathcal{M}$ and to decide according to a significance threshold α whether the pixel k belongs to a changed area or to a noisy area in the image.

The model $\mathcal{M}$ is defined based on two assumptions. First, each frame of the sequence is affected by an additive Gaussian noise. Second, all the pixels in η_k have changed because of noise and not because of image motion. The neighborhood η_k is generally chosen as a square window centered at the pixel position k. Under the above hypothesis the model $\mathcal{M}$ is simply described by a χ^2 distribution whose properties depend on the number of pixels in η_k and on the variance of the Gaussian noise affecting each frame of the sequence [42, page 41].

The performances of the proposed statistical change detection depends on the choice of the neighborhood η_k, the properties of the Gaussian noise affecting the images, and the threshold α of the significance test. These issues are discussed in the following items.

- The choice of the neighborhood η_k should satisfy the compromise between reliability of the statistical analysis in η_k and the validity of the hypothesis on which the method is based. In particular, by increasing the dimension of η_k, the statistics computed for the squared difference image is more reliable. This results in a reduced sensibility to noise for the computation of the changed mask. On the other hand, in this case the hypothesis that all the pixels in η_k have changed because of noise has a lower degree of confidence. This may lead to a wrong labeling along the edge of moving objects and to the corresponding blocking effects in the final mask (*residual background*). This characteristic is shown in Fig. 6.32. In our tests we defined η_k as a 3×3 window.

- The properties of the Gaussian noise affecting the images are defined by a single parameter: the variance σ. Increasing the value of σ makes the algorithm less sensible to the changes between the reference and the current frame. Unless this parameter can be deduced by the technical properties of the acquisition system, its online estimation is required. This is important in order to adapt the noise model to external changes of the illumination that may occur along the sequence. To avoid confusing changes due to motion with changes due to noise and vice versa, it is important to locally estimate the variance of the noise. In Aach [41], the estimation of σ is performed on a neighborhood η_k^* bigger than the neighborhood η_k. However, this solution does not take into

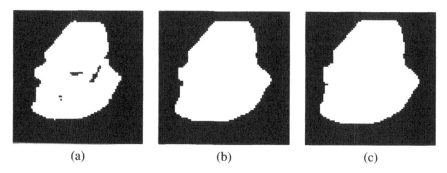

(a)   (b)   (c)

Fig. 6.32   Change detection masks from the same difference image but with different neighborhoods η_k. (a) η_k of 9 pixels (3×3 window). (b) η_k of 25 pixels (5×5 window). (c) η_k of 49 pixels (7×7 window).

account the fact that the statistics in η_k^* may combine both changes due to motion and changes due to noise. This may result in a wrong estimation of σ. To overcome this limitation, first an image motion detection is performed on the whole frame using as σ a value depending on the technical properties of the acquisition system. Then the areas that appear not affected by motion are used to evaluate a new value for σ. This new estimation takes into account the real noise affecting the actual frame. It is used to compute a new motion detection mask for the actual frame.

• The last required parameter is the significance threshold α that defines what is the probability of rejecting the hypothesis that no changes due to motion occur in η_k, although it is true. This is the reason why α should be as small as possible. Experimental results show that values ranging from 10^{-2} to 10^{-6} provide correct and similar masks.

The preceding discussions demonstrates that the suggested statistical approach does not require fine-tuning of any threshold and does not greatly increase the computational load compared to simple threshold techniques. The changing areas are generally detected with good spatial precision when the statistical change detection technique is applied to local information.

Besides its interesting properties, this method presents some limitations for surveillance applications. First it is not able to distinguish a moving object from its shadow or reflection since both are detected as changed areas. Next it only provides a binary description of the changed objects. This might be a limitation for those applications where details of the moving objects are required (face recognition in building surveillance, license plate extraction, etc.). In order to overcome these limits the binary mask generated by the statistical change detection is used to define the areas of interest in the scene. These can be analyzed by the segmentation tool described in Section 6.3.

6.7.4   Traffic Surveillance: Requirements and State of the Art

For some traffic surveillance applications, the detection of the areas where a change in the scene has occurred with respect to a reference frame could already be an appropriate solution. For instance, this is true when the video-based surveillance system is used as an alternative to physical detectors, such as magnetic loops, for an approximate vehicle counting or an estimation of traffic flow conditions. However, this solution is uneffective if a more precise analysis of the traffic conditions is required.

An advanced video-based surveillance system is required to detect and to monitor objects in a scene. In this case the different objects have to be tracked along time. Tracking moving objects is an important task in video surveillance, requiring the following characteristics: first the automatic segmentation of the vehicles from the background, then the ability of dealing with different traffic conditions and variety of objects. In the following, we present a brief overview of tracking methods that have been proposed in the framework of traffic surveillance.

Region-based tracking [34]   A region-based method tracks blobs of pixels, which roughly correspond to vehicles, and it relies on information provided by the entire region (motion, color, texture properties). This approach works properly only in case of free flow traffic and where the camera is placed in a high position with respect to the road. In congested traffic conditions, vehicles partially occlude each other, and this method groups together more vehicles in a unique large blob. In such a way the track of a single vehicle is lost.

Active contour-based tracking [43, 44]   This approach uses active contour models (snakes), and it relies on the information provided by the object boundaries. A contour-based representation can help in reducing the computational complexity, and it allows the tracking of both rigid and nonrigid objects. However, it is unable to track vehicles that are partially occluded. To overcome the problem of partial occlusions, in [45] a Kalman filtering approach and optical flow measurements are introduced in the active contour model.

3-D model-based tracking [46]   The problem of tracking vehicles that are partially occluded can be solved by considering their 3-D models. The definition of parameterized vehicle models makes it possible to exploit the a priori knowledge about the shape of typical objects in traffic scene. This approach is computationally intensive, and it presents two major drawbacks: first the need of object models with detailed geometry for all vehicles that might be found in the scene, and second, the lack of generality. This second drawback does not allow the system to detect objects different from vehicles, such as people or animals, that are interesting for interpreting the scene in case of dangerous situations.

Feature-based tracking [47]   Using a feature, instead of the entire object, its subparts are tracked. An example of these subparts is the corners of an objects, which allow the tracking despite partial occlusions. A major drawback is the problem of grouping the features by finding which features belong to the same object.

In considering these previous works, we propose a design that is a hybrid between the region-based and the feature-based techniques. It exploits the good characteristics of the two by considering first the object as an entity and then by tracking its subparts. In a first stage the motion detection algorithm identifies the objects from the background providing binary masks of the moving objects. In the second stage, objects are tracked by projecting and re-clustering the regions obtained with the tool proposed in Section 6.4. The details of this method are presented in the following section.

6.7.5   Segmentation Driven by Focus of Attention

The binary mask provided by the change detector is extended to guiding an additional step that analyzes the characteristics of the moving objects. These characteristics are extracted by a segmentation procedure that takes into account the spatial and temporal properties of the detected objects and defines the areas of interests in spatiotemporal regions. The entities that we consider are the "blobs," Fig. 6.33(c), defined by the binary change detection mask, and the subparts are the clusters obtained with the fuzzy C-means algorithm, which is applied on the multiple feature space presented in Section 6.3.

The attention-focusing approach consists in clustering only those pixels that are detected as changed and correspond to moving objects. This way each object is processed separately, and appears as a blob in the change detection mask. This results in a set of nonoverlapping clusters or regions, and each region is characterized by the values of its centroids. The information provided to the content-understanding step is in the form of regions and associated descriptors. The regions represent the different parts of the objects of interest, while the descriptors characterize their spatial and temporal properties: size, shape, color, texture, motion, trajectory, and so on. In order to link these properties to the correct object, a tracking procedure has to be defined, as presented in the following section.

6.7.6   Automatic Object Tracking

The tracking process integrates sequential information and provides a trajectory for each region: this may help in automatically recognizing alarm situations. One of the advantages of this approach is the ability to handle partial occlusions, and the appearance and disappearance of objects.

In considering only the areas of interest in images, some important issues may be raised in relation to to the management of the different video objects. The objects are divided into nonoverlapping regions by the clustering algorithm, and these regions must be assigned to the correct blob at each instant of time. In particular, the

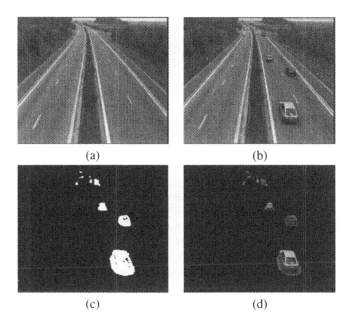

(a)   (b)

(c)   (d)

Fig. 6.33 Traffic surveillance sequence highway "n25w22" (courtesy of European ACTS project 304 Modest): (a) reference frame; (b) frame n.215; (c) corresponding change detection mask; (d) moving objects defined by the change detection mask. (The highway scenes are available in Postscript/Adobe Portable Document Format via anonymous FTP from `ftp://ftp.wiley.com/public/sci_tech_med/image_video/`.)

centroids assigned to an object in frame $n - 1$ have to be correctly transmitted to the same object that moved in frame n. The current change detection mask provides only binary information that is integrated with the motion information used for the tracking.

Thus the position and the motion information provided by the centroids are taken into account for the tracking (Fig. 6.34). In order to match objects in two successive frames, the center of mass of objects in frame n and the center of mass of the motion compensated (from frame $n - 1$) centroids are considered, as in Eqs. (6.23) and (6.26). Let $\mathbf{p}$ be the generic pixel position, n the frame number, and $T_b^{(n)}$ the total number of pixels in a generic blob $B_b^{(n)}$ in frame n with $b = 1, 2, \ldots, N^{(n)}$, where $N^{(n)}$ represents the total number of blobs in frame n. Thus the mean position $\mathbf{m}_b^{(n)}$ of the pixels belonging to each $B_b^{(n)}$ is given by

$$\mathbf{m}_b^{(n)} = \sum_{\mathbf{p} \in B_b^{(n)}} \frac{\mathbf{p}}{T_b^{(n)}}. \tag{6.23}$$

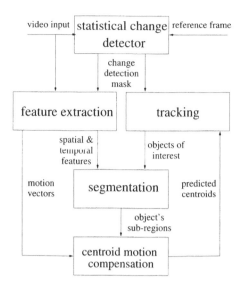

Fig. 6.34   Block diagram of the proposed tracking mechanism.

Now we have to compute the motion compensated position of the centroids in the new frame. Let $C_b^{(n)}$ be the number of centroids in $B_b^{(n)}$ and $\boldsymbol{\mu}_{k,b}^{(n)}$ the kth centroid $(k=1, 2, \ldots, C_b^{(n)})$ of $B_b^{(n)}$, with

$$\boldsymbol{\mu}_{k,b}^{(n)} = [\mathbf{p}_{k,b}^{(n)},\ \mathbf{d}_{k,b}^{(n)},\ \mathbf{c}_{k,b}^{(n)},\ t_{k,b}^{(n)}], \qquad (6.24)$$

where $\mathbf{p}_{k,b}^{(n)}$ is the position vector, $\mathbf{d}_{k,b}^{(n)}$ is the displacement vector, $\mathbf{c}_{k,b}^{(n)}$ is the color vector, and $t_{k,b}^{(n)}$ is the value of texture of centroid $\boldsymbol{\mu}_{k,b}^{(n)}$. We can compute the motion compensated prediction of the position of the centroids according to their motion information for each k in $B_b^{(n)}$:

$$\tilde{\mathbf{p}}_{k,b}^{(n)} = \mathbf{p}_{k,b}^{(n-1)} + \mathbf{d}_{k,b}^{(n-1)}. \qquad (6.25)$$

For each $B_b^{(n-1)}$ we find its mean projected position $\tilde{\mathbf{q}}_b^{(n)}$ in frame n:

$$\tilde{\mathbf{q}}_b^{(n)} = \sum_{k \in (1,\ldots,C_b^{(n)})} \frac{\tilde{\mathbf{p}}_{k,b}^{(n)}}{C_b^{(n-1)}}. \qquad (6.26)$$

To determine the correspondence between the objects in the current frame n and the centroids of the previous frame $n - 1$, the Euclidean distance D between all the objects is computed. This results in the following relationship between the mean predicted position of the centroids and the mean position of the blobs for each $B_b^{(n)}$:

$$D\left(\tilde{\mathbf{q}}_{b_1}^{(n)}, \mathbf{m}_b^{(n)}\right) = \|\tilde{\mathbf{q}}_{b_1}^{(n)} - \mathbf{m}_b^{(n)}\|_2. \tag{6.27}$$

The minimum distance for each $b_1 = 1, 2, \ldots, N^{(n-1)}$, $b = 1, 2, \ldots, N^{(n)}$ will allow us to assign the centroids to their corresponding object. Once the correspondence between the new detected objects and the old centroids is established, it is possible to begin the segmentation process in the new frame using the old projected centroids as initialization. The properties of the objects are therefore transmitted from frame to frame, allowing the automatic tracking of each single vehicle.

Figure 6.35 gives an example of the automatic object extraction and tracking capabilities of the proposed algorithm. Six frames taken from a highway sequence recorded during daylight by a stationary camera are presented. The vehicle is automatically extracted and tracked along the frames after the statistical change detector has provided the mask containing all the moving objects. The figure shows the tracking of a single vehicle from its appearing in the camera scope in frame 20, until it exits the scene in frame 130. In the same way all the other objects in the scene are automatically extracted and separately tracked along the frames, thus providing the content understanding module with segmented objects and their associated trajectories. This information will help this last module in describing events in the scene and in generating alarms in case of dangerous situations.

6.8   CONCLUSIONS

This chapter addresses an important and rather complex problem: segmentation and tracking of video objects. We discussed here a general architecture for segmentation and tracking of video information that we believe to be efficient and flexible enough to be used in a large number of applications including those addressed in the past, present and future, thanks to its modular structure. The basic idea of this architecture resides in the distinction between the notion of regions and the notion of objects. This distinction leads to an elegant separation of the problem of segmentation into two subproblems, in which one is more appropriately resolvable by computers (region extraction) and one more appropriately resolvable by a priori assumptions of the application or interaction of the user (object identification). To be concrete, this general architecture was then implemented making use of adequate components. To this end, a multi-feature clustering approach was developed to extract regions, taking into account a priori and a posteriori assumptions to improve such extraction. Other components such as a tracking mechanism and attention focusing were added to the scheme. They were adapted to interactive video editing and automatic segmentation, and tracking for surveillance applications. It is important to mention that the proposed architecture for segmentation and tracking remains general enough for a wide range of other applications such as object based video coding, computer vision, scene understanding, and content-based indexing and retrieval. Also depending on the

Fig. 6.35 Traffic surveillance sequence highway "n25w22" (Courtesy of European ACTS project 304 Modest). First and third row: Original frames n.20, n.50, n.80, n.110, n.120, and n.130 (available in Postscript/Adobe Portable Document Format via anonymous FTP from `ftp://ftp.wiley.com/public/sci_tech_med/image_video/`); second and forth row: example of object tracking: the vehicle has been automatically extracted and tracked.

constraints of the application, such as limitations on delay and complexity, each component used in this implementation can be replaced by a more adequate one without changing the general approach. This architecture is also compatible with a view of image analysis and scene understanding that divides the problem into a low-level stage and a high-level one. The low-level stage is equivalent to region-oriented processing and the high-level stage is related to object-level processing in the proposed architecture. Importantly this architecture also allows for interaction among low and high levels (e.g., region versus object). Understanding this interaction is one of the directions of future research.

6.9   ACKNOWLEDGMENTS

The authors are grateful to Dr. John Wang of the Hewlett Packard Laboratories in Palo Alto, California, and Dr. Philippe Schroeter of the Swiss Telecom for their contribution to the work presented in this chapter. They would like to thank Mr. Diego Santa Cruz, Mr. Massimiliano Lenardi, and Mr. Andrea Sodomaco for their contribution to the development of the software.

REFERENCES

1. R.A. Schowengerdt. *Techniques for Image Processing in Remote Sensing*. Academic Press, San Diego, CA, 1983.

2. A.P. Dhawan and L. Arata. Segentation of medical images through competivite learning. In *Proc. IEEE Int. Conf. Neural Networks*, Vol. 3, San Francisco, CA, March 1993, pp. 1277-1282.

3. B. Klaus and P. Horn. *Robot Vision*. MIT Press, Cambridge, 1986.

4. N. Negroponte. *Being Digital*. Hodder and Stoughton, London, 1995.

5. P. Correia and F. Pereira. The role of analysis in content-based video coding and indexing. *Signal Processing*, 66(2): 125-142, April 1998.

6. R. Castagno. *Video Segmentation Based on Multiple Features for Interactive and Automatic Multimedia Applications*. Ph.D. thesis. Swiss Federal Institute of Technology, Lausanne, Switzerland, 1998.

7. A.M. Bensaid, L.O. Hall, J.C. Bezdek, and L.P. Clarke. Partially supervised clustering for image segmentation. *Pattern Recognition*, 29(5): 859-871, May 1996.

8. V.M. Bove. Algorithms and systems for modeling moving scenes. In *Proc. 8th European Signal Processing Conf.*, Vol. 3, Trieste, Italy, September 1996, pp. 1685-1688.

9. E. Chalom. *Statistical Image Sequence Segmentation Using Multidimensional Attributes*. Ph.D. thesis. Massachusetts Institute of Technology, January 1998.

10. E. Chalom and V.M. Bove. Segmentation of an image sequence using multidimensional image attributes. In *Proc. 1996 IEEE Int. Conf. Image Processing*, Vol. 2, Lausanne, Switzerland, September 1996, pp. 525-528.

11. P. Salembier, F. Marques, M. Pardas, J.R. Morros, I. Corset, S. Jeannin, L. Bouchard, F. Meyer, and B. Marcotegui. Segmentation-based video coding system allowing the manipulation of objects. *IEEE Trans. Circuits and Systems for Video Technology*, 7(1): 60-74, February 1997.

12. *IEEE Trans. Circuits and Systems for Video Technology*, Vol. 9, December 1999. Special issue object-based video coding and description.

13. MPEG Video and SNHC Groups. Committee draft of MPEG-4, part 2, 14496-2. Technical Report ISO/IEC JTC/SC29/WG11/N1902, ISO/IEC, Fribourg, Switzerland, October 1997. Available http://drogo.cselt.stet.it/mpeg/.

14. E. Chalom and V.M. Bove. Segmentation of frames in a video sequence using motion and other attiributes. In *Proc. SPIE Digital Video Compression, Algorithms, and Technologies*. Vol. 2419, 1995, pp. 230-241.

15. O. Pichler, A. Teuner, and B.J. Hosticka. A comparison of texture feature extraction using adaptive gabor filtering, pyramidal and tree structured wavelet transforms. *Pattern Recognition*, 29(5): 773-742, 1996.

16. F. Moscheni. *Spatio-Temporal Segmentation and Object Tracking: An Application to Second Generation Video Coding*. Ph.D. thesis. Swiss Federal Institute of Technology, Lausanne, Switzerland, 1997.

17. F. Moscheni, S. Bhattacharjee, and M. Kunt. Spatiotemporal segmentation based on region merging. *IEEE Trans. Pattern Analysis and Machine Intelligence*, 20(9): 897-915, 1998.

18. P. Schroeter. *Unsupervised Two-Dimensional and Three-Dimensional Image Segmentation*. Ph.D. thesis. Swiss Federal Institute of Technology, Lausanne, Switzerland, 1996.

19. P. Schroeter and J. Bigun. Hierarchical image segmentation by multi-dimensional clustering and orientation-adaptive boundary refinement. *Pattern Recognition*, 28(5): 695-709, May 1995.

20. J.L. Barron, D.J. Fleet, and S.S. Beauchemin. Performance of optical flow techniques. *Int. J. Computer Vision*, 12(1): 43-77, February 1994.

21. B. Lucas and T. Kanade. An iterative image registration technique with an application to stereo vision. In *Proc. DARPA Image Understanding Workshop*, 1981, pp. 121-130.

22. E.P. Simoncelli, E.H. Adelson, and D.J. Heeger. Probability distribution of optical flow. In *Proc. IEEE Conf. Computer Vision and Pattern Recognition*, Maui, Hawaai, 1991, pp. 310-315.

23. J. Shi and C. Tomasi. Good features to track. In *Proc. IEEE Conf. Computer Vision and Pattern Recognition*, Seattle, WA, June 1994, pp. 593-600.

24. J.C. Bezdek. *Pattern Recognition with Fuzzy Objective Function Algorithm.* Plenum Press, New York, 1981.

25. J.C. Bezdek, R.J. Hathaway, M.J. Sabin, and W.T. Tucker. Convergence theory for Fuzzy C–Means: Counterexamples and repairs. *IEEE Trans. Systems, Men, and Cybernetics*, 17(5): 873-877, September-October 1987.

26. J.C. Bezdek. A convergence theorem for the fuzzy isodata clustering algorithms. *IEEE Trans. Pattern Analysis and Machine Intelligence*, 2(1): 1-8, January 1980.

27. R. Castagno, T. Ebrahimi, and M. Kunt. Video segmentation based on multiple features for interactive multimedia applications. *IEEE Trans. Circuits and Systems for Video Technology*, Special Issue on Image and Video Processing for Emerging Interactive Multimedia Services. 8(5): 562-572, September 1998.

28. J.C. Dunn. A fuzzy relative of the isodata process and its use in detecting compact well-separated clusters. *J. Cybernetics*, 3: 32-57, 1974.

29. http://drogo.cselt.stet.it/mpeg/.

30. C. Sacchi and C.S. Regazzoni. Multimedia communication techniques for remote cable-based video-surveillance systems. In *Proc. 10th Int. Conf. Image Analysis and Processing*, 1999, pp. 1100-1103.

31. F. Ziliani and A. Cavallaro. Image analysis for video surveillance based on spatial regularization of a statistical model-based change detection. In *Proc. 10th Int. Conf. Image Analysis and Processing*, 1999, pp. 1108-1111.

32. A. Mitiche and P. Bouthemy. Computation and analysis of image motion: A synopsis of current problems and methods. *Int. J. Computer Vision*, 19(1): 29-55, 1996.

33. G. W. Donohoe, D. R. Hush, and N. Ahmed. Change detection for target detection and classification in video sequences. In *Proc. 1988 IEEE Int. Conf. Acoustics, Speech, and Signal Processing*, New York, 1988, pp. 1084-1087.

34. K. P. Karmann, A. Brandt, and R. Gerl. Moving object segmentation based on adaptive reference images. In *Proc. 5th European Signal Processing Conf.*, Barcelona, 1992, pp. 951-954.

35. A. Makarov. Comparison of background extraction based intrusion detection algorithms. In *Proc. 1996 IEEE Int. Conf. Image Processing*, 1996, pp. 521-524.

36. P. L. Rosin. Thresholding for change detection. In *Proc. IEEE Int. Conf. Computer Vision*, 1998, pp. 274-279.

37. A. Teschioni and C.S. Regazzoni. Performance evaluation strategies of an image processing system for surveillance applications. In *Advanced Video Surveillance Systems*. Kluwer Academic, Boston, 1999, pp. 76-90.

38. M. Barni, F. Bartolini, V. Cappellini, F. Lombardi, and A. Piva. Fuzzy motion detection for highway traffic control. In *Advanced Video Surveillance Systems*. Kluwer Academic, Boston, 1999, pp. 58-66.

39. A. Anzalone and A. Machi. Video based management of traffic light at pedestrian crossing. In *Advanced Video Surveillance Systems*. Kluwer Academic, Boston, 1999, pp. 49-57.

40. T. Nakanishi and K. Ishii. Automatic vehicle image extraction based on spatio-temporal image analysis. In *Proc. 11th Int. Conf. Pattern Recognition*, 1992, pp. 500-504.

41. T. Aach, A. Kaup, and R. Mester. Statistical model-based change detection in moving video. *Signal Processing*, 31: 165-180, 1993.

42. S. G. Wilson. *Digital Modulation and Coding*. Prentice Hall, Englewood Cliffs, NJ, 1996.

43. D. Koller, J. Weber, and J. Malik. Robust multiple car tracking with occlusion reasoning. In *Proc. European Conf. Computer Vision*, 1994, pp. 189-196.

44. N. Paragios and R. Deriche. Geodesic active regions for motion estimation and tracking. In *Proc. 7th IEEE Int. Conf. Computer Vision*, 1999, pp. 688-694.

45. N. Peterfreund. Robust tracking of position and velocity with Kalman snakes. *IEEE Trans. Pattern Analysis and Machine Intelligence*, 21(6): 564-569, June 1999.

46. D. Koller, K. Danilidis, and H.H. Nagel. Model-based object tracking in monocular image sequences of road traffic scenes. *Int. J. Computer Vision*, 10(3): 257-281, 1993.

47. D. Koller, K. Danilidis, and H.H. Nagel. Model-based object tracking in monocular image sequences of road traffic scenes. In *Proc. IEEE Computer Vision and Pattern Recognition*, 1997, pp. 495-501.

7 Invariant Features in Pattern Recognition – Fundamentals and Applications

H. BURKHARDT and S. SIGGELKOW

Albert-Ludwigs-Universität Freiburg
Computer Science Department
Institute for Pattern Recognition and Image Processing[1]
79085 Freiburg, Germany

7.1 INTRODUCTION

The aim of invariant pattern recognition is to identify objects no matter from which position they are observed. All images of the same object represent the same object, so their position and orientation are irrelevant for the task to classify them. Mathematically speaking, patterns form an equivalence class with respect to a group action $g \in G$ describing the geometric coordinate transformation:

$$\mathbf{x}_1 \overset{G}{\sim} \mathbf{x}_2 \Leftrightarrow \exists g \in G : \mathbf{x}_2 = g(\mathbf{x}_1). \tag{7.1}$$

We consider image transformations of the following kind:

$$\mathbf{n} = (n_0, n_1) \tag{7.2}$$

be the coordinates of an image $\mathbf{x}(\mathbf{n})$ or a subset of it like contour or single points. We assume in the most general case an affine transformation of the image coordinates:

$$\mathbf{n}' = \mathbf{A}\mathbf{n} + \mathbf{t}. \tag{7.3}$$

---

[1] http://www.informatik.uni-freiburg.de/~lmb

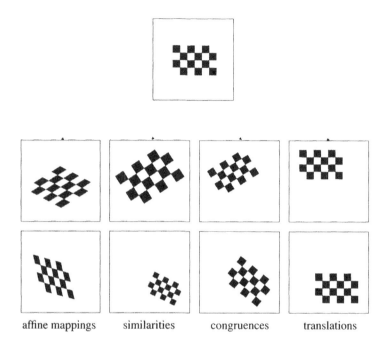

Fig. 7.1   Illustration of the transformation groups of Eq. (7.4). (The original color figure is available in Postscript/Adobe Portable Document Format via anonymous FTP from `ftp://ftp.wiley.com/public/sci_tech_med/image_video/`.)

The following special cases, which are illustrated in Fig. 7.1, will be discussed here:

$$
\begin{aligned}
\mathbf{A} = \mathbf{I} & \quad \text{group of translations } (p = 2 \text{ parameters}), \\
\mathbf{A}^T \mathbf{A} = \mathbf{I} & \quad \text{group of congruences } (p = 3 \text{ parameters}), \\
\mathbf{A}^T \mathbf{A} = k\mathbf{I} & \quad \text{group of similarities } (p = 4 \text{ parameters}), \\
\det(\mathbf{A}) = |\mathbf{A}| \neq 0 & \quad \text{group of affine mappings } (p = 6 \text{ parameters}).
\end{aligned}
\tag{7.4}
$$

The elements of these groups can also be expressed by a parametric formulation; that is, for each parameter vector $\boldsymbol{\lambda}$ there is an associated group element $g(\boldsymbol{\lambda})$. $p = \dim(\boldsymbol{\lambda})$ denotes the degrees of freedom in the parametric formulation. For example, the group of congruences (group of Euclidean motion) can be described by three parameters: 2 for translation in horizontal and vertical direction and 1 for rotation, as can be seen in Eq. (7.35).

In pattern recognition, a pattern in an unknown position is compared with different prototypes. The brute force and direct solution to this problem is to compare the images in all possible transformed positions and to find the optimal coincidence (direct matching). That is, using the Euclidean metric, one ends up finding the maximum of a high-order correlation function between the images, which is a rather time-consuming operation. The costs are growing exponentially with the degrees of freedom describing the coordinate transformation.

Invariant features are an elegant way to solve the problem. The idea is to find a mapping T that is able to extract the intrinsic features of an object, namely the features that stay unchanged if the object's position and/or orientation changes. Such a transformation T necessarily maps all images of an equivalence class into one point of the feature space:

$$\mathbf{x}_1 \overset{G}{\sim} \mathbf{x}_2 \quad \Rightarrow \quad T(\mathbf{x}_1) = T(\mathbf{x}_2). \tag{7.5}$$

Besides this property there are a number of additional requirements to make the invariants useful for real applications:

1. Measure of completeness (to avoid ambiguities).

2. Clustering properties (continuity with respect to a certain metric).

3. Discrimination performance (SNR in feature space).

4. Behavior with respect to systematic and stochastic errors.

5. Computational complexity and considerations for hard- and software implementation.

Completeness   The set of invariants of an image/object $\mathbf{x}$ with respect to a mapping T is given by all elements that are mapped by T into one point:

$$I_T(\mathbf{x}) := \{\mathbf{x}_i | T(\mathbf{x}_i) = T(\mathbf{x})\}. \tag{7.6}$$

Further we denote by $G(\mathbf{x})$ the set of objects/images within one equivalence class, that is, all images that can be generated from a prototype $\mathbf{x}$ by applying the transformation group G. This set is called *orbit*, namely

$$G(\mathbf{x}) := \{\mathbf{x}_i | \mathbf{x}_i \overset{G}{\sim} \mathbf{x}\}. \tag{7.7}$$

From Eq. (7.5) we can conclude that

$$G(\mathbf{x}) \subseteq I_T(\mathbf{x}). \tag{7.8}$$

The properties of an invariant transformation T are depicted in Fig. 7.2.

A mapping T that is invariant with respect to G is said to be *complete* if both sets are equal (sufficient condition):

$$T(\mathbf{x}_1) = T(\mathbf{x}_2) \quad \Rightarrow \quad \mathbf{x}_1 \overset{G}{\sim} \mathbf{x}_2. \tag{7.9}$$

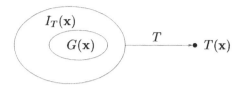

Fig. 7.2   Properties of an invariant transformation T.

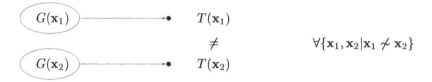

Fig. 7.3   Completeness of an invariant transformation T.

This property is illustrated in Fig. 7.3. So, if we meet the necessary conditions Eq. (7.5), as well as the sufficient condition, Eq. (7.9), we require that

$$T(\mathbf{x}_1) = T(\mathbf{x}_2) \quad \Leftrightarrow \quad \mathbf{x}_1 \overset{G}{\sim} \mathbf{x}_2. \tag{7.10}$$

As proved in [9], complete mappings are necessarily *nonlinear* mappings with contracting behavior. A linear mapping would have a nontrivial null-space in order to fulfill the necessary condition, Eq. (7.5). Given an element $\mathbf{n}$ of the null-space, all $\rho\mathbf{n}$ ($\rho \in \mathbb{R}$) would also be elements of the null-space. Therefore all objects $\mathbf{x} + \rho\mathbf{n}$ would be equivalent. This, however, means that nontrivial equivalence classes of linear systems are always unbounded ($\rho \to \infty$). So one cannot find complete *linear* invariants for compact transformation groups (against many statements in literature where, e.g., wavelets are used to characterize anisotropic textures).

Summarizing one can say:

Linear systems can discriminate between
"frequencies" (spectral components)

whereas

Nonlinear systems are able to discriminate between
"equivalence classes".

The advantage of complete mappings T is that there exist no ambiguities. With noncomplete transformations, instead, one can obtain the same features for different equivalence classes and thus cannot distinguish them. Therefore it is desirable to find transformations with a "high degree" of completeness; that is, $|I_T(\mathbf{x})|$ will come close to $|G(\mathbf{x})|$.

In practice, the separability of a fixed set of classes may be sufficient thus weakening the demand for completeness. However, in having completeness, one can easily exchange or add classes without getting ambiguities. This implies a high degree of generalization of the approach.

Clustering properties   We need good clustering properties of the elements in the feature space as a prerequisite for a simple classifier design. For small distortions in the image space we should also expect small changes in the feature space. Such changes may be characterized by the continuity of the used transformation with respect to a certain metric. Larger deviations within an area of the original space should also cluster in the feature space.

Discrimination performance   The classification performance is characterized by the distance of the object clusters in relation to the variance of the images due to noise. This signal-to-noise ratio should not be decreased by the mapping into the feature space.

Noise sensitivity   Images are disturbed by typical deterministic and stochastic errors like changes in brightness and contrast, and Gaussian noise. The chosen mapping T should produce easily interpretable results with respect to these characteristic disturbances.

Computational complexity   The computational complexity of the chosen transformation T should be as low as possible. A definite upper bound is given by the amount of computation that is necessary to calculate the brute force solution in form of a correlation in all parameters of the underlying group. A further possibility to accelerate an algorithm may be achieved by a computation on parallel hardware. Therefore the aspect of a parallel implementation of the transformation algorithm T is of fundamental interest [6, 8, 13]. Very often the computational complexity is linear. In a special case it can be even reduced to constant complexity [29].

7.2   PRINCIPLES FOR THE CONSTRUCTION OF INVARIANT FEATURES

In Eq. (7.5) we defined the key characteristics of invariant features. However, we have not discussed yet how to construct invariants systematically. To the best knowledge of the authors all principles to find invariants can be subdivided into three categories:

1. *Normalization.* Normalization methods utilize extreme points on orbits and normalize the object with respect to these. For example, Fourier descriptors normalize the object with respect to the main axes of ellipses [4, 1].

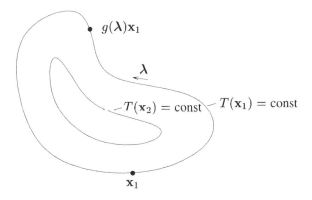

Fig. 7.4   Orbits of equivalent objects.

2. *Differential approach.* This approach considers invariant features obtained by solving partial differential equations (Lie theory) [33]. It is obvious that the features must be insensitive to infinitesimal variations of the parameters λ of the group; hence we get

$$\frac{\partial T(g(\lambda)\mathbf{x})}{\partial \lambda_i} \equiv 0, \quad i = 1, 2, \ldots, p. \tag{7.11}$$

3. *Integral approach.* The equivalence class of an object $\mathbf{x}$ forms an orbit in object space parameterized by λ (Fig. 7.4). The idea is to average arbitrary functions evaluated on the orbit (Haar integrals) [12, 14, 25]. It is obvious that the integral over the entire orbit is independent of the parameter λ characterizing the transformation, and therefore it is invariant to the transformation group:

$$T(\mathbf{x}) = \frac{1}{|G|} \int_G f(g\mathbf{x}) dg, \tag{7.12}$$

with $|G| = \int_G dg$ being the "power" of the group.

We emphasize that we consider only approaches that systematically try to bring $I_T(\mathbf{x})$ close to $G(\mathbf{x})$. Of course there exist other methods. However, these mappings have an invariant set $I_T(\mathbf{x})$ that includes $G(\mathbf{x})$ and in general is much larger than $G(\mathbf{x})$. As an example consider the histogram of an object $\mathbf{x} \in \mathbb{R}^N$. This, of course, is invariant to the symmetric group of all permutations and therefore also invariant to cyclic translations. However, the symmetric group has size $N!$, whereas the group of cyclic translations has size N only. So there is a large amount of ambiguities.

In the following, we will discuss examples of each of the three categories mentioned above. The third method-invariant integration-will be treated in more detail.

7.3   INVARIANTS BY NORMALIZATION

The goal of image normalization is to find a representative member of the equivalence class. We will consider object contours in this section and discuss two methodologies. The first performs an affine normalization [21, 20] up to remaining scaling, translation, and rotation; the second performs a normalization in the case of contour similarities by calculating Fourier descriptors [4]. The second method could therefore be used after the first method to solve the remaining normalization.

7.3.1   Affine Normalization

Let an object's contour be given as a polygon with N points $\mathbf{n}^i = (n_0^i, n_1^i)$, $i = 0, 1, \ldots, N$, and $\mathbf{n}^0 = \mathbf{n}^N$. This can be brought to normal form (minimal perimeter under the constraint of a fixed area ($\det \mathbf{A} = 1)^2$) by the following minimization:

$$\sum_{i=0}^{N-1} \left\| \begin{pmatrix} a & 0 \\ 0 & 1/a \end{pmatrix} \begin{pmatrix} \cos\varphi & -\sin\varphi \\ \sin\varphi & \cos\varphi \end{pmatrix} \begin{pmatrix} n_0^{i+1} - n_0^i \\ n_1^{i+1} - n_1^i \end{pmatrix} \right\|^2 \longrightarrow \min. \quad (7.13)$$

This leads to the following explicit solution: with $\Delta\mathbf{n}^i = \mathbf{n}^{i+1} - \mathbf{n}^i$, we obtain

$$a^4 = \frac{\sum_{i=0}^{N-1}(\Delta n_0^i \sin\varphi + \Delta n_1^i \cos\varphi)^2}{\sum_{i=0}^{N-1}(\Delta n_0^i \cos\varphi - \Delta n_1^i \sin\varphi)^2}. \quad (7.14)$$

Therein φ is given by

$$\tan(2\varphi) = -2\frac{\sum_{i=0}^{N-1} \Delta n_0^i \Delta n_1^i}{\sum_{i=0}^{N-1}((\Delta n_0^i)^2 + (\Delta n_1^i)^2)} \quad (7.15)$$

with a twofold ambiguity. However, the solutions are congruent, so one solution is sufficient to give a normalization up to remaining scaling, translation, and rotation. The case that the denominator of Eq. (7.14) equals zero can be neglected as it occurs only if the contour is one-dimensional.

The normalized coordinates can then be calculated as

$$T(\mathbf{n}^i) = \begin{pmatrix} a & 0 \\ 0 & 1/a \end{pmatrix} \begin{pmatrix} \cos\varphi & -\sin\varphi \\ \sin\varphi & \cos\varphi \end{pmatrix} \mathbf{n}^i. \quad (7.16)$$

The problem of the normalization given so far is that instead of the sum of distances the sum of squared distances is minimized (in order to obtain an explicit solution). But this can be solved by a suitable parameterization with an affine length

---

[2]To be exact, we minimize the sum of squared distances instead of the sum of distances to get an analytic solution.

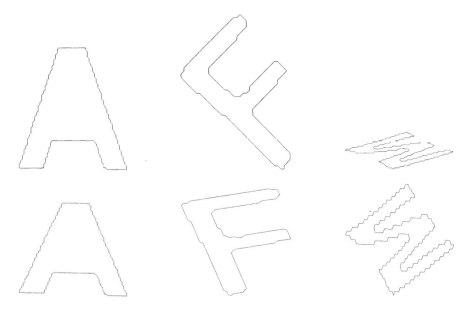

Fig. 7.5   Affine transformed contours before and after affine normalization.

[21]. One just has to replace Δn^i with

$$\Delta \mathbf{n}^i = \frac{\mathbf{n}^{i+1} - \mathbf{n}^i}{\sqrt{\lambda_{i+1} - \lambda_i}}, \tag{7.17}$$

where

$$\lambda_{i+1} = \lambda_i + \tfrac{1}{2} \left| \hat{n}_0^i \hat{n}_1^{i+1} - \hat{n}_0^{i+1} \hat{n}_1^i \right|, \quad \lambda_0 = 0, \tag{7.18}$$

$$\hat{\mathbf{n}}^i = \mathbf{n}^i - \mathbf{n}^a. \tag{7.19}$$

The contour's area center $\mathbf{n}^a$ is given by

$$n_0^a = \frac{1}{3} \frac{\sum_{i=0}^{N-1} (n_0^i n_1^{i+1} - n_0^{i+1} n_1^i)(n_0^{i+1} - n_0^i)}{\sum_{i=0}^{N-1} (n_0^i n_1^{i+1} - n_0^{i+1} n_1^i)}, \tag{7.20}$$

$$n_1^a = \frac{1}{3} \frac{\sum_{i=0}^{N-1} (n_0^i n_1^{i+1} - n_0^{i+1} n_1^i)(n_1^{i+1} - n_1^i)}{\sum_{i=0}^{N-1} (n_0^i n_1^{i+1} - n_0^{i+1} n_1^i)}. \tag{7.21}$$

Because of stability reasons it is useful not to use the contour itself for determining a and φ but to use its convex hull instead.

Examples of contours normalized this way are given in Fig. 7.5. Loosely speaking, the shape of the convex hull has become more similar to a circle after normalization.

The affine normalization alone is not sufficient to find an invariant description. The objects can still be translated, rotated, or scaled. These transformations have still to be eliminated from the description. One possibility is to use Fourier descriptors that will be treated now.

7.3.2 Fourier Descriptors

For easier presentation the points in 2-D space will be treated in complex notation[3], namely

$$n = n_0 + jn_1 \tag{7.22}$$

is a complex number here. We consider closed curves $n(\lambda)$ parameterized by λ and with perimeter T. The finite Fourier series expansion is given as

$$n(\lambda) = \sum_{k=-N/2}^{N/2} c_k e^{jk\omega\lambda}, \quad \omega = \frac{2\pi}{T}. \tag{7.23}$$

The Fourier coefficients c_k are given by

$$c_k = \frac{1}{T} \int_{\lambda=0}^{T} n(\lambda)e^{-jk\omega\lambda}d\lambda. \tag{7.24}$$

In the following we will use Fourier coefficients for objects with normalized perimeter $T = 2\pi$, namely, $\omega = 1$. We want to find invariant features for the case of similarity transformations. This means that the contour can be obtained from a reference contour n^0 by the following law:

$$n(\lambda) = sn^0(\lambda + \lambda_0)e^{j\varphi} + t. \tag{7.25}$$

Therein s is the scaling factor, t is a complex translation, φ is the rotation angle and λ_0 is a shift of the contour's starting point.

Such a transformation of the curve results in the following changes of the Fourier coefficients:

$$c_0 = c_0^0 se^{j\varphi} + t, \tag{7.26}$$
$$c_k = c_k^0 se^{j(\varphi+k\lambda_0)}, \quad k \neq 0. \tag{7.27}$$

c_0 just describes the center point and is invariant to a shift of the starting point. The other coefficients, which depend on scaling, rotation, and a shift of the starting point, however, are invariant to translation.

---

[3]We use j to denote $\sqrt{-1}$.

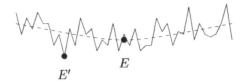

Fig. 7.6 Variation of an extreme point E caused by noise in case of an orbit with low curvature. The original orbit is given as dashed line, the noisy one as solid line.

According to [4] features that are invariant to all changes can be calculated by

$$T_k(n) := \frac{|c_k|}{|c_q|} e^{j(\varphi_k + a\varphi_r - b\varphi_q)}, \quad k = -\frac{N}{2}, \ldots, -1, 0, 1, \ldots, \frac{N}{2}, \qquad (7.28)$$

$$a = \frac{q - k}{r - q}, \quad b = \frac{r - k}{r - q}, \quad r = q + S, \quad q \in I\!N, \qquad (7.29)$$

with S being the degree of the contour's rotation symmetry (e.g., $S = 3$ for an equilateral triangle). These features are called Fourier descriptors. In practice, c_q and c_r have to be chosen of high absolute value in order to achieve a robust normalization.

From a geometrical point of view the Fourier descriptors do an orientation normalization of the main axes of ellipses. In [4] it is proved that the set of Fourier descriptors given by Eq. (7.28) is complete and minimal for contours that can be described by a finite Fourier expansion. It is also interesting to mention that the transformation parameters can be extracted from the Fourier descriptors; for details, we refer the leader to [4].

The weakness of normalization approaches is caused by the uncertainty of the localization of the extreme points that are used for normalization. This uncertainty can be caused by noise or not highly distinctive extremal points, as can be seen in Fig. 7.6.

7.4   DIFFERENTIAL INVARIANTS

For many transformation groups G the group elements $g \in G$ can be described by parameters λ so that the group product as well as the inverse elements continuously depend on the parameters (Lie groups). An easy example is the group of rotations that can be parameterized with an angle φ:

$$g(\varphi) = \begin{pmatrix} \cos\varphi & -\sin\varphi \\ \sin\varphi & \cos\varphi \end{pmatrix}, \quad \varphi \in [0, 2\pi]. \qquad (7.30)$$

By this parameterization a topology is introduced. Group elements are neighbors if their parameters λ are similar. The idea is that invariant features shall not change if an infinitesimal small transformation is performed:

$$\frac{\partial T(g(\lambda)\mathbf{x})}{\partial \lambda_i} \equiv 0, \quad i = 1, 2, \ldots, p. \tag{7.31}$$

For connected groups it is sufficient to solve this differential equation for fixed $\lambda = \mathbf{0}$ as each group element can be expressed as a product of many infinitesimal small transformations.

Let us consider the example of the rotation group again. A coordinate $\mathbf{n}$ after rotation is given by

$$g(\varphi)\mathbf{n} = (n_0 \cos \varphi - n_1 \sin \varphi, n_0 \sin \varphi + n_1 \cos \varphi). \tag{7.32}$$

We solve Eq. (7.31)

$$\left. \frac{\partial T(g(\varphi)\mathbf{n})}{\partial \varphi} \right|_{\varphi=0} = 0 \quad \Leftrightarrow \quad -n_1 \frac{\partial T}{\partial n_0} + n_0 \frac{\partial T}{\partial n_1} = 0 \tag{7.33}$$

which results in $T(\mathbf{n}) = f(n_0^2 + n_1^2)$ with arbitrary f, and thus obviously is invariant to rotations.

However, in practice, the differential equations become quite more complex and hard to solve. Therefore often a priori knowledge is used to reduce the number of necessary partial derivatives (semidifferential invariants) [3, 10, 33].

7.5   INTEGRAL INVARIANTS

The third possibility to construct invariant features is given by integrating over the transformation group. In the following we deal with gray scale images. In contrast to the notation used before (with subindexes like $\mathbf{X}_{n_0,n_1}$), we write them in the form $\mathbf{X} = \{\mathbf{X}(n_0, n_1)\}$, $0 \le n_0 < N_0$ and $0 \le n_1 < N_1$. This is due to the reason that we will have to treat noninteger indexes later. The value $\mathbf{X}(\mathbf{n})$ is called gray value at the pixel coordinate $\mathbf{n}$. In the following, it will be convenient to use both a continuous and a discrete formulation. In the continuous case the pixel coordinates (n_0, n_1) are real numbers in the range $0 \le n_0 < N_0$ and $0 \le n_1 < N_1$, respectively, whereas in the discrete case they are integers. Again there is a transformation group G with elements $g \in G$ acting on the images. For an image $\mathbf{X}$ and a group element $g \in G$ the transformed image is denoted by $g\mathbf{X}$.

We will consider the compact groups of translations and of congruences here (with cyclic boundary conditions). Let us shortly recall the coordinate transformation laws of these in parameterized form. For a group element g of the group of translations a translation vector $(t_0, t_1)^T \in I\!\!R^2$ exists, so that

$$(g\mathbf{X})(\mathbf{n}) = \mathbf{X}(\mathbf{n}'), \quad \text{with} \quad \mathbf{n}' = \mathbf{n} + (t_0, t_1)^T. \tag{7.34}$$

Similarly for the group of Euclidean motion (translation and rotation) additionally an angle $\varphi \in [0, 2\pi]$ exists, so that

$$(g\mathbf{X})(\mathbf{n}) = \mathbf{X}(\mathbf{n}'), \quad \text{with} \quad \mathbf{n}' = \begin{pmatrix} \cos\varphi & \sin\varphi \\ -\sin\varphi & \cos\varphi \end{pmatrix} \mathbf{n} + \begin{pmatrix} t_0 \\ t_1 \end{pmatrix}. \tag{7.35}$$

The transformation law states that an image transformation consists of a rotation around the rotation center followed by a translation. This rotation center is not known a priori and it does not necessarily coincide with the coordinate origin chosen in the image. However, by applying an appropriate translation, it is always possible to bring the coordinate origin to the rotation center, so that Eq. (7.35) is sufficient to describe also Euclidean motion with nonorigin rotation center.

As we deal with finite images all indexes are understood modulo the image dimensions N_0 and N_1, respectively. These periodic boundary conditions will be used throughout the section for all the index arithmetic. Note that by this convention the range of the translation vector $\mathbf{t} = (t_0, t_1) \in I\!\!R^2$ can be restricted to $0 \le t_0 < N_0$ and $0 \le t_1 < N_1$.

Obviously, if we are using the discrete formulation (i.e., the pixel coordinates n_0, n_1 are restricted to integers), an appropriate rounding or interpolation procedure must be applied, since the index vector $\mathbf{n}'$ in Eqs. (7.34) or (7.35) will have no integer values for many values of $\mathbf{t}$ and/or φ. We mention that theoretically rotations with cyclic boundary conditions do not satisfy the requirements for a group anymore (for the pure translation the requirements are still fulfilled). For example, two cyclic rotations cannot be represented by one cyclic rotation alone, but this is of no importance in practice.

7.5.1 Construction of Invariant Features by Integration over the Transformation Group

For a given gray scale image $\mathbf{X}$ and a complex-valued kernel function $f(\mathbf{X})$, it is possible to construct an invariant feature $T[f](\mathbf{X})$ by integrating $f(g\mathbf{X})$ over the transformation group G:

$$T[f](\mathbf{X}) := \frac{1}{|G|} \int_G f(g\mathbf{X}) dg. \tag{7.36}$$

The meaning of this formula is illustrated in Fig. 7.7. This averaging technique for constructing invariant features for general transformation groups is explained in detail in [24, 26, 25, 27].

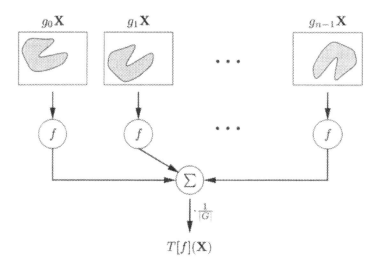

Fig. 7.7   Illustration of Eq. (7.36) for a group with n elements [18].

We consider especially monomials for the kernel function f, as it was proved that one can construct complete feature sets with these [15, 5, 25]. For patterns of dimensionality N and finite groups G with $|G|$ elements, the group averages of monomials with degree $\leq |G|$ form a generating system of the pattern space. The theoretical upper bound of monomials needed is given by $\binom{N+|G|}{N}$. But, in practice, the number of monomials needed falls far below this upper bound. In the following section we discuss the group of translations and the group of Euclidean motion separately.

7.5.2   Invariant Integration for the Group of Translations

We consider here the group of cyclic translations Eq. (7.34), both for one-dimensional signals as well as for images.

7.5.2.1   Invariant integration for translation in 1-D   For the one-dimensional signals $\mathbf{x}$ of length N the integral over the transformation group of cyclic translations, Eq. (7.36) can be rewritten as

$$T[f](\mathbf{x}) = \frac{1}{N} \int_{t=0}^{N} f(g(t)\mathbf{x})dt. \qquad (7.37)$$

Table 7.1   Orbits of binary signals that are equivalent with respect to cyclic translation.

Orbits	Members			
O_0	$(0,0,0,0)^T$			
O_1	$(0,0,0,1)^T,$	$(0,0,1,0)^T,$	$(0,1,0,0)^T,$	$(1,0,0,0)^T$
O_2	$(0,0,1,1)^T,$	$(0,1,1,0)^T,$	$(1,1,0,0)^T,$	$(1,0,0,1)^T$
O_3	$(0,1,0,1)^T,$	$(1,0,1,0)^T$		
O_4	$(0,1,1,1)^T,$	$(1,1,1,0)^T,$	$(1,1,0,1)^T,$	$(1,0,1,1)^T$
O_5	$(1,1,1,1)^T$			

In practice, the signals are not continuous, so that only integer coordinates and translations are regarded. Equation (7.37) then is given by

$$T[f](\mathbf{x}) = \frac{1}{N} \sum_{t=0}^{N-1} f(g(t)\mathbf{x}). \tag{7.38}$$

with the finite group $\{g(t)\} := G$.

Let us consider a concrete example. We consider a binary $\mathbf{x}$ of length 4. So our signals are $\mathbf{x} = (b_0, b_1, b_2, b_3)^T$, with $b_i \in \{0,1\}$. The orbits of equivalent signals are given in Table 7.1.

We calculate the group average of all monomials $x_0^{e_0} x_1^{e_1} x_2^{e_2} x_3^{e_3}$ with $\sum_{i=0}^{3} e_i \leq 4 = |G|$ in order to obtain a complete feature set. Because of the averaging over cyclic translation, we need not to consider monomials that just differ by a cyclic translation. As a second hint we know that for e_i positive,

$$x_i^{e_i} = x_i \quad x_i \in \{0,1\}, \quad i = 0,1,2,3.$$

Therefore we can restrict our attention to the following monomials:

$$
\begin{aligned}
f_0(\mathbf{x}) &= x_0, \\
f_1(\mathbf{x}) &= x_0 x_1, \\
f_2(\mathbf{x}) &= x_0 x_2, \\
f_3(\mathbf{x}) &= x_0 x_1 x_2, \\
f_4(\mathbf{x}) &= x_0 x_1 x_2 x_3
\end{aligned}
$$

(obviously $f_5(\mathbf{x}) = 1$ will be of no practical use). The group averages of these are

$$
\begin{aligned}
T[f_0](\mathbf{x}) &= \tfrac{1}{4}(x_0 + x_1 + x_2 + x_3), \\
T[f_1](\mathbf{x}) &= \tfrac{1}{4}(x_0 x_1 + x_1 x_2 + x_2 x_3 + x_3 x_0), \\
T[f_2](\mathbf{x}) &= \tfrac{1}{2}(x_0 x_2 + x_1 x_3),
\end{aligned}
$$

Table 7.2 Feature values for the orbits of Table 7.1.

Orbits	$T[f_0](\mathbf{x})$	$T[f_1](\mathbf{x})$	$T[f_2](\mathbf{x})$	$T[f_3](\mathbf{x})$	$T[f_4](\mathbf{x})$
O_0	0	0	0	0	0
O_1	1/4	0	0	0	0
O_2	1/2	1/4	0	0	0
O_3	1/2	0	1/2	0	0
O_4	3/4	1/2	1/2	1/4	0
O_5	1	1	1	1	1

$$T[f_3](\mathbf{x}) = \tfrac{1}{4}(x_0 x_1 x_2 + x_1 x_2 x_3 + x_2 x_3 x_0 + x_3 x_0 x_1),$$
$$T[f_4](\mathbf{x}) = x_0 x_1 x_2 x_3.$$

In Table 7.2 the resulting feature values for the different orbits (described in Table 7.1) are given. One can easily see that $T[f_0](\mathbf{x})$ and $T[f_1](\mathbf{x})$ are sufficient already to form a complete feature set.

7.5.2.2 Invariant integration for translation in 2-D Of course, the method is also valid for the two-dimensional case of images $\mathbf{X}$ of size $N_0 \times N_1$. The integral over the transformation group of cyclic translations, Eq. (7.36), then can be rewritten as

$$T[f](\mathbf{X}) = \frac{1}{N_0 N_1} \int_{t_0=0}^{N_0} \int_{t_1=0}^{N_1} f(g(t_0, t_1)\mathbf{X}) dt_1 dt_0. \tag{7.39}$$

We will not consider this case in detail, as it is very analogous to the case of Euclidean motion that will be treated in Section 7.5.3.

7.5.2.3 Relations with the invariant transformation class $\mathbb{C}T$ Interestingly these invariant features are strongly related to the invariants that can be obtained with the class of fast translation invariant transformations $\mathbb{C}T$ [34, 4, 7]. Those are defined by the following recursive transformation T with commutative operators $f_1(.,.)$, $f_2(.,.)$:

$$T(\mathbf{x}) = \left\{ \begin{array}{c} T(f_1(\mathbf{x}_{1|2}, \mathbf{x}_{2|2})) \\ T(f_2(\mathbf{x}_{1|2}, \mathbf{x}_{2|2})) \end{array} \right\}, \tag{7.40}$$

starting from scalar $T(x_i) = x_i$. Therein $\mathbf{x}_{1|2}$ denotes the first half of $\mathbf{x}$ and $\mathbf{x}_{2|2}$ the second half and $f_1(\mathbf{x})$ and $f_2(\mathbf{x})$ have to be evaluated on each component, respectively.

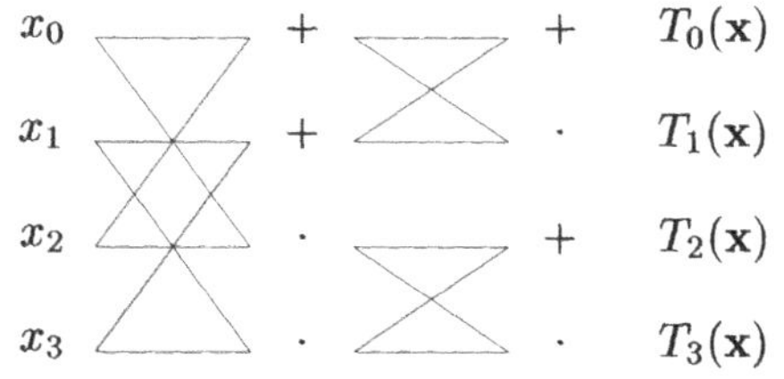

$$
\begin{array}{llll}
x_0 & + & + & T_0(\mathbf{x}) \\
x_1 & + & \cdot & T_1(\mathbf{x}) \\
x_2 & \cdot & + & T_2(\mathbf{x}) \\
x_3 & \cdot & \cdot & T_3(\mathbf{x})
\end{array}
$$

Fig. 7.8 Connection graph for invariant transformation class $\mathcal{C}T$ with $N = 4$.

We again consider the case $N = 4$ (e.g., as in Section 7.5.2.1) and use "+" and "·" as commutative functions. For illustration, the scheme is shown in Fig. 7.8 for this case. When solving the iteration, we obtain

$$
\begin{aligned}
T_0(\mathbf{x}) &= x_0 + x_1 + x_2 + x_3, \\
T_1(\mathbf{x}) &= x_0 x_1 + x_1 x_2 + x_2 x_3 + x_3 x_0, \\
T_2(\mathbf{x}) &= x_0 x_2 + x_1 x_3, \\
T_3(\mathbf{x}) &= x_0 x_1 x_2 x_3.
\end{aligned}
$$

These are nothing else than the invariant group averages $T[f_0](\mathbf{x}), T[f_1](\mathbf{x}), T[f_2](\mathbf{x})$, and $T[f_4](\mathbf{x})$ (up to a constant factor), that we obtained by invariant integration before, however calculated with a fast algorithm with complexity $\mathcal{O}(N \log_2 N)$.

7.5.3 Invariant Integration for the Group of Euclidean Motion

As mentioned before, we are considering image rotations and translations with cyclic boundary conditions, Eq. (7.35). In this case the integral over the transformation group can be written as

$$
T[f](\mathbf{X}) = \frac{1}{2\pi N_0 N_1} \int_{t_0=0}^{N_0} \int_{t_1=0}^{N_1} \int_{\varphi=0}^{2\pi} f\left(g(t_0, t_1, \varphi)\mathbf{X}\right) \, d\varphi \, dt_1 \, dt_0. \tag{7.41}
$$

$T[f]$ is called the group average of f. We want to mention that the method can be applied to 3-D data underlying 3-D Euclidean motion analogously. In this case the transformation group reads the same as the congruence group given in Eq. (7.4), but $\mathbf{n}$ is three-dimensional and $\mathbf{A}$ has size 3×3. This can be parameterized with a three-dimensional translation vector $\mathbf{t}$ and three Euler angles φ, θ, ψ. With the abbreviations $c_\alpha = \cos \alpha$ and $s_\alpha = \sin \alpha$, one obtains

$$
\mathbf{n}' = \begin{pmatrix} c_\psi & -s_\psi & 0 \\ s_\psi & c_\psi & 0 \\ 0 & 0 & 1 \end{pmatrix} \begin{pmatrix} 1 & 0 & 0 \\ 0 & c_\theta & -s_\theta \\ 0 & s_\theta & c_\theta \end{pmatrix} \begin{pmatrix} c_\varphi & -s_\varphi & 0 \\ s_\varphi & c_\varphi & 0 \\ 0 & 0 & 1 \end{pmatrix} \mathbf{n} +
$$
$$
\begin{pmatrix} t_0 \\ t_1 \\ t_2 \end{pmatrix}. \tag{7.42}
$$

But let us come back to the two-dimensional case. In order to give some intuitive insight in Eq. (7.41) we discuss several examples for constructing invariant image features:

- If the function $f(\mathbf{X})$ is already invariant, that is, $f(\mathbf{X}) = f(g\mathbf{X}); \forall g \in G$, then it is left unchanged by group averaging, that is, $T[f](\mathbf{X}) = f(\mathbf{X})$.

- For the function $f(\mathbf{X}) = \mathbf{X}(0,0)$ the group average is given by

$$
T[f](\mathbf{X}) = \frac{1}{N_0 N_1} \int\limits_{t_0=0}^{N_0} \int\limits_{t_1=0}^{N_1} \mathbf{X}(t_0, t_1) \; dt_1 dt_0. \tag{7.43}
$$

This is nothing but the average gray value of the image.

- For the function $f(\mathbf{X}) = \mathbf{X}(0,0) \cdot \mathbf{X}(0,1)$ the group average is given by

$$
T[f](\mathbf{X}) = \frac{1}{2\pi N_0 N_1} \int\limits_{t_0=0}^{N_0} \int\limits_{t_1=0}^{N_1} \int\limits_{\varphi=0}^{2\pi} \mathbf{X}(t_0, t_1)
$$
$$
\cdot \mathbf{X}(\sin \varphi + t_0, \cos \varphi + t_1) \; d\varphi dt_1 dt_0. \tag{7.44}
$$

- Finally we consider the function $f(\mathbf{X}) = \mathbf{X}(0,1) \cdot \mathbf{X}(2,0)$. Here the group average is given by

$$
T[f](\mathbf{X}) = \frac{1}{2\pi N_0 N_1} \int\limits_{t_0=0}^{N_0} \int\limits_{t_1=0}^{N_1} \int\limits_{\varphi=0}^{2\pi} \mathbf{X}(\sin \varphi + t_0, \cos \varphi + t_1)
$$
$$
\cdot \mathbf{X}(2\cos \varphi + t_0, -2\sin \varphi + t_1) \; d\varphi dt_1 dt_0
$$
$$
= \frac{1}{2\pi N_0 N_1} \int\limits_{t_0=0}^{N_0} \int\limits_{t_1=0}^{N_1} \int\limits_{\varphi=0}^{2\pi} \mathbf{X}(\sin \varphi + t_0, \cos \varphi + t_1)
$$
$$
\cdot \mathbf{X}\left(2\sin(\varphi + \pi/2) + t_0, 2\cos(\varphi + \pi/2) + t_1\right)
$$
$$
d\varphi dt_1 dt_0. \tag{7.45}
$$

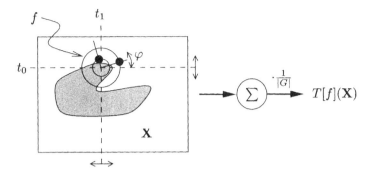

Fig. 7.9   Calculation strategy for the invariant integration in case of Euclidean motion.

Let us have a closer look at Eq. (7.45) in order to develop an efficient strategy for calculating the invariant features. We first consider the integral over the angle φ. The integrand in Eq. (7.45) is

$$\mathbf{X}(\sin\varphi + t_0, \cos\varphi + t_1) \cdot \mathbf{X}\,(2\sin(\varphi + \pi/2) + t_0, 2\cos(\varphi + \pi/2) + t_1) \quad (7.46)$$

Since $(k\sin\alpha + t_0, k\cos\alpha + t_1)$ describes for $0 \leq \alpha < 2\pi$ a circle of radius k around (t_0, t_1), this integral can be described as follows: Consider the pixel with coordinates (t_0, t_1), and determine all pixels which have distance 1 and 2 from this coordinate, respectively. The corresponding gray values of points on the circle with radius 1 and on the one with radius 2 have to be multiplied. In this case the corresponding coordinate points will have a phase shift of $\pi/2$ in polar representation. This procedure has to be done for all angles φ and for all shifts t_0, t_1, and all these results have to be averaged. This strategy for calculating invariant features is shown in Fig. 7.9.

We use kernel functions defined on a local domain. So we assume that only pixels from a small neighborhood of radius r are taken into account, that is, $f(\mathbf{X}(\mathbf{n})) \equiv 0$ for $\|\mathbf{n}\| \geq r$. For better distinction such functions of local support (FLS) will be denoted by $f_r(\mathbf{X})$ (Fig. 7.10), which means that, the radius r is added as subindex. With this definition the calculation of Eq. (7.41) can be reformulated by saying that we have to evaluate first for every pixel a function of local support (integration over φ). The local function is in general nonlinear. The second step is to average the results of the local computations (integration over t_0, t_1).

This interpretation of the group average given in Eq. (7.41) will be useful for determining the feature properties for scenes with multiple (possibly overlapping) objects and for articulated objects in Section 7.5.4. In the examples discussed so far, we only considered polynomials $f_r(\mathbf{X})$. It is also possible to determine invariant features by integrating other functions than polynomials over the transformation group. However, it is shown in [23, 25] that polynomials are sufficient to generate

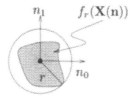

Fig. 7.10   Kernel function of local support f_r.

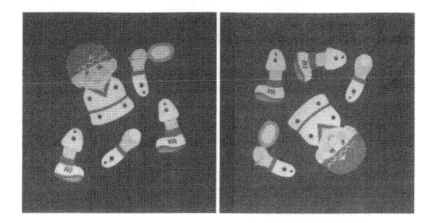

Fig. 7.11   Example with independent Euclidean motion of the objects.

complete feature sets. Therefore we consider only invariant features generated by averaging polynomials over the group here.

7.5.4   Additional Advantages of These Invariants

7.5.4.1   Independent motion and articulated objects   The method described before allows the calculation of features for gray scale images that are invariant under global image transformations. That is to say, it must be possible to find a translation vector **t** (and an angle φ respectively) to describe the transformation according to Eqs. (7.34) and (7.35). This is impossible for some applications where the image transformation is described by several local transformations with different parameters. We mention as examples scenes with several objects that are transformed independently and articulated objects that have joints. Figure 7.11 shows an example for several objects that are rotated and translated independently.

Another transformation of interest is if object parts have independent motion. In Fig. 7.12 an articulated object is given. The image transformation is described

Fig. 7.12   Example of an articulated object in front of a uniform background.

by different rotations and translations of the parts. By using appropriate functions, it is possible to use the group averaging for constructing invariant features that are not exactly invariant in the latter case but that vary only slightly. Such features are sometimes called *quasi-invariants* (cf. [2]).

To explain this, recall the two-step calculation strategy mentioned before. The invariant features could be calculated by first evaluating a function of local support for each pixel (integration over φ) and then averaging over all these results (integration over t_0, t_1).

Let us assume for the moment that the background intensity is zero and that the objects are separated. Furthermore we assume that the result of the local computation for a pixel equals zero if all gray values in the local neighborhood are zero (that is always true for the polynomials considered here). Under these assumptions the results of the local computations can only be different from zero for pixels belonging to an object or for pixels within a local neighborhood around the objects. The value of the invariant feature $T[f_r](\mathbf{X})$ is the sum of the results of the local computations and is therefore invariant even if the objects are transformed independently as long as the object separation in the scene is greater than r.

This reasoning is no longer valid if the background intensity is different from zero. However, there is one special case that can be treated along the same lines. The background is called homogeneous if for an image $\mathbf{X}$ without any objects the results of the local computations are identical (in practice it is sufficient that the variance is low enough). In this case the feature value $T[f_r](\mathbf{X})$ consists of terms determined by the objects and a local neighborhood around the objects, and a term determined by the background. If the background is homogeneous, this last term is independent from the positions of the objects and the feature is invariant even if the objects are transformed independently.

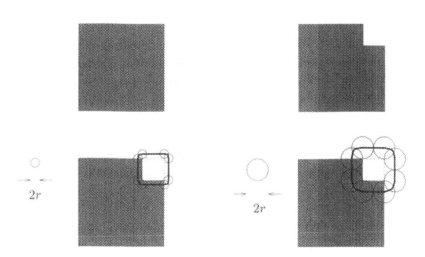

Fig. 7.13 Influence of distortions on the results of the local computations for two different support sizes. Upper left: undistorted pattern; upper right: distorted pattern; lower left: influence area of the distortion for a small support size of the rotated kernel function (given at the left); lower right: influence area of the distortion for a bigger support size of the rotated kernel function (given at the left).

The invariance is lost if the distance of the objects in the scene is smaller than r, or if the objects even overlap. However, the deviation only depends on the results of the local computations in the region where the overlap occurs. If this region is small (i.e., only moderate overlap), then the feature value varies only moderately.

The same reasoning applies for articulated objects. In this case, the feature value $T[f_r](\mathbf{X})$ consists of terms which are determined by the individual parts of the articulated object. The separate terms are not invariant if the parts move independently but the deviation is determined by the regions where the parts join. If the support size r of the rotated kernel function is small enough then only a small amount of all local computations is altered and therefore the feature after averaging stays nearly invariant. Figure 7.13 shows how local pattern distortions influence the results of the local computations for two different support sizes.

7.5.4.2 Robustness to topological deformations   Another useful property of the features given above is their robustness to topological deformations as shown in Fig. 7.14. The reasoning follows the same line as given above: if the deformations are moderate, only a small amount of local computations is altered so that the average of all local computations changes only slightly.

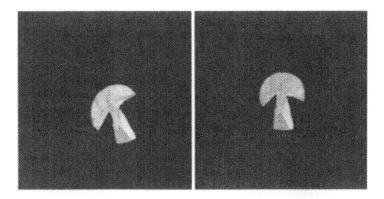

Fig. 7.14 Example of a deformable (rubber-type) object.

Fig. 7.15 Two objects with moderate overlap.

7.5.4.3 Additivity in feature space The reasoning in Section 7.5.4.1 for scenes with several objects allows an interesting extension to the recognition of two or more objects within one scene. These objects could even have a moderate overlap (so that segmentation attempts would fail) as in Fig. 7.15. Let T be an invariant transformation, and let $\mathbf{X}_1$, $\mathbf{X}_2$ be gray scale images that show a single object of type 1 and 2, respectively. Furthermore, let $\mathbf{X}$ be a gray scale image which shows these two objects together. We say that the feature T is additive in feature space if

$$T(\mathbf{X}) = T(\mathbf{X}_1) + T(\mathbf{X}_2). \tag{7.47}$$

It is obvious how to extend this definition to scenes with more than two objects. If there is moderate object overlap in $\mathbf{X}$ and T is continuous, then $T(\mathbf{X}) \approx T(\mathbf{X}_1) + T(\mathbf{X}_2)$; in other words, the deviation from the exact additivity is small. These concepts are visualized in Figs. 7.16 and 7.17.

7.5.5 Histogram of Local Features

The use of the features mentioned above is limited to ideal settings. In practical applications, however, one has to deal with greater object occlusion, partial changes

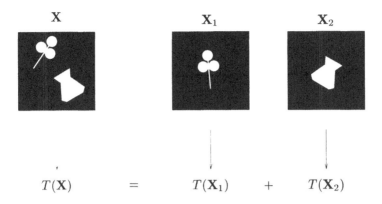

Fig. 7.16   Additivity in feature space.

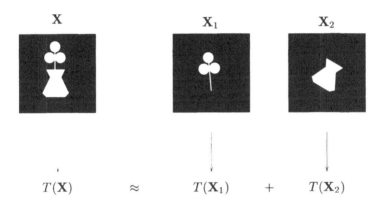

Fig. 7.17   Approximate additivity in feature space.

of the image, and so on. Because of global averaging in the calculation of the invariant features, these cases cannot be dealt with properly. We therefore construct features that preserve more of the local characteristics. As an example we again consider the transformation group of translation and rotation. Recall that the group integration could be written as Eq. (7.41).

We already mentioned that in practice, the integration is replaced by a sum and t_0, t_1, and φ are incremented in finite steps. In other words Eq. (7.41) is approximated by

$$T[f](\mathbf{X}) \approx \frac{1}{PN_0N_1} \sum_{t_0=0}^{N_0-1} \sum_{t_1=0}^{N_1-1} \sum_{p=0}^{P-1} f\left(g(t_0, t_1, \varphi = \frac{2\pi p}{P})\mathbf{X}\right). \qquad (7.48)$$

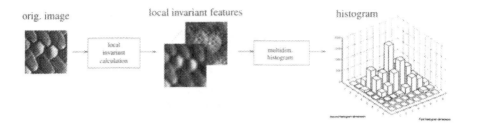

Fig. 7.18   Calculation of multidimensional feature histograms.

To get rid of the global averaging, one replaces the sums by a histogram operation, still keeping the invariance property (or one can keep the inner sum so that one obtains a histogram of local rotation invariant features). Also multiple features can be combined in histograms. Depending on the needs, one can build histograms for each kernel function separately, thus obtaining multiple histograms. A better (but more space-consuming) way is to combine the local results for different kernel functions f into multidimensional feature histograms. Thus the feature correspondence is given on pixel level.

Figure 7.18 illustrates the method for the case of histograms of locally invariant features; the two outer sums of Eq. (7.48) are replaced by histogram operations. Recall that one demand of an invariant transformation was its continuity in order to obtain clustered classes in feature space. The traditional histogram does not fulfill this requirement.

A traditional one-dimensional histogram might be considered as a set of assignment functions, one for each of K bins, that assign a feature value v (e.g., in the simplest case v is just a gray value) to an individual histogram bin (see Fig. 7.19):

$$\mathbf{h}(v) = \{h_i(v)\}, \quad i = 0, 1, \ldots, K - 1, \tag{7.49}$$

with

$$h_i(v) = \begin{cases} 0 & \text{for } v \text{ within a certain range} \\ 1 & \text{else} \end{cases} \tag{7.50}$$

and

$$h_i(v) \neq h_j(v) \quad \forall i \neq j. \tag{7.51}$$

The overall histogram of a set V of values v then is

$$\mathbf{h}(V) = \sum_{v \in V} \mathbf{h}(v). \tag{7.52}$$

The assignment of a feature value v to a specific histogram bin obviously is not a continuous function. As illustrated in Fig. 7.19, a small error ϵ can lead to a jump

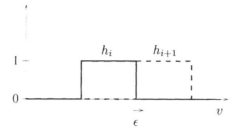

Fig. 7.19   Traditional histogram assignment function.

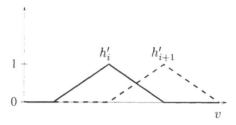

Fig. 7.20   Example of a fuzzy histogram assignment function.

in the assignment. We therefore propose a fuzzy version of the histogram that has a continuous bin assignment function.

The main idea is shown in Fig. 7.20. The fuzzy histogram assignment h'_i is a continuous function. It continuously increases from zero to one and decreases again. In addition there is an overlap of neighboring assignment functions, so the sum of them equals one. More general a fuzzy histogram will have the following properties:

1. *Continuity.* The histogram bin assignment function of each bin will be continuous.

2. *Mass preservation.* The sum of all bin assignment functions will equal one for fixed v,

$$\sum_i h'_i(v) = 1. \tag{7.53}$$

We now come to the multidimensional case and give a formal solution for a possible fuzzy multidimensional histogram. Let v_j denote the jth component of a N-dimensional feature vector $\mathbf{v}$. In the multidimensional case the assignment functions have a multidimensional index $\mathbf{i}$; For example, a 3-D feature vector $\mathbf{v}$ leads to an assignment function of kind $h_{(i_0,i_1,i_2)}$.

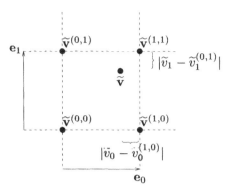

Fig. 7.21 Illustration of the fuzzy multidimensional histogram calculation for $N = 2$. The assignment function's value for bin $h_{\widetilde{\mathbf{v}}^{(0,0)}}$ is calculated as the product $|\widetilde{v}_0 - \widetilde{v}_0^{(1,0)}| \cdot |\widetilde{v}_1 - \widetilde{v}_1^{(0,1)}|$.

First we normalize the range of each dimension to $[0, K_j]$, with K_j being the desired number of histogram bins for dimension j of $\mathbf{v}$:

$$\widetilde{v}_j = (v_j - v_j^{min}) \frac{K_j}{v_j^{max}}, \quad j = 0, 1, \ldots, N - 1 \qquad (7.54)$$

with $[v_j^{min}, v_j^{max}]$ being the value range of v_j. In practice, it would be very inefficient to check all assignment functions $h_\mathbf{i}$ as most equal zero for a given $\mathbf{v}$ (especially for a large number of bins). We therefore do not regard all bin assignment functions $h_\mathbf{i}$ but restrict to calculating the value of the ones that do not equal zero for given $\mathbf{v}$.

Let $\mathbf{e}_j$ denote the jth unit vector. With

$$\widetilde{\mathbf{v}}^{(b_0, b_1, \ldots, b_{N-1})} = \sum_{j=0}^{N-1} \lfloor \widetilde{v}_j \rfloor \mathbf{e}_j + b_j \mathbf{e}_j, \quad b_j \in \{0, 1\}, \qquad (7.55)$$

we denote the 2^N surrounding bin centers, where $\widetilde{\mathbf{v}}$ lies in between. The nonzero assignment functions' values are

$$\begin{aligned} h_{\widetilde{\mathbf{v}}^{(b_0, b_1, \ldots, b_{N-1})}}(\widetilde{\mathbf{v}}) &= |\widetilde{v}_0 - \widetilde{v}_0^{(\overline{b_0}, 0, \ldots, 0)}| \, |\widetilde{v}_1 - \widetilde{v}_1^{(0, \overline{b_1}, 0, \ldots, 0)}| \\ &\quad \ldots |\widetilde{v}_{N-1} - \widetilde{v}_{N-1}^{(0, \ldots, 0, \overline{b_{N-1}})}|, \end{aligned} \qquad (7.56)$$

for all combinations of $b_j \in \{0, 1\}$. Therein $\overline{b_j}$ denotes the complement of b_j. For illustration, the procedure is shown for the case $N = 2$ dimensions in Fig. 7.21. The calculation of the weights is done like for bilinear interpolation. Note that the sum of all these increments resulting from one feature vector $\widetilde{\mathbf{v}}$ equals one-the property that we described as mass preservation before.

7.6   APPLICATIONS

Invariants are useful for several applications that deal with moving objects. We distinguish two different cases:

- The position and orientation of the objects are completely unimportant.

- The position and orientation are unimportant for the classification of the object but nevertheless one wants to know them, such as, for being able to pick the object up with a robot arm.

We will focus on the first case here. However, as mentioned before, there also exist invariant features that not only allow for classification but also for localization of objects, such as, the Fourier descriptors treated in Section 7.3.2.

7.6.1   Content-Based Image Retrieval

Besides other techniques, methods working fully automatically are of interest for image retrieval and of prime importance if the amount of data is too large to be handled manually or growing faster than manual annotation can be performed. Most current image retrieval systems use measures of similarity that require some match between full images or presegmented objects in images. Typical features used for that are the color histogram [11, 32], texture distribution [11], localized color [31, 30], or features derived from a local neighborhood of each point [22, 19]. Segmentation-based methods suffer from the uncertainty of the segmentation step, thus making these methods usable for special cases only.

Color histograms proved to be successful in automatic image retrieval. However, a major drawback is that all textural information is lost. As explained in Section 7.5.5, histograms can easily be extended by features that take the relations within a local pixel neighborhood into account. In contrast to the feature histograms of [19], the ones used here have the advantage that they stay unchanged in the case of translation and rotation. Thus we combine the advantage of an invariant description (e.g., we only need one histogram for a whole class of transformed images in the database) with the properties of histogram approaches, providing the possibility of also finding images by partial views, and vice versa, or to detect objects under occlusion. The disadvantage of histograms, their unsteady assignment at the bin boundaries, is removed by the fuzzy version of the histogram treated in Section 7.5.5.

Two different measures for comparing the histograms will be used. For simplicity we give the equations for the one-dimensional case here. Multidimensional histograms can be easily converted into one-dimensional histograms by ordering the dimensions after each other.

The χ^2-test [16] is a statistical method for determining whether two distributions differ. Given a query histogram $\mathbf{q}$ and a database histogram $\mathbf{h}$ the symmetric χ^2-test

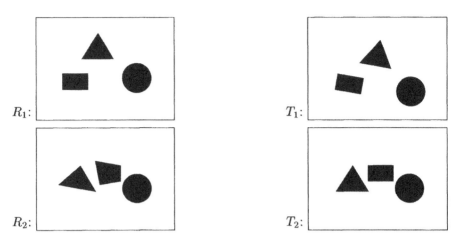

Fig. 7.22 Left images: two reference images (R_1, R_2). Right images: two test images (T_1, T_2). (The original color images are available in Postscript/Adobe Portable Document Format via anonymous FTP from ftp://ftp.wiley.com/public/sci_tech_med/image_video/.)

is given by

$$\chi^2(\mathbf{q}, \mathbf{h}) = \sum_i \frac{(q_i - h_i)^2}{q_i + h_i}. \qquad (7.57)$$

The intersection gives the common part of both histograms [32] and should be preferred if the search image is small compared to the database entries. We further normalize it with the search image size (which equals the sum of search histogram bins) in order to get an intuitive goodness of the match in percent of the search image

$$\cap(\mathbf{q}, \mathbf{h}) = \frac{\sum_i \min(q_i, h_i)}{\sum_i q_i}. \qquad (7.58)$$

7.6.1.1 Local versus global similarity In a first test we want to show the extended properties of our method compared to standard gray value histograms. According to the support size r of the kernel function, the similarity can be steered from local to more global similarity. An example is given in Fig. 7.22 and in Table 7.3 and Table 7.4.

Two test images are compared with two reference images. From a local point of view both test images are similar to the first reference image as they contain the same objects. However, from a global point of view the constellation of objects within the second test image is closer to the second reference image, although the objects are not exactly the same. Tables 7.3 and 7.4 reflect these two points of view: a 32 bin histogram simply based on the gray values classifies, both test images

Table 7.3   Results of histogram comparison for a kernel function using a simple gray value histogram, $f_0(\mathbf{X}) = \mathbf{X}(0,0)$

χ^2-test	R_1	R_2
T_1	0.0154	0.5450
T_2	0.0221	0.5863

Table 7.4   Same results as in Table 7.3 for a kernel function with bigger support, $f_{32}(\mathbf{X}) = \sqrt{\mathbf{X}(16,0) \cdot \mathbf{X}(0,32)}$

χ^2-test	R_1	R_2
T_1	9.1812	74.5491
T_2	76.8265	6.5463

being similar to the first reference, whereas a 32 bin histogram based on a kernel function of bigger support considers the second test image to be more similar to the second reference. So, in contrast to the simple gray value histogram, our method also considers structural information of different size, according to the support of the kernel functions chosen.

7.6.1.2   Combined color-texture histograms   Considering color information gives a significant relief in the task of finding visually similar images. Color histograms have proved to be successful in object recognition and image retrieval [32, 11]. However, for a huge image database, the color histogram may not be meaningful enough as all structural information contained in the image is lost. We therefore propose to add additional features based on the invariants given above.

Combined color-feature histograms, in fact, create a finer distinction: let each color bin h_i of a color histogram $\mathbf{h}$ be divided into K color-feature bins $h_{i,0}$, $h_{i,1}$, $\ldots$, $h_{i,K-1}$. Then the intersection of two color histograms $\mathbf{q}, \mathbf{h}$ is given by

$$\sum_i \min \left(\sum_{j=0}^{K-1} q_{i,j}, \sum_{j=0}^{K-1} h_{i,j} \right). \tag{7.59}$$

The intersection of the corresponding color-feature histograms,

$$\sum_i \sum_{j=0}^{K-1} \min \left(q_{i,j}, h_{i,j} \right), \tag{7.60}$$

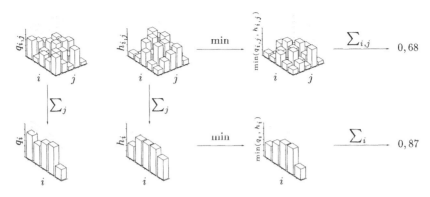

Fig. 7.23 Illustration of Eqs. (7.60) (upper row) and (7.59) (lower row). For abbreviation we define $q_i := \sum_j q_{i,j}, h_i := \sum_j h_{i,j}$.

is always less than or equal to Eq. (7.59). The denominator of Eq. (7.58) could be neglected here, as it is the same for both versions. Equations (7.59) and (7.60) are illustrated in Fig. 7.23.

For our tests we used a database of 438 color photograph images[4]. In addition to the normal $8 \times 8 \times 8$ color histogram, we calculated a second histogram extended by an invariant texture feature based on the gray value information, which normally contains most structural information. The kernel function f chosen for the following experiments was $f_8(\mathbf{X}) = \sqrt{\mathbf{X}(4,0) \cdot \mathbf{X}(0,8)}$ whose support diameter was about an eighth of the image width for square images.

In a first test we used original images from the database as search templates. In Fig. 7.24, an example is given for the combined color-texture histogram. While both methods were powerful enough to discriminate between all 438 images of this database, this might not be the case for a larger database.

Instead of increasing the database, we verified the improvement by our method searching with small cuts from the database entries as shown in Fig. 7.25, as this causes similar difficulties, such as, matching a constant red image of size 50×50 by color histogram matching against a database image. An intersection of 100% merely states that there are at least 2500 red pixels in the database image-located anywhere.

We compared both methods on the same image database, taking a rotated square cut from the center of each image for searching (about 15% of the image area). Because of the above-mentioned lack of spatial information within the histogram, we often obtained *several* images matching the search image with an intersection of 100%. Therefore we considered the worst rank in cases of multiple 100% matches. As a result we obtained an average rank of 3.2 for the color histogram, whereas for the

---

[4]The images were kindly provided by IBM in its QBIC demo and are courtesy of IBM Corporation.

Fig. 7.24 Search by an original image from the database (given at the top left). The results given at the right are ordered row by row. The corresponding intersection matches in percent are given below each image. (The original color images are available in Postscript/Adobe Portable Document Format via anonymous FTP from ftp://ftp.wiley.com/public/sci_tech_med/image_video/.)

Fig. 7.25 Search by rotated image detail (given zoomed at the top left). The results given at the right are ordered row by row. The corresponding intersection matches in percent are given below each image. (The original color images are available in Postscript/Adobe Portable Document Format via anonymous FTP from ftp://ftp.wiley.com/public/sci_tech_med/image_video/.)

combined color-texture histogram the average rank was improved to 1.7. A typical example showing the improvement is given in Figs. 7.26 and 7.27. In Fig. 7.26 the result of the normal color histogram comparison is given. As can be seen, several 100% matches were found, so the correct results cannot be uniquely identified. In Fig. 7.27 the result for the combined color-texture histogram is given instead. The correct results are clearly identified. We note that we do not achieve a full 100% match because the search template was rotated, and therefore small errors occurred.

7.6.1.3 Robustness to scaling   Scaling cannot be treated with the features given above; obviously the features do not stay invariant. Another, quite similar method treating scaling is presented in [28, 27]. However, this requires segmentation. Still another possibility is to consider features calculated at different scales. Figure 7.28 shows again the recognition results for the database of 438 images. Instead of color-feature histograms, the two feature kernel functions, $f_2(\mathbf{X}) = \sqrt{\mathbf{X}(1,0) \cdot \mathbf{X}(0,2)}$ and $f_4(\mathbf{X}) = \sqrt{\mathbf{X}(2,0) \cdot \mathbf{X}(0,4)}$ based on the gray value information only, are combined, as color would have made the task too simple. The images were then scaled by factors of 0.7 to 2.4 (i.e., roughly 0.5 to 5.75 in image area) and compared with a database containing the features for scales 1.3 and 1.7 (indicated in Fig. 7.28). Very good recognition accuracy is obtained between scale 1 and 2.2 (4.8 in image area). Note that whole images are scaled in Fig. 7.28. The histograms therefore could be scaled to have sum one. A comparison with scaled cuts from an image would be more difficult, but, to our knowledge this has not been treated yet in the literature on image retrieval (except for segmentation-based methods that sometimes implicitly are able to treat scaling, e.g., just describing an object by its color and its shape).

7.6.2   Analysis of 3-D Objects

A typical field dealing with 3-D objects is medicine. In medical applications the use of volume images generated by computer tomography (X-ray or magnetic resonance imaging) is a standard noninvasive diagnostic tool for physicians. Normally more than one scan of a body-part is made for control and documentation of medical treatment of diseases. The comparison of two or more datasets of one patient, scanned at different times, is difficult. The problems of the analysis of multiple-volume images are the nonuniform orientation and position of the body-parts of interest (unknown pose). In order to be able to compare and describe a process that causes morphological or metabolical differences, conventional techniques use registration. The presented method to construct invariant features for object recognition allows a comparison of multiple monomodal scans of objects. The method is able to compare multiple volume image scans from one body-part of a patient without any registration, segmentation, or other preprocessing. This can be done by computation of the invariants for each 3-D dataset, followed by a comparison of the invariants. It is possible to define metrics of the invariant features of different volume images and make a statement about morphological processes, that indicates the success or failure of a medical treatment. Furthermore the invariant features allow a content-based canonical coding of the

Fig. 7.26 Search by a rotated image detail (given zoomed at the top left) applying a simple color histogram. The results given at the right are ordered row by row. The corresponding intersection matches in percent below the images indicate that the expected results could not be identified uniquely. (The original color images are available in Postscript/Adobe Portable Document Format via anonymous FTP from ftp://ftp.wiley.com/public/sci_tech_med/image_video/.)

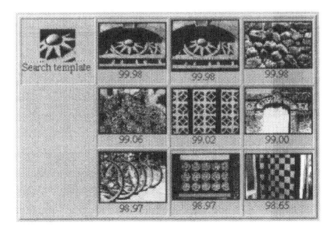

Fig. 7.27 Search by the same rotated image detail as in Fig. 7.26 (given zoomed at the top left) applying the combined color-texture histogram. The results given at the right are ordered row by row. The corresponding intersection matches in percent below the images show that the correct matches could be clearly identified, in contrast to the example given in Fig. 7.26. (The original color images are available in Postscript/Adobe Portable Document Format via anonymous FTP from ftp://ftp.wiley.com/public/sci_tech_med/image_video/.)

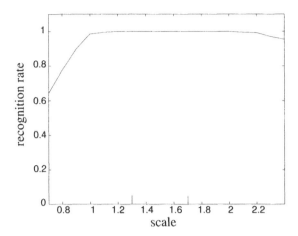

Fig. 7.28   Robustness to scaling.

images for effective database administration and retrieval [17]. Beside medicine there are various fields that can make use of 3-D invariant features as well.

7.7   CONCLUSION

In this chapter we presented the possibility to distinguish between similar objects by means of invariant features. After a formal description of what is considered similar, we mentioned three general principles to construct invariant features: image normalization, differential methods, and the integration over the transformation group.

We gave examples for the two former principles and discussed one in more detail because of its advantageous features: The main advantage is that these features do not need any preprocessing like segmentation or extraction of extreme points. Furthermore they are robust to local changes, thus making them suitable also for articulated objects. In restricted applications one can even use the additivity of the features to classify multiple objects within one scene simultaneously.

In order to achieve an even better robustness with regard to bigger occlusions, the features were been modified to histograms of local features. To preserve the continuous behavior of the features, a fuzzy histogram was created to replace the uncontinuous traditional histogram. Finally we presented examples from real applications that emphasize the practical use of the invariant features in image query applications.

REFERENCES

1. K. Arbter, W. Snyder, H. Burkhardt, and G. Hirzinger. Application of affine-invariant Fourier descriptors to 3-D objects. *IEEE Trans. Pattern Analysis and Machine Intelligence*, 12(7): 640-647, July 1990.

2. T. O. Binford. Inferring surfaces from images. *Artifical Intelligence*, 17: 205-244, 1981.

3. M.H. Brill, E.B. Barrett, and P.M. Payton. Projective invariants for curves in two and three dimensions. In J. L. Mundy and A. Zisserman, eds., *Geometric Invariance in Computer Vision*. MIT Press, Cambridge, 1992, pp. 193-214.

4. H. Burkhardt. *Transformationen zur lageinvarianten Merkmalgewinnung*. Habilitationsschrift, Universität Karlsruhe, 1979. Ersch. als Fortschrittbericht (Reihe 10, Nr. 7) der VDI-Zeitschriften, VDI-Verlag.

5. H. Burkhardt, A. Fenske, and H. Schulz-Mirbach. Invariants for the recognition of planar contour and gray-scale images. In *Proc. ESPRIT Basic Research Workshop at the ECCV 92*, Santa Margherita Ligure, Italy, May 1992, pp. 1-26.

6. H. Burkhardt, B. Lang, and M. Nölle. Aspects of parallel image processing algorithms and structures. In H. Burkhardt, Y. Neuvo, and J.C. Simon, eds., *From Pixels to Features II: Parallelism in Image Processing*. ESPRIT BRA 3035 Workshop, Bonas, France, August 1990, North-Holland, 1990, pp. 65-84.

7. H. Burkhardt and X. Müller. On invariant sets of a certain class of fast translation-invariant transforms. *IEEE Trans. Acoustics, Speech, and Signal Processing*, 28(5):517-523, October 1980.

8. H. Burkhardt, Y. Neuvo, and J.C. Simon, eds. *From Pixels to Features II: Parallelism in Image Processing*. ESPRIT BRA 3035 Workshop, Bonas, France, August 1990, North-Holland, 1990.

9. H. Burkhardt and H. Schulz-Mirbach. A contribution to nonlinear system theory. In *Proc. IEEE Workshop Nonlinear Signal and Image Processing*, Halkidiki, Greece, June 1995, pp. 823-826.

10. N. Canterakis. Complete projective semi-differential invariants. In *Proc. Int. Workshop on Computer Vision and Applied Geometry*, Nordfjordeid, Norway, August 1995.

11. J.J. Ashley, R. Barber, M.D. Flickner, J.L. Hafner, D. Lee, W. Niblack, and D. Petkovic. Automatic and semi-automatic methods for image annotation and retrieval in QBIC. In *Proc. SPIE Storage and Retrieval for Image and Video Databases III*, Vol. 2420, 1995, pp. 24-35.

12. A. Hurwitz. Über die Erzeugung der Invarianten durch Integration. In *Nachr. Akad. Wiss. Göttingen*: pp. 71-89, 1897.

13. B. Lang and H. Burkhardt. A parallel transputer system with fast pipeline interconnection. In *Proc. SPIE Applications of Digital Image Processing XIII*, Vol. 1349, San Diego, CA, July 1990, pp. 313-322.

14. T.G. Newman. A group theoretic approach to invariance in pattern recognition. In *Pattern*. Chicago, 1979, pp. 407-412.

15. E. Noether. Der Endlichkeitssatz der Invarianten endlicher Gruppen. *Math. Ann.*, 77: 89-92, 1916.

16. W.H. Press, S.A. Teukolsky, W.T. Vetterling, and B.P. Flannery. *Numerical Recipes in C*, 2nd ed. Cambridge University Press, Cambridge, 1992.

17. M. Schael and S. Siggelkow. Invariant greyscale features for 3D sensordata. In *Proc. Int. Conf. Pattern Recognition*, Barcelona, Spain, September 2000.

18. M. Schael and H. Burkhardt. Automatic detection of errors on textures using invariant grey scale features and polynomial classifiers. In M. Pietikäinen and H. Kauppinen, eds., *Proc. Workshop Texture Analysis in Machine Vision*, Oulu, Finland, June 1999, pp. 45-51.

19. B. Schiele and J.L. Crowley. Object recognition using multidimensional receptive field histograms. In B. Buxton and R. Cipolla, eds., *Computer Vision: ECCV'96*. Vol. I, Springer, Berlin, 1996, pp. 610-619.

20. R. Schiller. Normalization by optimization. In *Proc. 4th European Conf. Computer Vision*, Cambridge, England, April 1996. Lecture Notes in Computer Science, 1064, Springer, Berlin, 1996, pp. 620-629.

21. R. Schiller. *Konturbasierte Verfahren in der lageinvarianten Mustererkennung*. Ph.D. thesis. Technische Universität Hamburg-Harburg, May 1998.

22. C. Schmid and R. Mohr. Local gray value invariants for image retrieval. *IEEE Trans. Pattern Analysis and Machine Intelligence*, 19(5): 530-534, May 1997.

23. H. Schulz-Mirbach. Algorithms for the construction of invariant features. In W.G. Kropatsch and H. Bischof, eds., *16. DAGM–Symp. "Mustererkennung,"* Vienna, September 1994. Reihe Informatik XPress, Vienna, 1994, pp. 324-332.

24. H. Schulz-Mirbach. Constructing invariant features by averaging techniques. In *Proc. 12th Int. Conf. Pattern Recognition*, Vol. II, Jerusalem, October 1994, pp. 387-390.

25. H. Schulz-Mirbach. *Anwendung von Invarianzprinzipien zur Merkmalgewinnung in der Mustererkennung*. Ph.D. thesis. Technische Universität Hamburg-Harburg, February 1995. Reihe 10, Nr. 372, VDI-Verlag.

26. H. Schulz-Mirbach. Invariant features for gray scale images. In G. Sagerer, S. Posch, and F. Kummert, eds., *17. DAGM–Symp. "Mustererkennung"*, Bielefeld, 1995. Reihe Informatik aktuell, Springer, Berlin, 1995, pp. 1-14.

27. H. Schulz-Mirbach. Invariant gray scale features. Internal Report 8/96, Technische Informatik I, Technische Universität Hamburg-Harburg, 1996.

28. H. Schulz-Mirbach. The Volterra theory of nonlinear systems and algorithms for the construction of invariant image features. Internal Report 7/96, Technische Informatik I, Technische Universität Hamburg-Harburg, 1996.

29. S. Siggelkow and M. Schael. Fast estimation of invariant features. In W. Förstner, J.M. Buhmann, A. Faber, and P. Faber, eds., *Mustererkennung, DAGM 1999*, Informatik aktuell, Bonn, September 1999. Springer, Berlin, 1999.

30. J.R. Smith. *Integrated Spatial and Feature Image Systems: Retrieval, Analysis and Compression*. Ph.D. thesis. Columbia University, 1997.

31. M. Stricker and A. Dimai. Color indexing with weak spatial constraints. In *Proc. SPIE Storage and Retrieval for Image and Video Databases IV*, Vol. 2420, 1996, pp. 29-40.

32. M.J. Swain and D.H. Ballard. Color indexing. *Int. J. Computer Vision*, 7(1): 11-32, 1991.

33. L.J. van Gool, T. Moons, E. Pauwels, and A. Oesterlinck. Semi-differential invariants. In J. L. Mundy and A. Zisserman, eds., *Geometric Invariance in Computer Vision*. MIT Press, Cambridge, 1992, pp. 157-192.

34. M.D. Wagh and S.V. Kanetkar. A class of translation invariant transforms. *IEEE Trans. Acoustics, Speech, and Signal Processing*, 25: 203-205, April 1977.

8 Image Models for Facial Feature Tracking

D. SHAH[†] and S. MARSHALL

University of Strathclyde, Department of Electronic & Electrical Engineering,
204 George Street, Glasgow G1 1XW, Scotland, United Kingdom

8.1 INTRODUCTION

Locating and tracking facial features, especially the mouth region in a head and shoulder image sequence, is an important stage in the robust performance of telecommunication and recognition systems. Accurate extraction and analysis is a difficult problem due to the large number of varying parameters and conditions. In some systems these variations must be measured and classified; in others they must be ignored. From psychological studies it has been observed that the use of visual information can dramatically improve the performance of speech recognition systems. One of the major problems encountered while integrating visual speech into such a system, is to determine which features are important and how to represent them and extract them automatically. One of the main tools of feature location and extraction is the use of robust but flexible image models. A range of models exists from geometrical to statistical. The models can incorporate gray level image variations, such as in principal component analysis (PCA), or they can be confined to feature boundaries, as is the case with active contour models or snakes.

In this chapter the various models available will be reviewed and their characteristics outlined. These are given in Section 8.2. A model is of no use unless it accurately represents the underlying data. Fitting a model to image data frequently involves a compromise between preserving the a priori form of the model and obtaining a close fit to the data. Many models require some form of iteration or optimization technique, and for image sequence tracking they may incorporate information from preceding frames. The different techniques available for adapting image models to data will be

[†]Presently with the University of Cambridge, Computer Laboratory, New Museums Site, Pembroke Street, Cambridge CB2 3QG, United Kingdom

described in Section 8.3. Section 8.4 summarizes the development of a number of new image models and gives examples of this work [52, 53]. Some of these models have made use of the accompanying audio data to quantify the statistical link between mouth shape parameters and the corresponding speech phonemes–an example of this can be found in [53]. This link may be exploited to achieve lip synchronization and for low bit rate coding. Finally, conclusions are given in Section 8.5.

8.2   IMAGE MODELS

Since it is known that accurate extraction and analysis of facial features is a difficult problem due to the large number of varying parameters and conditions, it is necessary to adopt robust techniques that are based on prior information. This also helps in eliminating the effects of reflective color on the lips or patterned illuminations on the face. One of the many ways to track the features is to use image models that are able to exploit the prior knowledge of the image in different modes, and so on. This section deals with different image models and their corresponding applications.

8.2.1   Intensity Models for Face Detection and Face Recognition

Much work on processing facial features using intensity models has been carried out in the last decade. The first such work was by Sirovich and Kirby [54]. They presented a method that showed that any human face, whether cropped or full, could be described in terms of an optimum coordinate system. In their approach, a series of facial images is stored as a one-dimensional (1-D) array by concatenating rows of pixels. The 1-D vectors are then manipulated to obtain a coordinate system whose dimensionality is smaller than the total image space. They showed that a number of subimages (which they called eigenpictures or eigenfaces) need to represent the picture within some error bound that increases as the number of features (i.e., hair, background) increase.

Later Kirby and Sirovich made an improvement by exploiting the natural symmetry of the human face [31]. The original ensemble of faces was extended to include reflections about the mid-line of the face: this helped speed the method toward convergence. Shortly afterward, Turk and Pentland developed a real-time system that locates a subject's head and recognizes a person by comparing characteristics of their face to those of known individuals [58]. They defined an individual's face as the weighted sum of eigenpictures from the principal components. A particular face was recognized by comparing the weights to those of known individuals. Shackleton and Welsh used the same method to categorize an image in terms of principal components [51]. They discussed a process in which the geometric description and textural content of facial features are independently characterized for recognition purposes. The method employed deformable templates to isolate the facial features and principal component analysis (PCA) to obtain coefficients describing the textural content.

Craw and Cameron proposed further improvements based on purely theoretical considerations [16]. They suggested that if the image could be preprocessed using a priori knowledge of the scene content, then it would make the PCA method more robust and practical. For example, if two face images are summed up, the resultant image will be a face; that is, the faces could form a linear subspace. This property holds true if the faces are normalized (i.e., standardized) before subjecting them to the technique. The method given by Craw and Cameron utilizes a triangular mesh to standardize the facial image before applying the PCA method to reduce the vector space. Moghaddam and Pentland described an approach based on Karhunen-Loève (K-L) expansion to construct eigenfeatures and used them to find facial features [40]. Their approach for tracking features is carried out by performing an exhaustive search over the entire image at different scales, and then a local measure is used to evaluate target saliency.

Yuille et al. developed the deformable template method used for feature extraction and person identification [68]. The method is based on simple geometric shapes that can deform and move under the influence of image evidence in an attempt to minimize an energy function. Petajan used geometric features of the mouth, such as height and width, to track and extract the other features [43]. A simple thresholding technique is used to find the mouth opening and a linear time-warping model is then applied to the images and the training templates. Brunelli and Poggio employed a correlation method for locating and tracking facial features. Their technique uses templates of different scales to find the maximum correlation with respect to normalized images [13]. Wolff et al. described a method based on edge detection, thresholding, and triangulation of the eyes with respect to the mouth [66]. Later they incorporated motion information obtained from the difference images after preprocessing. Mak and Allen described an approach for the localization of the mouth as a pre-processing step to lip tracking [36]. Their method is based on morphological filters, difference images, thresholding, and cluster analysis.

Belhumeur et al. worked on developing a face recognition algorithm that is insensitive to large variation in lighting direction and facial expression [9]. They considered each pixel in an image as a coordinate in a high-dimensional space. They took advantage of the observation that the images of a particular face, under varying illumination but fixed pose, lie in a three-dimensional (3-D) linear subspace of the high-dimensional image space. They projected the image space to a low-dimensional subspace using the Fisherface method (based on Fisher's Linear Discriminant). The method produced well-separated classes even under severe variation in lighting and facial expressions. They compared this method with the eigenface technique and demonstrated that their Fisherface method obtained lower error rates than those of the eigenface technique for tests on the Harvard and Yale face databases.

Another group of researchers provided a fast algorithm to perform image-based tracking [20]. The method relies on the selective integration of a small subset of pixels that contain a lot of information about the state variables to be estimated. They found that the resulting decrease in the number of pixels for processing resulted in a substantial speedup of the basic tracking algorithm.

Bala et al. developed an algorithm for detection and tracking of the face and facial features in video image sequences [5]. They located the face in the image sequences by utilizing both color and scene background information. In the next stage, the eyes were detected within the face region, and both the features were tracked constantly over time. This technique was integrated in an interactive computer system, in which it serves as a preprocessing step for the determination of the 3-D eye positions.

8.2.2   Statistical Distribution Models for Feature Localization and Tracking

There has been a lot of research into the use of probabilistic models for tracking features. Bayesian image analysis has been applied to develop deformable templates, for example, by Grenander et al. [26] and Phillips and Smith [44]. The main assumption in this work is that knowledge of the object's shape is known beforehand. For example, in agricultural image analysis, the shape of animals and plants are known.

Cootes et al. devised a method for building deformable templates in which the constraints on the flexibility of the model were determined from the available data [15]. Their method is described as follows: The object under consideration is represented by a set of labeled points. The position of these points is determined (by hand) for a collection of training samples. Statistics relating to the position and possible direction of movement of these points in the model are calculated using principal component analysis. A point distribution model is produced, comprising the set of points describing the mean shape of the examples, together with a small number of linearly independent parameters representing the variations of the shape from this mean. The template is deformed by allowing the model to be attracted to the edges within the image. At each point in the model, the gray level profile normal to the boundary is examined, and the point is propagated toward the strongest edge. The size of the adjustment depends on the variability parameters, as the new template must lie within the specified range of allowable shapes.

Grenander et al. defined a template by a piecewise linear closed contour [26]. They used a Bayesian approach to fit the template to the image object by deforming the template to maximize an appropriate posterior distribution. Their algorithm consists of a posterior distribution made up of a prior term and a data term. The data term describes the fidelity of the template reconstruction to the data. It measures the discrepancy between the original image and a binary image representing the two regions defined by the template outline. The posterior distribution is a cyclic Markov random field model in which the conditional distribution of the position of a vertex is dependent only on the two neighboring vertex positions. To determine the template deformation, each vertex is visited in turn, and its new position is chosen to maximize the posterior probability function. This process is repeated iteratively until no further changes occur.

Another approach for feature localization was described by Lanitis et al. in [32]. This approach is based on statistical models that are derived from a training set of face images. Shape deformation is captured by a point distribution model, and an active shape model search is used to locate the face and its features within an image.

Luettin et al. developed a technique for locating and tracking lips using active shape models where the deformation of the lip model as well as image search is based on a priori knowledge derived from the training set [35]. They used a database consisting of a broad variety of speakers and lighting conditions to demonstrate the robustness and accuracy of the technique.

Baumberg et al. worked with techniques of generating a point distribution model (PDM) automatically from real image data [8]. Traditionally a PDM is derived by analyzing the modes of variation of a set of training examples. This process is laborious and may lead to a nonrepresentative set of training examples being used. Baumberg et al. used a cubic B-spline as the shape vector for training the model and observed that the resulting modes of variation showed the potential of the model for labeling and tracking of objects in real work scenes.

8.2.3   Wireframe Models

The use of wireframe models has arose because of the method's ability to exploit much more of the redundancy present in an image sequence compared to other methods. The idea is to reconstruct image scenes through detailed analysis of the input image and to create plausible models. This may seem fairly simple, but the feasibility of extracting redundancy depends on the complexity of objects and the background. For example, the head can be reasonably modeled by a rigid ellipsoid, whereas the face, being nonrigid, is more complicated. It can be coded by mapping the subject's image onto a wire-frame model and transmitting the movements as parameters of the model [63]. Figure 8.1 shows Welsh's wireframe model for a head and shoulder image: details of the model can be found in [64, 65].

The earliest work on wireframe models was carried out by Parke [41, 42]. His work was based on face synthesis and animation using models. The method adopted is as follows: Facial parameters containing structural information and expression information are obtained through something called facial action coding system (FACS) action units [22]. For synthesizing the face model, the parameter set and an interconnected polygon network are employed. It has been shown that a sequence of facial images can be created using a limited amount of information (a set of parameters obtained from the model), and this has demonstrated the potential of the technique in low-bandwidth visual communications.

Platt et al. worked on a physically based muscle-controlled facial expression model [45]. They developed a set of facial parameters dependent on the underlying structure of the face that creates facial expressions. Shortly afterward, Forchheimer worked on displaying a moving wireframe model of faces through animation for low bit rate coding [23]. The final system generated real-time moving images of a wireframe face. The wireframe model adopted was the CANDIDE model which consisted of 100 triangles distributed according to the curvature of the human face [49]. During the same period Aizawa [1, 2] proposed a model-based analysis/synthesis image coding system (MBASIC) that utilized a 3-D wireframe model of the object, composed of

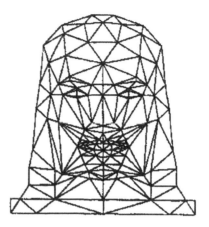

Fig. 8.1   Welsh's wireframe model.

about 400 triangular tessellations. He analyzed the input image and reconstructed the output image using the information of the 3-D model.

Recently a group of researchers described a facial image retrieval system based on verbal feature description and a neural network program [37]. They collected facial photographs and scanned them into a computer system where the images were converted into wireframe models. The geometrical feature values for each face were calculated using the wireframe models. Personal subjective impressions on each face were also collected via a questionnaire, and from the results the average impressions for each face were calculated. These impressions, which are related to verbal feature descriptions, were then converted to numerical values and used as indexes for retrieval. These researchers observed that the method showed a greatly improved retrieval ratio in comparison with the systems without background neural network learning.

Karunaratne et al. worked on an abstract muscle model capable of deforming a 3-D facial mesh of a synthetic human face to generate emotional expressions [29]. They used a triangulated facial mesh with smooth shading for the face. About 12 muscles on each side of the face were modeled by means of muscle abstractions derived from both facial anatomy and the FACS [22].

Strom et al. developed a real-time system for tracking and modeling of faces using an analysis-by-synthesis approach [55]. They mapped the texture of a 3-D face model with a head-on view of the face. The system then selected a set of feature points in the face-texture based on second derivatives of the surface. These selected points of the rendered image were tracked in the incoming video using normalized correlation, and the result was fed into an extended Kalman filter to recover camera geometry, head pose, and structure from motion. This information was used to rigidly move the face model to render the next image needed for tracking.

8.2.4   Active Contour Models: Snakes

The development of active contour models has resulted from the work of Kass et al. [30]. The contours or snakes are defined as energy-minimizing splines, and their operation can be understood as a special case of a more general technique of matching a deformable model to an image by means of minimizing an energy function. One of the factors notable is that they are dependent on the prior information of the object's shape and the likelihood position of the object and thus cannot find the contours independently. The work by Kass has shown that these models are able to establish the fine contour details of an object, but they have difficulty in incorporating shape constraints. Thus Kass et al. calculated a factor that quantifies the amount of compromise allowable between the shape elasticity and the resolution factor of fine contour lines.

Further research has been carried out on improving the active contour models for tracking [3]. Improvements to the original method in eliminating numerical instability was done by Berger et al. through an idea of snake growing [10]. This technique allowed the primary snake to divide itself into smaller pieces. The strength of this approach was that at each stage, good convergence conditions were realized. Work was reported by Bregler et al. on constraining the active contour models for tracking [11]. Their work describes an outer lip contour that is constrained to lie in a subspace learned from a training set. The distance of the contour to the subspace is considered when the internal energy is found to be minimal; that is, the contour follows a straight line. Later on Cohen used the finite element method to minimize the energy, this proved to be more numerically stable than other methods and worked at a higher efficiency [14]. This was shown to be useful when the contours were closed or nearly closed. An approach using splines was reported in [6] where shape constraints were imposed on the deforming template by limiting the number of degrees of freedom. They applied the approach for image tracking by searching for high-contrast edges in images.

Ramos Sanchez et al. combined a spline-based lip model with a color-based image search [48]. Their work describes a lip boundary search formula as a two-class classification problem. They showed that as the lip intensity is not adequately represented by an unimodal distribution, use of individual intensity models was adopted for each subject. A similar approach on using color information and shape modeling based on Bezier curves is described in [61].

The papers reviewed in this section have applications for facial feature tracking. For more details, the interested reader may refer to the cited papers. In addition to these papers, there are a great number of other publications on facial tracking. A representative cross section of papers drawn from international journals and conferences is included in the references. For example, papers placing a great deal of emphasis on real-time aspects of facial feature tracking include [50, 34, 46, 62, 17, 27, 33, 56, 67].

Papers using facial tracking for telecommunications and image coding applications include [60, 59, 24, 25]. Integration of speech with image data is addressed in [39, 38, 57]. Further applications include pose estimation [7, 18] and database retrieval [28]. Other papers in this area are [47, 19].

8.3 FITTING MODELS TO DATA

A general description of fitting a model to an image follows. Given a model and an image containing an example of the object modeled, the interpretation involves choosing optimum values for all the model parameters so as to best fit the model to the image according to some criteria. These models are able to capture the natural variability within a class of shapes and can be used in image search and tracking to find examples of the object modeled. In this section a description of techniques used for model fitting is given.

8.3.1 Statistical Technique

Statistical methods for fitting a model to an object are often represented by sets of points systematically marked on each object shape. For example: it is possible to represent the objects under study in a parameterized form, such as a template $\underline{T}$. The parameters of the template are assigned a probability distribution $\pi(\underline{T})$ which is able to model the variations τ in $\underline{T}$. Hence τ is a random variable following the distribution and representing all possible template deformations. Through this prior density it is possible to express the knowledge of the shape of the object under study. Thus the number of parameters representing the template/object belong to the finite space $\underline{\theta} = (\theta_1, \theta_2, \dots, \theta_n)$.

To employ Bayes's rule, it is necessary to know what the likelihood function should be so that the posterior density found makes it possible to infer the distribution and statistics of the object under study. The image model or full description of the image is the likelihood. For example, let the observed image $\underline{I}$ be considered as a matrix of gray levels, $i(t_j)$, $t_j \in \mathbb{R}$, $j = 1, \dots, N^2$, where the size of the image is $N \times N$ and $\mathbb{R}$ denotes the set of real numbers. The likelihood is the joint probability density function $L(\underline{I}/\underline{T})$ of the gray levels of $\underline{I}$ given the templates T. This quantity expresses the dependence of the observed image on the deformed template.

By Bayes's theorem, the posterior density of the deformed template τ given the observed image $\underline{I}$ is proportional to

$$\pi(\underline{T}|\underline{I}) \propto L(\underline{I}|\underline{T})\pi(\underline{T}). \tag{8.1}$$

The solution maximizing the expression with respect to T is the maximum a posteriori estimate of the true scene. It has been found empirically that Monte Carlo Markov chain algorithms are efficient for simulating templates from any arbitrary posterior density.

8.3.2   Optimization Techniques

In image recognition the best image representation is sought, meaning the best match between the image and the model is required, and the best in this understanding is the goal. Whatever is the "best," an objective function of goodness must be available.

Thus an optimization function problem can be described as follows: Given some finite space D and a function $f : D \to \mathbb{R}$, find the "best" value in D under f. This is basically understood as finding a value $x \in D$ that is the minimum of the function f:

$$f_{min}(x) = \min_{x \in D} f(x). \tag{8.2}$$

The function f is the objective function and represents the distance between the model and data. Therefore its minimization is the desired goal.

It should be noted that the optimization algorithms cannot guarantee finding a good solution if the objective function does not reflect the true nature of the problem. Design of the objective function is the key factor in the performance of an optimization problem. Many conventional direct methods (e.g., calculus-based) as well as sophisticated iterative methods (e.g., genetic algorithms) exist for optimization.

8.3.3   Heuristic Methods

Heuristics are criteria, methods, or principles for deciding which among several alternative courses of action promises to be the most effective in order to achieve some goal. They represent a compromise between two requirements, namely the need to make such criteria simple and at the same time the desire to see them discriminate correctly between good and bad. It is the nature of good heuristics that they provide a simple means of indicating which among several courses of action is to be preferred, and that they are not necessarily guaranteed to identify the most effective course of action but do so sufficiently often.

Many heuristic methods for matching data to models exist [21] that have modeled uncertainty in imaging applications. One such theory is the possibility theory, an extension of fuzzy theory that can be used for data modeling purposes. The theory assumes the conditional interdependence of pieces of evidence in support of a hypothesis. This theory can be said to be an extension of fuzzy theory. A fuzzy set is a set of objects having a continuous degree of membership function that relates each set element to a real number in the range [0,1] representing its degree of membership in the set.

For a heuristic method to be used for tracking, the deformable prototype gives a practical estimation of a membership function [12]. The idea behind the method is quite simple. If P is a prototype model that can be deformed by manipulating parameters $(p_1, \ldots, p_n)$, then given an object, it is possible to deform the prototype model so that a maximal matching is obtained. The dissimilarity D between the object x and the model depends on the minimal distance between them and the distortion "energy" of the deformation. Formally, this can be written as

$$D(x) = \min_{p_1,\dots,p_n} \left(m(x; p_1, \dots, p_n) + w\, \delta(p_1, \dots, p_n) \right), \tag{8.3}$$

where m is a distance function between x and the model and δ is a distortion function weighted by w. A membership function γ_p can then be defined as

$$\gamma_p(x) = 1 - \frac{D(x)}{\sup D}, \tag{8.4}$$

where $\sup D$ is the least upper bound of D.

8.4 LOWER LIP EXTRACTION AND TRACKING

In order to classify the parameterized mouth and lip shape, detection and extraction of the appropriate region of the face is necessary. In this section two methods for lip tracking and extraction are discussed. The first is based on the amplitude projections of an image, and the second is based on the Bayesian paradigm, which involves designing parametric template models for describing the shape of the lips.

8.4.1 Lip Extraction and Tracking Through Amplitude Projection Method

The method described here uses two-dimensional amplitude projections of the image and a set of filters to extract the feature points (Fig. 8.2). In the initial stage, a subimage containing the mouth area is passed through a zero memory gray scale reversal and thresholding filter, where the threshold value is set at the minimum point of the histogram between its bimodal peaks. This process results in a binary image, Fig. 8.2(c), from which amplitude projections along its rows and columns are obtained as depicted in Fig. 8.2(d),

$$C(k) = \sum_{j=1}^{N} I(j, k), \tag{8.5}$$

$$R(j) = \sum_{k=1}^{N} I(j, k), \tag{8.6}$$

where $\underline{I}$ is the binary image of size $N \times N$, and $\underline{R}$ and $\underline{C}$ are the resulting column and row intensity vectors obtained from Eqs. 8.5 and 8.6. For simplicity, only the column vector $\underline{C}$ is considered in the remaining calculations.

The next step in the process is to transform $\underline{C}$ such that the function W is true:

$$W = \begin{cases} e & \text{if } C(k-1) < C(k) \text{ for } k = 2, \dots, N/2 \\ 0 & \text{otherwise,} \end{cases} \tag{8.7}$$

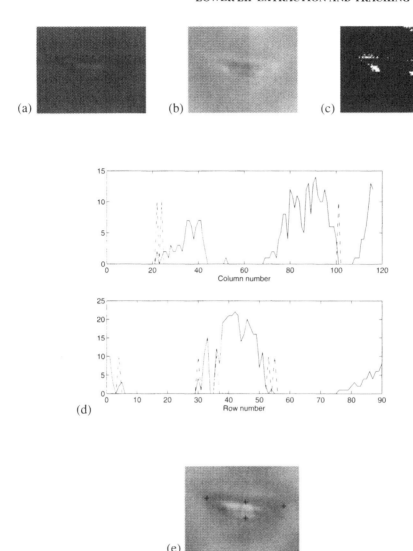

Fig. 8.2 Extraction of mouth image coordinates. (a) Original image; (b) negated image; (c) thresholded image; (d) column and row amplitude projections; (e) located feature points.

where e is any arbitrary constant. The function W detects areas that are of a strictly increasing intensity profile. The operation above is carried out for the first half of $\underline{C}$; for the remaining half the function W will be slightly changed and is left to the reader's interest to derive. The new vector, say $\underline{H}$, obtained from the above is convolved with a window V of width 5 whose components have the value 0, f, or g, creating a transformed column vector $\underline{H}^*$ (dotted spikes in top plot of Fig. 8.2(d)).

The window V is defined by

$$V = \begin{cases} f & \text{if } H(k) > \{H(k-1), \ldots, H(k-4)\} \text{ for } 5 < k \leq N/2 \\ g & \text{if } H(k-4) > \{H(k-3), \ldots, H(k)\} \text{ for } N/2 < k \leq N \\ 0 & \text{otherwise.} \end{cases} \qquad (8.8)$$

From $\underline{H}^*$, the column coordinates of the lip corners are obtained by isolating the component indexes and labeling them with a constant value f or g. The corresponding row coordinates situated along the final column coordinate (i.e., the chosen column number with the spike) is found by selecting the row index number whose amplitude is a maximum. In a similar way the coordinates for the top and bottom lip are found. Figure 8.2(e) shows the coordinates obtained using the method. From the coordinates, subimages with constant resolution were obtained with the mouth positioned at the center of the frame.

8.4.2   Lip Extraction and Tracking Through a Parametric Method

The accuracy of the foregoing method depends on the quality of the image as it deals with the intensity matrix, and it is therefore possible to obtain incorrect estimates for the position of the feature due to the lighting variations. The use of Bayesian models to locate and track features in an image has been widely investigated. Work in this area has shown that it is possible to track objects in an image space [44, 68, 4]. This technique can be used to model the shapes in an image by their boundary and the texture content. The key to using Bayesian analysis lies in the specification of prior distributions. Methods exist that can be used to obtain the prior [26].

8.4.2.1   The lower lip template   For this work a geometrical template of the lower lip is used, described by parametric components such as line segments on arcs. For open and closed mouths the most noticeable features are the groove separating the teeth from the lower lip and the valley where the lips meet. It is known that while talking, the lower lip actually moves in making any utterances as the skull stays relatively still and jaw moves up and down. Thus the lower lip only is modeled as described below.

A simple two-dimensional (2-D) template (Fig. 8.3) of the lower lip is considered where (x, y) is the right-hand edge/tip of the lower lip and a is the distance separating the two edges. The distance between the central line (dotted line) and the upper edge of the lower lip is b, the depth of the lower lip is c, and the rotation of the main axis from the horizontal is ϕ. Thus it is possible to represent the two edges of the lower lip by parabolas. In Fig. 8.3, let A be the origin with AD as the x-axis, then ABD and ACD are two parabolas with apex B and C, respectively. Since B is the apex at $(a/2, b)$ and passes through $(0, 0)$, then the equation of parabola is given by

$$y = -b + k_1 \left(x - \frac{a}{2} \right)^2, \qquad (8.9)$$

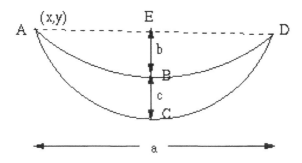

Fig. 8.3 Lower lip template: Rotation not shown.

where $k_1 = 4b/a^2$ and for ACD,

$$y = -(b + c) + k_2(x - \frac{a}{2})^2 \qquad (8.10)$$

with $k_2 = 4(b + c)/a^2$. Thus the coordinates of A, B, C and D are as follows: $A = (0, 0)$, $B = (a/2, b)$, $C = (a/2, b + c)$, and $D = (a, 0)$. Any point p on the parabola ABD will have coordinates $(a - p, -b + k_1(a - p - a/2)^2)$.

8.4.2.2   The prior function   From experiments it is found that for head and shoulder images of size 576×720, in the training set, in a neutral mouth position, the mean values of the parameters are

$$a = 79, \qquad b = \frac{a}{4}, \qquad c = \frac{a}{5}. \qquad (8.11)$$

Changes to the template can be achieved by adjusting the length a and altering the remaining parameters in turn. Thus from the training data the following multivariate densities were assigned to describe the variations,

$$a \sim \mathcal{N}(79, 49), \qquad b|a \sim \mathcal{N}(\frac{a}{4}, 4), \qquad c|a \sim \mathcal{N}(\frac{a}{5}, 9). \qquad (8.12)$$

Simulations for this set of parameters are shown in the results section. Considering the prior with $\phi = 0$ and (x, y) as uniform on the image, the joint prior of (x, y) and a are obtained using the pdf of the normal distribution with mean and variance as above:

$$f(\theta_x, \theta_y, \theta_a) = \begin{cases} \exp(-\frac{1}{2*49}(\theta_a - 79)^2) & \text{for } 0 < \theta_x < M \text{ and } 0 < \theta_y < N \\ 0 & \text{otherwise.} \end{cases}$$

$$(8.13)$$

8.4.2.3 The likelihood/texture function To relate the template to the observed data, a likelihood function is devised that depends on the different factors, such as the method of obtaining the images, or lighting conditions. Using simple probabilistic tools, it is possible to model the underlying pixel intensities of an image $\underline{I}$ given the template $\underline{\theta}$. The statistical form of the model above is

$$L(\underline{I}|\underline{\theta}) = \frac{1}{Z} \exp\left(-\frac{1}{2}\sum_k E_k\right),\tag{8.14}$$

where E_k is the energy associated with the kth region in the image related to the template and Z is the normalizing constant. The relevant energy function is given by

$$E_k = \frac{1}{\sigma_k}\sum_{j\in k}(I_j - \mu_k)^2,\tag{8.15}$$

where (μ_k, σ_k) are the parameters that model the underlying pixel intensities (i.e., the texture).

8.4.2.4 The overall function Finally, inference on a scene can made by obtaining the posterior distribution for the lip template, therefore using the Bayes's rules,

$$\pi(\theta|\underline{I}) \propto \exp\left(-\frac{1}{2}(E_k + f(\theta_x, \theta_y, \theta_a))\right).\tag{8.16}$$

Using the posterior density, the lower lip template can be sampled using Markov Chain Monte Carlo methods such as the Hastings-Metropolis algorithm [44].

8.4.2.5 Results For the purpose of this work, the parameters (μ_k, σ_k) were estimated using the original subimage of resolution 84×116, and the initial values for the Hasting-Metropolis algorithm were chosen to be $\mathcal{N}((M/4, N/4, M/2),$ $\text{diag}(10, 10, 49))$, where $M \times N$ is the image size.

The starting template was positioned randomly onto the subimage, Fig. 8.4(b). Lower lip templates were sampled from the density given in Eq. 8.16 using the Hastings-Metropolis algorithm. In this way new values were generated from a three-dimensional normal distribution with mean equal to the current parametric values and variance kept at the same constant value.

Figure 8.5 displays the convergence graphs for the parameters (x, y, a). Note that the difference between the original and estimated parametric values gradually decreases and after a number of iterations reaches a steady-state. The same is observed subjectively where after a few iterations, no significant change in the template position is obtained. This can be seen in Fig. 8.4(c) through (h), where a sequence of templates are overlaid onto the same image.

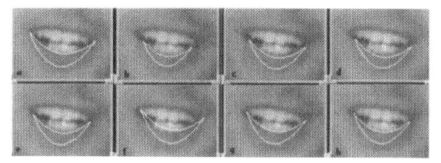

Fig. 8.4 Convergence of the lower lip template to the lower lip image feature. (a) Template created using original parametric value placed onto the image. (b) Randomly placed template (obtained using initial parametric values). (c)-(h) Template position at different iterations (3rd, 8th, 75th, 150th, 225th, and 600th).

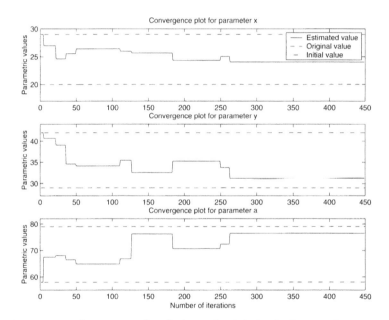

Fig. 8.5 Convergence of parameters (x, y): lower lip right-hand edge coordinate and a: distance between the edges.

From the final parameters of the template, the edge coordinates of the mouth were obtained (Fig. 8.6). The parameters at the initial and final stages of the simulation are given below:

- Original parametric values: $x = 20, y = 29, a = 79, b = 19, c = 15$.

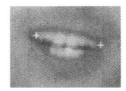

Fig. 8.6   Corner coordinates of the mouth edges.

- Initial parametric values: $x = 29, y = 42, a = 58, b = 15, c = 12$.

- Final estimated parametric values: $x = 25, y = 33, a = 72, b = 17, c = 14$.

8.5   CONCLUSIONS

This chapter has sought to provide the reader with a crosssection of the models that may be used to track facial features over a video sequence. A description of existing techniques was provided along with a summary of some the work in the field. The robust tracking of facial features in image sequences is an important goal in image analysis. A reliable system can be beneficial in a range of applications, including very low bit rate coding for telecommunications systems, and enhanced systems for the handicapped and improved speech recognition.

All faces are the same to some extent in that they contain eyes, nose, and mouth. There are only subtle variations from person to person that need to determined. Nevertheless, the use of models in describing pixel data invariably leads to a trade-off between model and data constraints. Not surprisingly Bayesian techniques are frequently employed to provide the balance between sometimes conflicting inputs.

Many of the techniques are well established and work consistently in controlled lighting conditions. There are, however, opportunities for future development in this area. One area is the use of three-dimensional models in head and shoulder type sequences to capture a more realistic description of the speaker and allow a greater range of movement without model distortion. Another area of research is in the integration of speech with the image signal to allow improved speech recognition and even low bit rate coding.

REFERENCES

1. K. Aizawa et. al. Model-based image coding–construction of a 3-D model of a person's face. In *Proc. Picture Coding Symp.*, Stockholm, Sweden, 1987, pp. 3-11.

2. K. Aizawa and H. Harashima. Model-based analysis-synthesis image coding (MBASIC) system for a person's face. *Image Communication*, 1(2): 139-152, October 1989.

3. T.E. Amini, A.A. Weymouth, and R.C. Jain. Using dynamic programming for solving variational problems in vision. *IEEE Trans. Pattern Analysis and Machine Intelligence*, 12(9): 855-867, September 1990.

4. A.J. Baddeley and M.N.M. van Lieshout. Stochastic geometry models in high-level vision. In K. V. Mardia and G. K. Kanjii, eds., *Advances in Applied Statistics*, Vol. 1, Carfax Publishing, Abingdon, 1994, pp. 235-256.

5. L.R. Bala, K. Talmi, and J. Liu. Automatic detection and tracking of faces and facial features in video sequence. In *Proc. Picture Coding Symp.*, Berlin, Germany, 1997.

6. R. Bartels, J. Beatty, and B. Barsky. *An Introduction to Splines for Use in Computer Graphics and Geometry Modeling*. Morgan Kaufmann, San Francisco, CA, 1987.

7. B. Bascle and A. Blake. Separability of pose and expression in facial tracking and animation. In *Proc. 6th Int. Conf. Computer Vision*, Bombay, India, 1998, pp. 323-328.

8. A.M. Baumberg and D.C. Hogg. Learning flexible models from image sequences. Technical Report 93.36. School of Computer Studies, University of Leeds, 1993.

9. P.N. Belhumeur, J.P. Hespanha, and D.J. Kriegman. Eigenfaces vs. fisherfaces: Recognition using class specific linear projection. *IEEE Trans. Pattern Analysis and Machine Intelligence*, 19(7): 711-720, July 1997.

10. M-O. Berger and R. Mohr. Towards autonomy in active contour models. In *Proc. 10th Int. Conf. Pattern Recognition*, Atlantic City, NJ, June 1990, pp. 847-851.

11. C. Bregler and S.M. Omohundro. Nonlinear manifold learning for visual speech recognition. In *Proc. Fifth IEEE Int. Conf. Computer Vision*, 1995, pp. 494-499.

12. H. Bremermann. Pattern recognition. In *Systems Theory in the Social Sciences*. Birkhaeuser, 1976, pp. 116-159.

13. R. Brunelli and T. Poggio. Face recognition: Features verses templates. *IEEE Trans. Pattern Analysis and Machine Intelligence*, 15(10): 1042-1052, October, 1993.

14. L.D. Cohen and R. Kimmel. Global minimum for active contour models: A minimum path approach. In *Proc. Computer Vision and Pattern Recognition Conf.*, 1996, pp. 666-673.

15. T.F. Cootes, C.J. Taylor, D.H. Cooper, and J. Graham. Training models of shape from sets of examples. In *Proc. British Machine Vision Conf.*. Springer, Berlin, 1992, pp. 9-18.

16. I. Craw and P. Cameron. Parameterising images for recognition and reconstruction. In *Proc. British Machine Vision Conf.*, Springer, Berlin, 1991, pp. 367-370.

17. T. Darrell, G. Gordon, J. Woodfill, and M. Harville. A virtual mirror interface using real-time robust face tracking. In *Proc. 3rd IEEE Int. Conf. Automatic Face and Gesture Recognition*, Nara, Japan, 1998, pp. 616-621.

18. T. Darrell, B. Moghaddam, and A.P. Pentland. Active face tracking and pose estimation in in interactive room. In *Proc. IEEE Computer Society Conf. Computer Vision And Pattern Recognition*, San Francisco, CA, 1996, pp. 67-72.

19. L.C. De-Silva, K. Aizawa, and M. Hatori. Detection and tracking of facial features. In *Proc. SPIE Mobile Robots X*, Vol. 2501, 1995, pp. 1161-1172.

20. F. Dellaert and R. Collins. Fast image-based tracking by selective pixel integration. In *Proc. ICCV 99 Workshop on Frame-Rate Vision*, Corfu, Greece, September 1999.

21. D. Dubois and H. Prade. *Fuzzy Sets and Systems: Theory and Applications*. Academic Press, San Diego, CA, 1980.

22. P. Ekman and W.V. Friesen. *Facial Action Coding System*. Consulting Psychologists Press, Palo Alto, CA, 1977.

23. R. Forchheimer and O. Fahlander. Low bit-rate coding through animation. In *Proc. Picture Coding Symposium*, March 1983, p. 13.5.

24. Y. Fujino, T. Ogura, and T. Tsuchiya. Facial image tracking system architecture utilizing real-time labeling (TV telephones and conferencing). In *Proc. SPIE Visual Communications and Image Processing '93*, Vol. 2094, 1993, pp. 2-11.

25. B. Girod. Image sequence coding using 3-D scene models. In *Proc. SPIE Visual Communications and Image Processing '94*, Vol. 2308, 1994, pp. 1576-1591.

26. U. Grenander and D.M. Keenan. Towards automated image understanding. *J. Applied Statistics*, 16: 207-221, 1989.

27. J. Heinzmann and A. Zelinsky. Robust real-time face tracking and gesture recognition. In *Proc. 15th Int. Joint Conf. Artificial Intelligence*, Nagoya, Japan, 1997, pp. 1525-1530.

28. N. Herodotou, K.N. Plataniotis, and A.N. Venetsanopoulos. A content-based storage and retrieval scheme for image and video databases. In *Proc. SPIE Visual Communications and Image Processing '98*, Vol. 3309, 1998, pp. 697-708.

29. S.K. Karunaratne and H. Yan. An abstract muscle model to generate facial expressions on a synthetic 3d human face. In *Proc. Inaurgural Conf. Victorian Chapter of the IEEE Engineering in Medicine and Biology Society: Biomedical Research in the 3rd Millennium*, 1998.

30. M. Kass, A. Witkin, and D. Terzopoulos. Snakes: Active contour models. *Int. J. Computer Vision*, 1: 321-331, 1988.

31. M. Kirby and L. Sirovich. Application of the K-L procedure for the characterization of human faces. *IEEE Trans. Pattern Analysis and Machine Intelligence*, 12(1): 103-108, 1990.

32. A. Lanitis, T.F. Taylor, and C.J. Cootes. Automatic interpretation and coding of face images using flexible models. *IEEE Trans. Pattern Analysis and Machine Intelligence*, 19(7): 743-756, 1997.

33. C.W. Lee, A. Tsukamoto, K. Hirota, and S. Tsuji. A visual interaction system using real-time face tracking. In *Proc. 28th Asilomar Conf. Signals, Systems and Computers*, Pacific Grove, CA, 1994, pp. 1282-1286.

34. Y.F. Liang and J. Wilder. Real-time face tracking. In *Proc. Conf. Machine Vision Systems for Inspection and Metrology*, Boston, MA, 1998, pp. 149-156.

35. J. Luettin, N.A. Thacker and S.W. Beet. Active shape models for visual speech feature extraction. In D. G. Storck and M. E. Hennecke, eds., *Speechreading by Humans and Machines*, Vol. 150, NATO ASI Series, Series F: Computer and Systems Sciences. Springer, Berlin, 1996, pp. 383-390.

36. M.W. Mak and W.G. Allen. Lip-tracking system based on morphological processing and block matching techniques. *Signal Processing: Image Communication*, 6(4): 335-348, 1994.

37. T. Matsushima and S. Hashimoto. Facial image retrieval by the feature description words. In `http://jazz.ma.is.sci.toho-u.ac.jp/~ matusima/kao/`.

38. H. Mizoguchi, T. Shigehara, Y. Goto, M. Teshiba, and T. Mishima. Virtual messenger to whisper in a person's ear by integrating real-time face tracking and speakers array. In *Proc. 4th Int. Conf. Virtual Systems and MultiMedia (VSMM 98)-Future Fusion, Real Applications for the Virtual Age*, Gifu, Japan, 1998, pp. 90-95.

39. H. Mizoguchi, T. Shigehara, M. Yokoyama, and T. Mishima. Virtual wireless microphone—A novel human robot interface utilizing face tracking and sound signal processing. In *Proc. 8th Int. Symp. Measurement and control in Robotics*, Czech Technical University, Prague, Czech Republic, 1998, pp. 437-442.

40. B. Moghaddam and A. Pentland. Maximum likelihood detection of faces and hands. In *Proc. Int. Workshop Automatic Face and Gesture Recognition*, Zurich, June 1995, pp. 122-128.

41. F.I. Parke. Computer generated animation of faces. *ACM Nat. Conf.*, 1: 451-457, August 1972.

42. F.I. Parke. Parameterized models for facial animation. *IEEE Computer Graphics and Applications*, 2(9): 61-68, 1982.

43. E.D. Petajan, B. Bischoff, and D. Bodoff. An improved automatic lipreading system to enhance speech recognition. In *Proc. ACM SIGCHI*, 1988, pp. 19-25.

44. D.B. Phillips and A.F.M. Smith. Dynamic image analysis using Bayesian shape and texture models. In *Statistics and Images*, Vol. 1, Carfax, Oxford, 1994, pp. 299-322.

45. S.M. Platt and N.I. Badler. Animating facial expressions. *Computer Graphics*, 15(3): 242-252, 1981.

46. R.J. Qian, M.I. Sezan, and K.E. Matthews. A robust real-time face tracking algorithm. In *Proc. 1998 IEEE Int. Conf. Image Processing*, Chicago, IL, 1998, pp. 131-135.

47. M.J.T. Reinders, F.A. Odijk, J.C.A. Van-Der-Lubbe, and J.J. Gerbrands. Tracking of global motion and facial expressions of a human face in image sequences. In *Proc. SPIE Visual Communications and Image Processing '93*, Vol. 2094, 1993, pp. 1516-1527.

48. M.U. Ramos Sanchez, J. Matas, and J. Kittler. Statistical chromaticity models for lip tracking with B-splines. In J. Bigün, G. Chollet, and G. Borgefors, eds., *Audio- and Video-Based Biometric Person Authentication*. Springer, Berlin, 1997, pp. 69-76.

49. M. Rydfalk. CANDIDE: A parameterised face. Technical report, Dept. Electrical Engineering, Linkoping University, S-581 83 Linkoping, Sweden, 1987.

50. H. Sako, M. Whitehouse, A. Smith, and A. Sutherland. Real-time facial-feature tracking based on matching techniques and its applications. In *Proc. 12th IAPR Int. Conf. Pattern Recognition*, Jerusalem, Israel, 1994, pp. 320-324.

51. M. Shackleton and W. Welsh. Report on classification of facial features for recognition. Technical report. British Telecom, RT4343/FACE008, 1990.

52. D. Shah and S. Marshall. Parametric method for tracking and analysing lip movements. In *Proc. Noblesse Workshop Non-linear Model Based Image Analysis(NMBIA98)*, Springer, Berlin, 1998.

53. D. Shah and S. Marshall. Processing of audio and visual speech for telecommunicaiton systems. *J. Electronic Imaging*, 8(3): 263-269, 1999.

54. L. Sirovich and M. Kirby. Low-dimensional procedure for characterization of human faces. *J. Optical Society America A*, 4(3): 519-524, 1987.

55. J. Strom, T. Jebara, S. Basu, and A. Pentland. Real time tracking and modeling of faces: An EKF-based analysis by synthesis approach. In *Proc. ICCV'99 Workshop Modeling People*, Corfu, Greece, September 1999.

56. W.Y. Tae and S.O. Il. A method for real-time face region tracking based on color histogram. In *Proc. SPIE Acquisition, Tracking, and Pointing X*, Vol. 2739, 1996, pp. 361-365.

57. C. Tsuhan and R.R. Rao. Audio-visual integration in multimodal communication. *Proc. IEEE*, 86(5): 837-852, 1998.

58. M. Turk and A. Pentland. Representing faces for recognition. Technical report. MIT Media Lab Vision and Modeling Group 132, 1990.

59. S. Valente and J.L. Dugelay. Face tracking and realistic animations for telecommunicant clones. *IEEE Multimedia*, 7(1): 34-43, January-March 2000.

60. S. Valente and J.L. Dugelay. Face tracking and realistic animations for telecommunicant clones. In *Proc. 6th Int. Conf. Multimedia Computing and Systems*, Florence, Italy, 1999, pp. 678-683.

61. M. Vogt. Fast matching of a dynamic lip model to color video sequences under regular illumination conditions. In *Speechreading by Humans and Machines*, Vol. 150, Springer, Berlin, 1996, pp. 399-408.

62. C. Wang and M.S. Brandstein. A hybrid real-time face tracking system. In *Proc. 1998 IEEE Int. Conf. Acoustics, Speech, and Signal Processing*, Seattle, WA, 1998, pp. 3737-3740.

63. W.J. Welsh. Model-based coding of moving images at very low bit-rates. In *Proc. Picture Coding Symp.*, 1987, page 3.9.

64. W.J. Welsh. Model-based image coding. *British Telecom Technology J.*, 8(3): 94-106, 1990.

65. W.J. Welsh. *Model-based coding of images*. Ph.D. thesis, Dept. Electronic Sytems Engineering, University of Essex, January 1991.

66. G.J. Wolff, K.V. Prasad, D.G. Stork, and M.E. Hennecke. Lipreading by neural networks: Visual prepocessing, learning and sensory integration. In *Advances in Neural Information Processing Systems*, Vol. 6, MIT Press, Cambridge, 1994.

67. L. Yufeng and J. Wilder. Real-time face tracking. In *Proc. SPIE Machine Vision Systems for Inspection and Metrology VII*, Vol. 3521, 1998, pp. 149-156.

68. A.L. Yuille, P. Hallinan, and D. Cohen. Feature extraction from faces using deformable template. *Int. J. Computer Vision*, 8(2): 99-111, 1992.

Index

3-D autoregressive model, 106

Active contour models, 305
Active shape models, 303
Additivity in feature space, 282
Affine normalization, 267
Affine order statistic filters, 190
   affinity function, 191
Amplitude projection method, 308
Analysis-by-synthesis approach, 304
Analysis bank, 169
Analysis order statistics bank, 171
α-stable distribution, 193
Audiovisual
   object, 215
   scene, 215
Automatic Restoration System for Animated
      Sequences (SARSA), 94

Backpropagation algorithm, 166
Bayes rule, 306
Bayesian
   restoration, 106
   techniques for vertical scratches restoration, 102
Block matching, 106, 111
Boolean
   filters, 18
   Positive Boolean functions, 19
   reference Boolean filter, 16

CALIC system, 41
CANDIDE wireframe model, 303
Cauchy distribution, 193
Centroids, 229
Change detection, 248
   neighborhood, 249
   temporal, 247
   texture, 248
Characteristic exponent, 193
CIF image sequence, 79
Clustering properties, 265
Clustering

class centers, 229
   fuzzy, 232
   fuzzy C-means algorithm, 216, 233
   introduction, 229
   ISODATA, 229
   techniques, 229
Coding
   content-based, 215
Color space
   Lab, 223
   Luv, 223
   RGB, 223
   YUV, 223
Combination filter (C-filter), 125
Complete mappings, 264
Completeness, 263
Compression methods
   Differential Pulse Code Modulation (DPCM),
      35, 39
   JPEG compression, 31
   lossless, 30, 46
   lossy, 30
   spatial redundancy, 30
Computational complexity, 265
Confidence measure, 227
Connected operators, 108
Constrained MSE minimization by the algorithm
      of Frost, 154
Constraint, spatial, 237
Content-based
   coding, 215
   description, 215
   image retrieval, 287
   manipulation, 215
Context
   energy, 46
   texture, 46
Contextual modeling of prediction errors, 46
Correction
   flicker, 96
   jitter, 101
Cost coefficients, 21

Printed and bound by CPI Group (UK) Ltd, Croydon, CR0 4YY

27/10/2024

14580330-0001